Advances in Bioclimatology 4

Springer

Berlin
Heidelberg
New York
Barcelona
Budapest
Hong Kong
London
Milan
Paris
Santa Clara
Singapure
Tokyo

Volumes in the series

Vol. 1: R.L. Desjardins, R.M. Gifford, T. Nilson, E.A.N. Greenwood (1992)

Vol. 2: J.D. Kalma, G.P. Laughlin, J.M. Caprio, P.J.C. Hamer
The Bioclimatology of Frost.
Its Occurrence, Impact and Protection (1992)

Vol. 3: Y. Cohen, J.M. Elwood, M.G. Holmes, V.A. Kanevski, S.M. Kochubey,
J. Ross, T. Shadchina, E. Simensen, F.I. Woodward (1994)

Vol. 4: D.J. Beerling, P.W. Brown, K.G. Hubbard, J.D. Kalma, S.J. Meyer,
M. Nunez, O.D. Sirotenko, R.L. Snyder, G. Stange, F.J. Woodward (1996)

Advances in Bioclimatology_4

With Contributions by

D.J. Beerling P.W. Brown K.G. Hubbard
J.D. Kalma S.J. Meyer M. Nunez O.D. Sirotenko
R.L. Snyder G. Stange F.J. Woodward

With 51 Figures

Springer

Edited by
Dr. GERALD STANHILL
Agricultural Research Organization
The Volcani Center
Institute of Soils and Water
P.O. Box 6
Bet Dagan, 50250
Israel

ISBN-13: 978-3-642-64709-3 e-ISBN-13: 978-3-642-61132-2
DOI: 10.1007/978-3-642-61132-2

Softcover reprint of the hardcover 1st edition 1996

Typesetting: Best-set Typesetter Ltd., Hong Kong
SPIN 10508319 31/3137 – 5 4 3 2 1 0 – Printed on acid-free paper

Preface to the Series

Advances in Bioclimatology – the study of the relations between the physical environment and the form and function of living organisms – have been spectacular during the last third of this century.

Before this period, the subject, having slowly emerged from its classical origins as a branch of natural history, had reached the stage of a collection of largely empirical, statistical relationships between standardized but often inappropriate climatological and biological measurements.

Since then, research into the basic physical and physiological mechanisms involved has used the latest techniques of measurement and analysis to develop various bioclimatic relations which have contributed much to improving crop and animal production and optimizing the human environment.

Recently, some of these relationships have been incorporated into larger models of climate-ecosystem interactions. Such models are being used to assess the often unintended effects of human activity on various elements of the biosphere.

However, the advances described have been very unevenly spread through the vast field of interest encompassed by bioclimatology; the fields of plant, animal and human climatology have largely advanced in independent fashions and even within each biological province different techniques of analysis and measurement have developed for different time and space scales of organization.

One of the major aims of this new review series is to overcome this separate development by providing a common forum for those wishing to obtain an authoritative review of the latest developments in bioclimatology.

The emphasis will be on advances which are soundly based on biological and physical principles rather than those describing empirical relationships. Reviews will also deal with the latest techniques of measurement and analysis which are of relevance to bioclimatology and to those describing broader ecological studies in which bioclimatology provides a major element.

Although most of the reviews to be published will be commissioned, the editors would welcome suggestions from individuals interested in contributing a review of the type described, as well as for ideas on major topics of wide interest around which a number of individual reviews could be centered.

Bet Dagan, Israel

G. Stanhill
Editor

List of Editors

Those interested in contributing a review to this series are invited to contact Dr. Gerald Stanhill or one of the Associate Editors. Proposals should outline briefly the review's aims and scope.

Contents

A Guide to Automated Weather Station Networks in North America

R.L. SNYDER[1], P.W. BROWN[2], K.G. HUBBARD[3], and S.J. MEYER[3]

1 Introduction

Automated weather station (AWS) networks are increasingly being used to gather weather data for agricultural and other bioclimatic applications. Many manually read conventional weather station (CWS) networks are being replaced with AWS networks. The use of AWS networks increased rapidly during the 1980s because of improvements in battery-powered dataloggers and computer communications. Although CWS networks continue to provide the basic climate information, AWS networks have the ability to gather and disseminate greater quantities of data at more frequent intervals than CWS networks.

1.1 Advantages and Disadvantages

Since all data gathering, transfer, and archiving in AWS networks are done through electronic means, the processes are faster and generally more accurate than for CWS networks. This is a definite advantage for applications such as freeze, fire, and flood forecasting. The timeliness of data is important, although some lag time can be tolerated, for estimating daily potential evapotranspiration (ET_p) or growing degree days (°D). However, an AWS network that provides timely, reliable ET_p and °D information has promotional advantages for agencies attempting to encourage the adoption and use of the information.

A major advantage offered by AWS networks is that the more frequent gathering of data has the potential to improve estimates of ET_p, pest, and crop development, etc. However, the ability to efficiently gather large amounts of weather data has improved much faster than the knowledge on how to convert the said data into usable information. Current weather driven models (e.g.,

[1] University of California, Department of Land, Air and Water Resources, Davis, California 95616, USA
[2] University of Arizona, College of Agriculture, Tucson, Arizona 85721, USA
[3] High Plains Climate Center, Department of Agricultural Meteorology, University of Nebraska-Lincoln, Lincoln, Nebraska 68583-0728, USA

calculation of ET_p and °D) are mainly based on daily data and assumptions about the diurnal trends in radiation, temperature, humidity, etc. Using measured rather than estimated diurnal weather data should provide more accurate estimates of ET_p and °D. Also, the addition of new parameters (e.g., hourly wind speed, wind direction, leaf wetness, etc.) may further improve other application models. As scientists develop better weather-driven models, agricultural and environmental production, protection, and conservation will also improve.

Another advantage of AWS over CWS networks is that digital data transfer from an AWS to a central computer eliminates human errors that sometimes occur in CWS networks. Because data transfer is electronic, the testing of incoming data quality is considerably faster and easier than for CWS networks. A major advantage of AWS networks is that labor is normally not needed on weekends.

The choice of which type of network to develop depends on many factors including (1) the network purpose, (2) the proximity of weather stations, (3) the availability of trained employees and employee time, and (4) short- and long-term funding.

1.2 Types of AWS Networks

AWS networks generally fall into one of three categories including (1) real-time data or event-recording networks, (2) climatic data networks, and (3) agricultural (near-real-time data) networks. The type of network to establish depends on the purpose for which the data are required.

The most common type of AWS network is the real-time data or event-recording network. These networks are developed to provide warnings for the effects of severe weather such as (1) freezing temperatures, (2) floods, (3) fire conditions, and (4) avalanches. Real-time networks typically recover information via radio telemetry or satellite data transfer.

In climatic data networks, real-time or near-real-time data recovery is unnecessary and data recovery by cassette tape or solid-state memory is often sufficient. The purpose of a climatic data network is to gather the data needed to characterize a climatic region. Therefore, weather stations are designed to gather a standard set of data from sites that are typical of the climate being assessed. For example, in a desert climate, the station should be located over a typical surface with sparse vegetation.

The third type of AWS network is the agricultural network requiring dissemination of data the day after recovery. Data are collected typically by telephone modem, radio telemetry, infrared telemetry, or satellite transfer. Using AWS rather than CWS networks facilitates data recovery and dissemination to users. Agricultural data are most often used for evapotranspiration estimation and pest management, so locating stations in areas representative of agriculture is an important concern.

1.3 Purpose of This Review

Considerable growth in the use of AWS networks has occurred, but usage is still small on a worldwide basis. Several networks in the USA and Canada have successfully operated for 10 to 15 years, and many more are currently in the planning phase. Establishing and operating an AWS network is a costly and time-consuming process and many pitfalls can be avoided by using the experience of others.

This review focuses on this experience, documenting the planning, component and operation of AWS networks in North America, mainly for agricultural applications. Since some forecasting networks and climatic networks are also used for agricultural applications, information and input from these multipurpose networks are also discussed although the review is in the main addressed to the non-climatologist reader.

The authors have considerable experience with the California Irrigation Management Information System (CIMIS), the Arizona Meteorological Network (AZMET), and the Automated Weather Data Network (AWDN) for the High Plains (Nebraska and surrounding States). Experiences with these USA networks will be frequently mentioned in this review. All references to costs will be given in USA 1992 dollars ($) unless otherwise indicated.

2 Planning

During the first phase of designing an AWS network, planners should prepare a statement of the need and the major use envisioned for data collected from the network. The intended use of the data will help, in later phases of planning, to quantify the desired accuracy of measurements and the required spatial density of network sites.

One should be as specific as possible in outlining the need for a network. For instance, if the network is being established to provide estimates of ET_p using a Penman-type equation then it is important to consider the effect of measurement accuracy on the resulting output. After selecting the magnitude of error that can be tolerated in the ET_p estimate, a sensitivity analysis (Meyer et al. 1989) can be used to assign limits to the measurement error for, in this case, solar radiation, temperature, wind speed, and humidity.

2.1 Spatial Density

The optimum spatial density of automated weather stations (AWSs) in a network can be determined by examining the variability associated with the quantities being measured and comparing this to a geographic or spatial monitoring goal. The goal might be to represent the spatial, statistical moment (s) (for example an area average) for quantities of interest over a specific region. Supposedly, if one were to start with only one station in the region and continue adding stations, eventually one would reach a point where the calculated average temperature over the region would not change significantly due to the addition of more stations. We can define the minimum station density as the number of stations needed to produce a spatial average that is not significantly different from the average that would be obtained from an unlimited number of stations in that region. Practically speaking, one never encounters a situation where station numbers are not limited; therefore, estimates of the number of stations required to adequately sample for a statistical moment must be inferred by using data from available stations and such techniques as Monte Carlo simulation. Some statistical procedures for examining network representativeness have been described (Gandin 1970). If the region of interest does not have enough conventional weather stations to form the basis of a network density study, then the planner is forced to estimate needs using data from other existing networks. A network from an area with similar climate and topography offers the best opportunity of obtaining reasonable and useful information. A higher station density is required to stabilize the desired moment for short time scales, such as an hour, than is required for longer time scales, such as a day.

Often, weather variables must be estimated for points within the network region where monitoring stations do not exist. Some work has been done to determine the relationship between the accuracy of the estimate and the distance from the surrounding stations (Hubbard 1994), but the relationships produced are specific to the area studied and cannot be extrapolated to other climates. These relationships cannot be derived until a network already exists in the region. Planners who use the results of other studies as a guide to determining station density should definitely examine the spatial resolution of their own network once it has been established.

The spatial variation of some weather variables is greater than for others. Based on analysis of climate data from the High Plains of the USA (Hubbard 1994), a 60-km spacing is required to explain greater than 90% of the variation between sites for maximum daily temperature data. About 30-km spacing is needed for minimum temperature, relative humidity, solar radiation, and ET_p. Spacing for soil temperature is 20 km and for wind it is 10 km. Precipitation requires 5 km spacing. Spacing requirements may also vary with the analysis region and the time of year.

A recent paper (Hubbard 1994) discusses the use of the coefficient of variation (r^2) between stations to develop variograms, variomaps, and annual variograms for illustrating the reliability of data from a weather network. The methodology is useful for determining the spacing of stations needed to characterize selected weather parameters and for identifying the area represented by individual AWS stations.

2.2 Siting Criteria

Siting criteria may differ depending on the network purpose, e.g., ET_p estimation, freeze forecasts, and pest management. For instance, if the data are used exclusively for one crop, the station(s) probably should be located inside fields of that crop. A general network with all stations located over the same type of common underlying surface serves to standardize the network.

The distribution of stations over a region depends to a large extent on the overall goal of the network. For example, the goal of the AWDN is to characterize the macroclimate of the High Plains; therefore, each AWS is located in a rain-fed pasture away from areas such as irrigated fields, which pose microclimatic problems (Hubbard et al. 1987). In contrast, the CIMIS program in California chose to place AWSs over irrigated lawns and pastures because those surfaces most closely match the grass reference used to calibrate the CIMIS version of the Penman equation (Snyder et al. 1985a; Snyder and Pruitt 1992).

2.2.1 Underlying Surface

The underlying surface in the AWDN network is rain-fed grass. The underlying surface for stations in CIMIS and AZMET is irrigated grass. Site qualifications for AWDN are listed in Appendix A. In all ecosystems (forests, deserts, etc.), it is important to choose a reference surface that is representative of the area. Regardless of the reference surface chosen, fetch (upwind distance of uniform surface) is an important consideration. There should be sufficient fetch in any direction that air passing by the station will be in equilibrium with the underlying surface. For climate studies in arid lands, it is not recommended that grass be irrigated because irrigation will change the water balance and in turn the energy balance, thereby altering the microclimate. Even so, if the goal is to estimate ET_p in an arid environment, an irrigated surface may be best if adequate irrigated fetch is provided to establish equilibrium after the dry-to-wet transition zone.

2.2.2 Suitable Surroundings

The general recommendation for siting agricultural weather stations is to locate the station in an open, relatively level location representative of the local agricultural environment (Newman et al. 1959; Doorenbos 1976). Locating the station a considerable distance from major landscape obstructions (e.g., buildings and trees) that would disrupt wind flow and/or shade the area is of primary importance. A horizontal distance of two to three times the height of small obstacles (e.g., small shrubs) and seven to ten times the height of large obstacles (e.g., large buildings and trees is recommended (Newman et al.

1959). Placement near roads should be avoided to minimize problems with dust and vandalism, and the site should be fenced to prevent large animals from coming in contact with equipment. The site should be at least 10 × 10 m and preferably 50 × 50 m (Doorenbos 1976). A well-maintained grass surface has been recommended for the site (Newman et al. 1959; Doorenbos 1976).

The above siting recommendations have come under close scrutiny in recent years, particularly in arid and semiarid regions where well-watered crops create microclimatic conditions significantly different from those observed in natural areas. The recommended area for an AWS site is insufficient to prevent the influence of surfaces beyond the site boundary (Brown et al. 1990). In many locations, these surfaces undergo pronounced seasonal changes (e.g., bare ground to full canopy). The standard fetch to instrument-height ratio of 100 : 1 would mandate AWS sites of 300 × 300 m for instruments mounted 1.5 m above the surface and even larger sites for higher sensor exposures. Large pasture areas would therefore seem best for AWS sites. Alfalfa has been suggested as an alternative surface for AWSs in arid regions where grass pastures are not routinely available. However, the cycle of cutting and regrowth as well as crop rotation an effect the resulting data (Brown et al. 1990).

Avoid potential AWS sites with high levels of electromagnetic radiation, e.g., microwave, radiowave, or television towers or power lines as these can affect the performance of sensors and other electronic equipment.

2.3 Datalogger Selection

The datalogger is chosen based on a number of general criteria including cost, reliability, and maintenance, while other more specific criteria such as accuracy, sensor compatibility, data storage and retrieval options, operating temperature range, and power supply should also be examined (Tanner 1990). As some of these criteria may be difficult to ascertain, demonstrations by the manufacturer should be arranged and other users of the equipment contacted.

Each datalogger type has advantages and disadvantages. For this reason our purpose here is to present information on dataloggers that are currently available and relevant to AWS networks and to provide information on their capabilities. Variables often measured by AWSs include those presented in Table 1.

Frequent sensor readings are recommended. If maximum wind speed gusts are to be recorded, a maximum sampling interval of 10 s or less is recommended to avoid underestimation. Generally, hourly samples, averages, or totals of the readings are stored for immediate or later transfer to a computer. Using hourly data greatly improves our ability to understand agricultural processes such as pest growth and development, crop growth and development, and ET_p. Storage of more frequent data seems unnecessary since models using these date are not readily available.

Table 1. Frequently measured AWS parameters

1.	Temperature
2.	Humidity
3.	Solar radiation
4.	Photosynthetically active radiation
5.	Leaf wetness
6.	Wind speed
7.	Wind direction
8.	Rainfall
9.	Soil temperature
10.	Pan evaporation

Hourly averages produce a statistically representative value of the variable. Humidity expression such as vapor pressure or dew-point temperature can be calculated in the datalogger from temperature and relative humidity samples and the hourly average values of those expressions recorded. Often total daily accumulation is needed for precipitation and irradiance. Selection of the maximum or minimum value during a time interval can be used for some purposes, e.g., maximum wind speed gust, air temperatures.

2.4 Sensor Selection

Prior to selecting a sensor or datalogger it is advisable to look at performance characteristics in three broad categories. Accuracy, and static characteristics describe the performance of the sensor in one or more steady-state environments, whereas dynamic characteristics describe sensor performance in a changing environment. The sensors necessary to measure the variables listed in Table 1 are offered by numerous manufacturers (Appendix B is a partial listing).

One critical performance characteristic to assess is the resolution of the sensors and the dataloggers. Here, resolution is defined as the smallest change in the environmental variable that will result in a detectable change in the electronic signal. The effective resolution of the sensors and the datalogger used in combination will be the poorest (largest) of the two resolutions.

The range of the sensor – the upper and lower limit of detection – should encompass all conditions that are anticipated in the region of the network. Readers should refer to texts for more discussion of performance characteristics (e.g., Schwerdtfeger 1976; Fritchen and Gay 1979; Wang and Felton 1983 and Brock 1984) and more discussion of instrumentation. As the more accurate sensors usually cost more to acquire, to some extent the monitoring goal of the network will determine its overall cost.

The American Association of State Climatologists (1985) has published a resolution giving general guidance on the accuracy of sensors and the conditions of their installation to standardize measurements for climate analyses. This resolution is shown in Appendix C.

2.5 Computer Selection for the Data Processing Center (DPC)

Rapidly changing computer technology opens many options to an AWS network planner. Software and data-storage capability are two factors determining the choice for a central computer. The software that allows communication with the AWS and the software for the management of the data base and subsequent analysis of the weather variables are of equal importance. Storage media for meeting the requirements of the weather data base can include magnetic tape, hard disk, floppy disk, WORM (write once read many times) devices, etc. One guideline for determining total storage capacity is to allow enough storage so that at least 1 year of hourly data can be on-line. It is critical that the operation of a network includes plans to backup the system on a frequent basis.

2.6 Data Retrieval Method Selection

A true AWS network requires a DPC that receives data from the stations via transfer by telephone, radio frequency (RF) communication (surface or satellite), or direct wire connection. A modem at the AWS site and at the DPC is required for telephone communication. Initial costs are usually low but telephone charges and long-distance charges are ongoing. With RF communication a user can own all the RF transmitters, receivers, and repeaters. However, stations must have line of sight for surface RF communication and generally must be no farther than about 25 km apart for transmitters rated below 10 W. The stations can be either self-timed (transmitting data on a fixed interval) or interrogated by the central computer. The development of satellite communications in the private sector and of new equipment (transmitters, etc.) for use with the weather station dataloggers make this a promising option for reliable, inexpensive, communication. Satellite RF communication does not require surface line of sight and the additional costs of the equipment can usually be justified at remote locations where conventional telecommunications are not possible. An important consideration is that communication via RF communication is only in the direction from the AWS to the DPC. When telephone communication is used, remote programming of the AWS is possible. The ability to remotely reset the AWS clock and to reprogram can save considerable time and money over the long term.

Often, AWS manufacturers and distributors offer software packages to retrieve data for various telecommunications methods and such packages are usually cost-effective. However, the vendor should be asked to provide demonstrations of the software being used. In some cases,specific hardware may be necessary to use the commercial data retrieval software and the data must be passed to a separate computer. If this is the case, software must be available to transfer between the data retrieval hardware and the main computer.

For AWS networks, data retrieval software should permit automatic, regular interrogation of the AWSs. If contact with the AWS is not accomplished on the first attempt, the software should automatically attempt to call again. The computer should alert the operator that a data transfer problem exists if the data retrieval cannot be accomplished in two or three attempts.

The data retrieval software should automatically retrieve data from an AWS beginning with the time period following the last set of transferred data. After retrieval, the software should perform a quality control (QC) test, calculate additional variables, and archive the data.

2.7 Network Management

A network established and maintained by only one agency has a desirable feature – involved parties are all responsible to the same organization, thus keeping the number of interfaces involved with administration of the network to a minimum. Likewise, the monitoring goal may be much narrower if a single agency is involved. The temptation to cut measurements and data to "the bone" because a narrow monitoring goal has been specified must be weighed against any needs that may develop later to compare data from the network to data from other networks. A single agency having a nationwide mission can provide the type of wide-area network necessary for homogenous measurements and this in turn allows for the possibility of comparing weather impacts from state to state. If an agency or group is limited geographically to a smaller area, such as a county or township, then the possibility of combining with other groups and agencies to form a combined network should be explored.

Small groups who work together to form a network can increase the effectiveness of the available funding. By pooling their maintenance, calibration, data dissemination, and other activities a multiple-agency network can reduce the technical expertise, sensor inventory, calibration facilities, etc. needed. This decided financial advantage has convinced some agency personnel of the need for cooperation and, in some cases, has resulted in a network where operational expenses are shared among the interested parties. As the administration of a network becomes more challenging as the number of involved parties increases it is probably best in such cases to strike written agreements that spell out the responsibilities and duties of the parties involved so that the agreement provides continuity for the network as an agency's personnel change.

2.8 Cost of Establishment

Establishing a network requires an initial investment to procure the equipment. The cost of the equipment including sensors, dataloggers, computers, etc. will depend upon the associated monitoring goals. Other costs such as installation of phone lines at AWS sites (if needed), fencing for AWS sites, and leasing of site property will vary with local conditions.

Capital costs associated with developing an AWS network largely depend on network size, types of sensors employed at the AWSs, and procedures used to retrieve data. The purchase price for most AWSs run between $4000 and $8000, in 1992 dollars. Sensor quality and the cost of data retrieval equipment account for the wide price variation. In Arizona, the purchase price of an AWS configured for phone modem data retrieval and equipped with sensors for measuring air and soil temperature, relative humidity, wind speed and direction, solar radiation, and precipitation is approximately $4000. In contrast, the AgriMet AWS (used in the USA Pacific Northwest), which is configured for satellite retrieval and equipped with a slightly different array of sensors relative to Arizona, cost about $8000 (A. Powers, pers. comm.).

Additional funds must be budgeted for site preparation (e.g., irrigation, phone, fence, grass) and AWS installation. These costs vary with the remoteness of the site and the type of station installed. Networks using RF systems for data retrieval must consider the additional capital costs of purchasing and installing repeater stations.

2.9 Operating Costs

Labor is generally the largest operating expense, and often consumes more than 75% of the total operating budget. Networks consisting of more than ten AWSs typically have a minimum of two employees.

Non-labor operating expenditure depends on the type of telecommunications employed for data retrieval, the distance between the DPC and AWSs, and the frequency of maintenance. For example, networks using telephone data retrieval find phone charges to be the next largest operating cost after labor. Networks using ground or satellite RF telemetry list travel and replacement parts as the next largest cost category. Annual expenditures for repair and replacement parts such as batteries, bearings, rheostats, etc. are predictable. Catastrophic events such as lightning strikes or vandalism can cause

Table 2. Non-labor annual operating cost (1991) of an AWS in the AZMET program of Arizona and AWDN of Nebraska

Operation	AZMET	AWDN
Telecommunications (phone)	$ 640	$ 480
Travel (mileage and lodging)	350	240
Replacement parts and repair	360	350
Computer supplies (DPC)	80	50
Total	$1430	$1120

major disruption of the repair and replacement budget. Most networks report non-labor operating costs for an AWS in the range of \$1000 to \$2000 per year (Table 2).

3 Network Components

3.1 Sensors

Several factors must be considered when selecting sensors for use in AWS networks. The most important is to select sensors that are compatible with the AWS datalogger. Other factors to consider include accuracy, resolution, and response time of the sensor. Accuracy refers to how closely the sensor will measure the environmental parameter in question; resolution refers to the smallest change in the parameter that can be detected by the sensor; and response time indicates how quickly the sensor responds to a changing environment. Sensor cost generally increases with improvements in accuracy, resolution, and response time. Therefore, careful assessment of these requirements should precede the purchase of both sensors and dataloggers. Most applications that make use of AWS data do not require sensors with exceptional resolution nor fast response times. The measurement of air temperature (T_a) provides a good example.

Most AWS networks measure T_a with a thermistor probe placed in a radiation shield that is aspirated by natural wind movement. Radiation shields are rated no better than ±0.5 °C (Gill 1983), so it makes little sense to select a temperature sensor that can resolve to 0.01 °C. Similarly, many AWS dataloggers scan sensors every 60 s and store data as hourly averages and totals. In such circumstances, expensive, fast response sensors are unnecessary.

Durability is equally important when selecting AWS sensors. Sensors constructed from corrodible metals or materials sensitive to ultraviolet light should be avoided. Poor sensor durability will increase maintenance and equipment costs.

3.1.1 Placement Height

Recommendations for sensor placement height have been suggested and published by several organizations, but a standard set has not been accepted by agricultural meteorologists. Complete conformity with respect to sensor placement is unlikely because many AWSs are installed for a specific application which may dictate sensor placement. Where possible, however, an effort should be made to maintain uniform sensor installations.

The United Nations Food and Agriculture Organization (FAO), Doorenbos 1976) and the AASC (American Association of State Climatologists

1985) have published recommendations for sensor placement (Table 3). The FAO recommendations are for agricultural weather stations, while the AASC recommendations are for climatic stations. WMO (1983) also lists recommended heights for sensors. The major discrepancy among the recommendations rests in the heights of anemometers and rain gauges. The FAO recommendation for anemometers results from widespread use of the Penman equation for estimating ET_p (Penman 1948) which requires wind runs at a height of 2 m. The 3-m recommendation from the AASC has no real historical precedence. According to the AASC, the 3-m height represents "a reasonable compromise between the expense of setting up the tall towers required for higher heights and the problems of exposure that increase at lower heights." The exposure standard for anemometers may be of little consequence in applications such as the Penman equation. Bausch (1990) found that wind speed (U) measured at 3 and 2 m differed by about 9%, and that U obtained at 3 m could be adjusted to within 5% of wind speed measured at 2 m using the standard power function.

The height of a rain gauge affects the total catch of precipitation (PPT). Gauges placed at higher elevations are subjected to higher wind speeds which produce a greater deformity of airflow around the gauge. Surrounding terrain and landscape features also affect airflow. FAO recommends a rain-gauge height of 0.3 m, whereas AASC recommends 1 m. The general recommendation from the World Meteorological Organization (WMO 1983) is to place the gauge at a height that eliminates the possibility of splash-in from the surrounding area. Also, potential for snow drifting should be considered when selecting a gauge height.

It is unlikely that the discrepancy in recommended rain-gauge height between the FAO and the AASC will make a large (i.e., >10%) difference in measured PPT. The AASC recommendation does minimize problems with splash-in and snow drifting and is probably the better general recommenda-

Table 3. Exposure heights for instrumentation used to measure environmental parameters for use in agriculture as recommended by the United Nation Food and Agriculture Organization (FAO) and the American Association of State Climatologists (AASC)

Parameter	FAO	AASC
Air temperature	1.5–2 m	1.5 m
Relative humidity	1.5 m	1.5 m
Soil temperature	5, 10, 20, 50, 100 cm	10 cm
Wind speed	2 m	3 m[a]
Precipitation	30 cm[b]	1 m[b]
Wind direction	2 m	3 m
Solar radiation	2 m	

[a] 2 m acceptable for specific applications such as estimation of ET_p.
[b] Higher levels acceptable in areas with significant snowfall.

tion. Perhaps of greater importance, however, is ensuring that all rain gauges within a network are installed at a common height.

The FAO recommends a temperature (T_a) sensor height in the range of 1.5 to 2 m. Bausch (1990) reported that T_a at 1.5 and 2 m can differ significantly. He found that daily maximum and minimum T_a at the two levels differed by as much as 1.8 and 2.4 °C, respectively. Differences of this magnitude are significant for climatological as well as certain agricultural purposes, such as frost and °D models, and should be considered when selecting the exposure height of the thermometer.

Tanner (1990) suggests that temperature, humidity, and wind sensors all be placed at some common level since AWS data are frequently used to model heat and vapor fluxes. Such models generally require wind, temperature, and humidity data from the same level. WMO recommends the placement of temperature, humidity, and wind sensors at 2-m height for agricultural stations.

3.1.2 Sensor Fundamentals

The following section provides a general overview of sensor fundamentals. More extensive discussions on sensors are given in Fritschen and Gay (1979), Rosenberg et al. (1983), Tanner (1990), and WMO (1983).

3.1.2.1 Temperature

Thermistors and thermocouples are the most commonly used temperature sensors in AWS networks. Both are relatively inexpensive, sufficiently accurate, and easy to monitor by remote battery-powered dataloggers. Acceptable T_a data are provided by either sensor if properly designed, monitored, and shielded from radiation. The theory and use of electronic temperature sensors are discussed by Fritschen and Gay (1979) and Rosenberg et al. (1983). Special considerations regarding the use of thermistors and thermocouples with dataloggers are explained by Tanner (1990).

Thermistors are materials whose electrical resistance decreases with an increase in temperature in a non-linear manner. However, this relationship is linearized numerically in modern dataloggers.

Thermocouples are formed when wires of dissimilar metals are connected together to form two junctions – the measurement junction and the reference junction. A temperature difference between the measurement and reference junctions produces a voltage in the thermocouple circuit, the magnitude of which varies in a nearly linear fashion over the environmental temperature range.

Many different types of thermocouples are available, the choice depends on the application. Tanner (1990) notes that many dataloggers measure the temperature of the input terminals to obtain the reference temperature, and he cautions that temperature gradients across the input terminals should be minimized to reduce errors. Tanner suggests that chromel-constantan (type E) thermocouples have an advantage over copper-constantan (type T) thermocouples because their lower thermal conductivity reduces errors due to heat conduction along the wire.

3.1.2.2 *Humidity*

The water vapor content of air is sometimes measured using a chilled-mirror dew-point temperature (T_d) hygrometer. The sensor works well for remote T_d measurements if it employs a low-voltage, direct-current power source. The current drain (3 A) for a hygrometer is great enough for a solar panel and rechargeable power supply to be recommended. Chilled-mirror hygrometers are accurate when properly maintained; however, they are expensive and frequent visits to clean dust from the mirror may be needed. Consequently, chilled-mirror hygrometers are not widely used in AWS networks.

Relative humidity (RH) sensors are the most common instrument for measuring humidity in AWS networks. RH sensors are less expensive than dew-point hygrometers, require less electrical power, and provide sufficiently accurate measurements when properly maintained. Filters are commonly included with these sensors to protect them from pollutants. RH sensors, like temperature sensors, must be shielded from radiation. Most RH sensors absorb faster than they desorb water vapor, producing a tendency to overestimate RH as the air dries. An evaluation of 20 different humidity sensors is provided by Skaar et al. (1989).

3.1.2.3 *Shields for Temperature and Humidity Sensors*

Thermocouples and thermistors measure the temperature of their environment, not T_a. To ensure the measured temperature is close to T_a, the sensor must be shielded from gains or losses in heat due to radiation. Some radiation shields may impede air flow and cause air to stagnate within the shield. This causes differences between T_a inside the shield (near the sensor) and the external ambient T_a. Stagnant air within a shield can also slow the response time of the sensor to changes in the ambient T_a. Either a Stevenson Screen, or a Gill-type electronic sensor shield (Gill 1983) or the standard Cotton Region shelter, will minimize radiation-induced temperature errors when wind speeds exceed approximately 2 m s^{-1}. Forced ventilation of temperature sensors may be required in regions with high radiant load and low wind speeds.

3.1.2.4 *Radiation*

Three types of radiation sensors are commonly used on AWSs. Pyranometers measure total global solar radiation (R_s), net radiometers measure radiation balance or net radiation (R_n) (the net amount of short and long-wave radiation absorbed at the surface) and quantum sensors measure photosynthetically active radiation (PAR). Depending on the purpose of the network, one or more of these sensors may be used.

Thermopile pyranometers and silicon photocells are commonly used to measure R_s. High-quality thermopile pyranometers typically measure in the 200 to 4500 nm waveband, nearly equivalent to the solar waveband (150 to 4000 nm). Silicon photocells typically measure only in the 400 to 1100 nm waveband and require calibration to estimate the total solar irradiance.

Thermopile pyranometers measure the difference in temperature between a black radiation-absorbing surface and a reference surface using a series of thermoelectric junctions. Radiation received is proportional to the temperature difference of the two surfaces. Black and reference surfaces are enclosed by a glass dome(s) which uniformly transmits shortwave radiation but protects the absorbing and reference surfaces from convective and latent heat exchange, i.e., wind, rain, and dew. Pyranometer models that meet the WMO standards for signal linearity, temperature compensated outputs, stability, and cosine response (WMO 1983) should be selected.

Thermopile pyranometers are temperature-dependent and computational or electronic corrections are required to correct for this error. Thermopile pyranometers can measure surface albedo if mounted upside down.

Silicon photocell pyranometers generate an electrical current that is proportional to the shortwave radiation impinging on the sensor. Often a resistor is placed in series with the silicon cell and the sensor output current is measured as a voltage drop across the resistor. Silicon pyranometers are calibrated against first-class thermopile pyranometers and are accurate only when used in the same exposure for which they were calibrated. They should not be used to measure light penetration through plant canopies or reflected radiation (albedo) because the resulting transmitted or reflected radiation differs spectrally from incident radiation, invalidating the calibration.

R_n is measured with a net radiometer in situations where ET_p or freeze forecasts are needed. A net radiometer uses an upper and lower thermopile consisting of a series of thermocouples imbedded within a black plate. Two hemispheric plastic domes are used to encase the thermopile. The domes transmit radiation in both the solar and terrestrial wavelengths 300–60 000 nm while protecting the thermopile from ventilation, wetting, and contaminants. The temperature difference between the upper and lower surfaces of the radiometer determines the thermopile electrical signal which is linearly related to R_n.

R_n measurements are affected by the surface under the instrument. When used to estimate ET_p, the instrument should be mounted over a surface cover of thick, green, growing grass. The grass area must be large enough such that it alone is viewed by the radiometer. Setting the sensor 1 m above the grass gives acceptable results.

Net radiometers give incorrect readings if condensation forms on the inside or outside surface of the domes. A small air pump can be used to blow air through a desiccant tube into the net radiometer to prevent condensation on the inside. Battery-powered aquarium pumps switched on for a short interval every 5 to 10 min work well as an intermittent aspirator. Silica gel is often used as the desiccant.

Condensation on the outside of the net radiometer domes can be prevented by aspirating the outside of the domes. This requires continuous fan operation and a source of power, and is usually not possible in remote locations. In many areas, the frequency of condensation on exposed surfaces is much higher at night, leading to measurement problems in early morning.

Net radiometers require frequent inspection and repair. Domes are often damaged by birds and other animals and the domes will yellow and crack with age. The black plates also discolor with time. Even when not discolored, the polyethylene domes need frequent replacement. Ultraviolet light decreases the transmissivity of the domes to long-waveband more than short-waveband radiation, so the calibration can be correct for short but not for long-waveband radiation. It is best to return the radiometer to the manufacturer for frequent repairs and recalibration. Calibration issues are addressed in Field et al. (1992). A spare net radiometer to use for testing AWS net radiometers is reommended.

PAR sensors are used to measure or estimate the amount of energy available to fix carbon dioxide in photosynthesis. Only radiation in the 400–700 nm waveband is effective for photosynthesis and this range comprises only 40–50% of the solar waveband. One mole of photons in the PAR waveband corresponds to approximately 0.172 MJ of energy (Rosenberg et al. 1983). Calibrated silicon sensors are typically used to estimate PAR. The silicon cell is filtered to match the photosynthetic response of a leaf.

3.1.2.5 Soil Heat Flux Density

Soil heat flux plates are used to measure the soil heat flux density (G) needed for energy balance calculations. Heat flux plates use a differential thermopile (like net radiometers) to measure the difference in temperature between the upper and lower surfaces of a plate. The plate must have a low heat capacity to ensure a rapid response to changes in G. The biggest problem is water-phase changes on either surface of the plate that lead to measurement errors.

When soil heat flux plates are placed in good contact with the soil they provide reliable results. Inserting a series of sensors at several locations helps to account for soil variability and the spatial variability of heat flux attributable to differential shading of the soil surface.

Heat flux plates ar buried sufficiently deep to minimize the region of vapor transfer and heat divergence and convergence. In cracking soils, plates too near the surface may be exposed to direct radiation as the soil dries. Tanner (1990) recommends placing heat flux plates at between 5 and 10 cm below the soil surface, and measuring soil temperature midway between plate depth and the soil surface to estimate the heat flux in the top layer of soil.

3.1.2.6 *Wind Speed*

Rotating horizontal-cup anemometers or propeller-type aerovanes are typically used to measure wind speed in AWS networks, providing sufficient accuracy for most agricultural applications and being more durable and stable than other sensors. Aerovanes point into the wind, so they measure the wind component that parallels the axis of propeller rotation.

The rotation of a mechanical anemometer is transformed into an electrical signal read by a datalogger. For photochopped anemometers, external power is required to power a light beam that is chopped by a rotating slotted disk. Other anemometers do not require external power, relying on contact closure or direct current generators to generate an electrical signal. Any mechanical anemometer meeting WMO standards will provide acceptable data for agricultural applications.

Wind speed is directly proportional to the turning rate of most rotational anemometers, but the linear regression of wind speed vs. turning rate does not pass through the origin (Baynton 1976). The regression is needed to obtain accuracy at high wind speeds, but a small wind speed equal to the regression offset will be recorded when the cups are not turning. A desirable datalogger feature is the capability to zero the wind speed when the cups are not turning. This is especially useful if statistics on the wind speed and direction will be computed by the datalogger.

3.1.2.7 *Wind Direction*

Wind vanes are used to measure wind direction and come in two basic types: (1) those combined with propeller anemometers and (2) stand-alone vanes. Wind vanes require an excitation voltage to measure the resistance of a circular potentiometer. Wind-direction instruments are usually programmed to record the wind direction on a standard compass scale of 0° to 360°, where 0° is true north.

3.1.2.8 *Precipitation*

Precipitation (rainfall) sensors (or rain gauges) are often included in AWS networks to measure site-specific precipitation amounts. Tipping bucket rain gauges are typically used because they are simple, inexpensive, and measure both rainfall duration and intensity. Alternatively, weighing rain gauges have been used which are considerably more expensive, but allow the measurement of snow when antifreeze is added to the bucket.

The size and shape of rain-gauge orifices and funnels can affect accuracy. These factors are discussed by Tanner (1990). Rain gauges require regular calibration to ensure accuracy. Manufacturer calibrations are sometimes inaccurate and should always be checked following installation.

3.1.2.9 Evaporation Pans

Evaporation pans are sometimes employed in AWS networks. A standard Class "A" evaporation pan with an automatic refilling and reading device has been employed in the USA Department of Agriculture weather stations included in the CIMIS network in California. The automatic evaporation pan depth recorder and refilling device is described by Phene and Campbell (1975). The device is programmed to measure water depth on a regular basis and automatically maintains the water level within 25 to 50 mm below the pan rim. The estimation of ET_0 (reference evapotranspiration) with evaporation pan, fetch, humidity, and wind speed data are discussed in Doorenbos and Pruitt (1977). Equations to estimate ET_0 from pan and weather data are given in Jensen et al. (1990).

3.2 AWS Dataloggers

Battery-powered dataloggers are typically used for AWS agricultural networks due to the unavailability of alternating current (AC) power. Two basic types of dataloggers are used in AWS networks, the general purpose datalogger and the turn-key datalogger. General purpose dataloggers are used for a wide array of datalogging needs and therefore offer a great degree of flexibility to the user. Most general purpose dataloggers offer extensive on-site processing and programming capabilities, and provide a wide choice of signal conditioning options. Consequently, general purpose dataloggers can accommodate most new sensors with just a programming change.

Turn-key dataloggers are configured in the factory using specifications supplied by the end user. These systems are set up for a specific array of sensors and specific on-site processing needs. Turn-key dataloggers offer simplicity and generally require less programming on the part of the user. The weakness of turn-key systems is the loss of flexibility after purchase. Often the dataloggers must be returned to the factory if a new sensor or processing capability is required.

It is rare that an AWS datalogger is purchased as a separate component. Many manufacturers provide AWS packages that include dataloggers, sensors, mounting hardware, and data retrieval equipment helping to ensure that all AWS components are compatible. A non-exhaustive listing of the major suppliers of dataloggers is provided in Appendix B.

The selection of the AWS datalogger is of paramount importance since it controls the entire operation of the AWS. Desirable datalogger features include (1) high reliability, (2) ease of programming, (3) high measurement frequency, (4) support functions for sensor compatibility, (5) sensor interfaces, (6) ease of wiring, (7) low power requirements, (8) high storage capacity, and (8) low cost.

Sensor compatibility is a primary consideration when selecting a datalogger. The manufacturer's specifications for *both* the datalogger and sensor must be examined to determine the compatibility of a datalogger. For example, a sensor with a 0–10 V DC output would require a resistive divider to use with a datalogger possessing a 0–5 V full-scale range. A datalogger with a resolution of ±1 mV would be unacceptable for measuring thermocouples with a sensitivity of 40 μV/°C because temperature would resolve to only ±25 °C.

A large proportion of AWS networks use battery power to avoid (1) the cost of establishing electrical service and (2) problems associated with lightning-induced power surges in AC lines. Dataloggers and sensors with high power demands require larger battery packs, solar battery chargers, or frequent battery service.

Much of the power consumed by a datalogger is used during brief periods when analog measurement, internal processing, and data transfer take place. The remainder of the time the datalogger waits in a "quiescent" mode that requires very little power (e.g., 0.5–2.0 mA). When comparing dataloggers, the equivalent continuous power requirement must be computed for the application. Higher scan rates increase the power drain. Dataloggers with selectable scan rates allow users to sample and process sensor inputs at frequencies that provide representative measurements and maximum battery life.

A datalogger that provides several data retrieval methods [e.g., phone or radio frequency (RF) by line of sight or satellite] is best since it affords flexibility should a particular method prove more convenient or cost effective in the future. The cost and power requirements of peripheral equipment required for data retrieval should also be considered.

On-site processing capabilities allow the manipulation and conversion of data at the AWS site and are an important feature of modern dataloggers. Processing capabilities range from the simple conversion of input signals to engineering units, to powerful programming commands. The computation of saturation vapor pressure and the linearization of thermistor outputs represent two simple examples of processing functions available on many dataloggers. Data reduction (forming averages, summations, etc.) is essential for some measurements. Data reduction and manipulation at the AWS level reduces the time required to retrieve, handle, and process data at the network data processing center.

Adequate on-site data storage (memory) to ensure no data are lost (or overwritten) between data retrieval events is an important requirement. Networks that retrieve data automatically should have at least 4 days of on-site data storage at each AWS to minimize data loss should the retrieval system fail.

A final and very important consideration when selecting a datalogger is the working environment. Datalogger specifications for thermal and humidity tolerance must be examined. It is imperative to select a datalogger that can operate over a range of climatic conditions that encompass the historical or expected climatic conditions at the AWS site. Some problems associated with high temperature environments can be reduced by proper use of protective enclosures and radiation shields. Many dataloggers are capable of operating at

−30 °C without heating. The environmental specifications for data storage and transfer peripherals must also meet the operating conditions of the site.

3.3 Data Processing Center (DPC)

A variety of methods are used to transfer data from AWSs to the DPC. The decision of which method to use depends largely on the type of network operation and the costs and/or availability of the various retrieval options. Real- or near-real-time networks typically use fully automated means of retrieving data such as radio, satellite, or telephone lines (e.g., Hubbard et al. 1983; Ley 1988; Powers 1988). Networks collecting general climate data can employ less automated retrieval procedures which include using datalogger memory, attached solid-state memory modules or audio tape as a temporary storage medium. In the case of the latter two methods, the storage medium (module or tape) is typically transported to the DPC where data are transferred to computer for processing (e.g., Dugas et al. 1984).

Solid-state memory modules are now the predominant mechanism for on-site data transfer with "credit card" storage use increasing. Memory modules are rugged, durable, vary in their capacity for data storage, and they last for many years. Like cassette tapes, memory modules must be retrieved from each station and this is an obvious drawback.

The most common method of retrieving AWS data is with modems and standard voice-grade telephone lines. This method offers several advantages over other data retrieval systems, including (1) access at any desired time or calling interval, (2) relatively low capital investment in equipment, (3) reliability, (4) ability to fully automate the retrieval process, and (5) the ability to communicate in two directions. Modems can be attached to nearly all AWS dataloggers manufactured today, and most AWS manufacturers sell communication software specifically designed to retrieve data from remote AWSs. With telephone communications it is possible to remotely reprogram the datalogger and reset the clock when needed. This is also possible with RF or satellite communications, but the costs for a two-way system may be double or triple that of a one-way system.

The main problems associated with using telephone communications are: (1) absence of a telephone service, (2) poor quality service, and (3) expensive installation. If there is any question about the quality of the telephone line, communications should be tested with components borrowed from the manufacturer or a colleague. Since deregulation of the phone industry, line establishment costs have increased dramatically in the USA, occasionally exceeding $1000 in sparsely populated rural areas.

Another disadvantage of phone retrieval is the day-to-day operational cost. Local line charges at the AWS site vary significantly from one location to another. In Arizona, local AWS line charges range from $378 to $672 annually.

Costs for long-distance data retrieval can vary due to local rates, calling frequency, and call length but generally amount to $70 to $120 annually per AWS for daily retrieval.

In areas where a phone service is available nearby, but not at the AWS site, a phone-to-short-haul modem extension can be used. A standard two or four pair cable is buried between the two short-haul modems to provide the connection. Short-haul modems work over relatively short distances. Tanner (1990) indicates the working range varies from 3 to 10.5 km.

Radio frequency (RF) transmission is used to retrieve data where phone service is either unaffordable, or where real-time access is required. Telephone charges are prohibitively expensive when real-time access is required and telephone retrieval may not provide the desired transmission speed. Radio systems are generally efficient and reliable, but require a larger capital investment and a longer lead time to establish. An RF base station must be established at the DPC and each AWS must be equipped with a transceiver which costs $800–$1200 per station (Tanner 1990). RF telemetry requires line-of-sight transmission, and repeaters may be necessary depending upon the local terrain features and distance. Licensing of a transmission frequency can be a lengthy process in many countries. Where possible, it is advisable to use existing RF systems operated by state and federal agencies such as the USA National Weather Service (NWS) or a local emergency broadcast system. The Public Access Agricultural Weather System (PAWS) in Washington, the largest agricultural weather network utilizing RF telemetry, developed a cooperative agreement with the NWS and uses a UHF frequency reserved for NWS purposes (Ley 1988).

Retrieval of AWS data by satellite is often the only feasible method in mountainous terrain or very remote areas. Geostationary satellite systems operated by the USA (GOES), European Space Agency (Meteosat), Japan (GMS), and India (INSAT) are available to transfer data from ground-based collection platforms (DCP). Each satellite system provides a data collection system (DCS) consisting of a transponder on the satellite and a ground-based DPC (Baldrige et al. 1983). The GOES satellite is available free of charge to government, non-profit organizations, and under special circumstances, to private users. Access requires the DCP owner to enter into a memorandum of agreement with the National Environmental Satellite Data and Information Service (NESDIS). Inside the USA, the DCP must also be certified by NESDIS (MacCallum and Nestlebush 1983). The authors are unfamiliar with the availability and cost for using satellite systems outside of the USA.

From an establishment perspective, satellite retrieval is by far the most costly data-retrieval method. Each AWS must have a transmitter costing $1800–$2500. Users of the GOES DCS can retrieve data directly from the satellite provided an Earth station is available locally. If no Earth station is available, data can be retrieved via phone/modem link (300–2400 bits/s) with the Central Data Distribution Facility (CDDF) computer located in Camp

Springs, Maryland, USA. The user must pay all applicable long-distance phone charges for data retrieval from the CDDF (NOAA/NESDIS 1990). Another cost consideration when using satellites is personnel and travel costs required to service the DCPs. Satellite managers require DCP owners to repair malfunctioning DCPs as soon as a problem is identified (MacCallum and Nestlebush 1983).

The DCSs transmit at 100 bits/s and the signal is burdened with overhead bytes (Tanner 1990; MacCallum and Nestlebush 1983). This limits the volume of data that can be transferred during the 1-min transmission window allotted each user. Users collecting hourly weather data can expect to transmit six to eight times a day (Warren and Vance 1981; Powers et al. 1988). The short transmission window and slow transmission rate also limit the usefulness of satellites for transmitting data in real time. Accommodations are made for random reporting via GOES, which is used to warn of extreme conditions such as heavy rains or floods (NESDIS 1985). Obtaining the data from NESDIS in a timely fashion is the greatest limitation to real-time data collection. Otherwise, the user must have their own Earth station.

The specific goals of the AWS network and the method of data retrieval dictate the frequency of data transfer between AWS and DPC. True real-time networks may retrieve data at intervals of 1 h or less (Ley 1988) whereas climate networks and networks with less-automated retrieval systems may transfer data at intervals approaching 1 week (Robbins and Titlow 1989; Dugas et al. 1984). Daily data retrieval is sufficient for most agricultural uses, including estimation of ET_p and operational use of crop and pest models. Consequently, the vast majority of networks retrieve AWS data daily (Hubbard et al. 1983; Snyder et al. 1985b; Brown 1987). Where data are retrieved via geostationary satellites, transfer proceeds every 3–4 h according to schedules set forth by the satellite manager (Warren and Vance 1981; MacCallum and Nestlebush 1983; Powers et al. 1988).

The DPC should have sufficient memory capacity to store at least 1 year of hourly data as well as hardware for long-term data archiving. There should be a sufficient number of dial-in ports for users to access data from the DPC without having to wait long periods.

3.4 Software

Selecting the proper software for a DPC is critical for smooth operation. Software to (1) retrieve data from the AWS, (2) manage a sizable database, (3) generate reports as needed, and (4) perform data quality control testing are needed. At this time there is no single software package available to accomplish all of these tasks. When the first networks (e.g., CIMIS and AWDN) were established, network software was unavailable and much of the software had to be developed by combining software packages and writing new computer codes. However, this method is very expensive and purchasing commercial software is now more cost effective.

3.4.1 Database Management

Database management software is needed to process and archive data for later access by users. The choice of database software (e.g., DBASE, RBASE, etc.) depends largely on the network size and the DPC. CIMIS used a large mainframe computer and a database management program was developed by necessity. AWDN originally used a mainframe computer but switched in 1987 to an IBM PC compatible computer with a locally developed database management system. Smaller and newer networks have purchased database management software. For example, AZMET uses FIDO bulletin board software to manage and disseminate data using an IBM compatible computer. CLICOM software was written by the USA National Climatic Data Center (NOAA 1988) to quality control (QC) and archive weather data and it is being used by several networks (Titlow and Robbins 1988; Ley and Evans 1990). CLICOM software and a training program are available for developing nations through the World Meteorological Organization (WMO).

3.4.2 Report Generators

The output of data from an AWS network can take several forms. When CIMIS is accessed by computer dial-up, the user can select data in either a daily or an hourly report format or non-report format. In the report format, the data are displayed on the screen in a tabular format with the rows and columns with headings, units, and QC flags indicated. In the non-report format, data are listed sequentially in ASCII files and specific variables can be requested. In either format, the starting and stopping date and time can be specified. In general, software capable of generating both report and non-report formats is recommended. The report format is better for dissemination to the general public, but the non-report format is more useful for research purposes and operational use of crop/pest models.

3.4.3 Data Quality Control

Most AWS networks employ a QC program for data testing. For example, the CLICOM program (NOAA 1988) contains quality-control features. A quality-control program typically tests current data against historical averages, nearby stations, or calculated estimates of data limits. A discussion of the CIMIS data quality-control program was given by Snyder et al. (1985a). Although quality-control programs identify most major data quality errors, regular human inspection of data are still recommended. Quality-control programs have limitations and can sometimes overlook errors a human will notice. The AWDN employs a daily mapping of data from the region to help technicians spot problems. The danger is that network operators often assume the quality-control program will identity all errors. In reality, it is not possible for errors such as low R_n values to be identified using a quality-control program. Frequent maintenance and calibration are needed to avoid such problems.

3.5 DPC Costs

The costs associated with purchasing a DPC computer, peripheral equipment, and software vary tremendously depending on the size of the AWS network, data applications, dissemination goals, etc. When CIMIS was being developed between 1982 and 1985, the DPC and peripherals cost approximately $500 000 including the room, electrical modifications, and the data retrieval and dissemination equipment. Fortran software was purchased for the DPC computer and two full-time programmers were employed for 3 years to develop data-transfer, QC, database, and information dissemination software. The cost for personnel and software was approximately $300 000. A microcomputer with data retrieval software was purchased to retrieve data from the AWSs. Computer software has decreased in cost since CIMIS was developed and there is considerably more commercially available database software for processing AWS data (e.g., CLICOM, DBASE, RBASE, SQL, etc.). The environmental requirements for mainframe computers are also now less stringent.

4 Network Operations

4.1 Maintenance and Calibration

Descriptions of the developmental research and associated operational experiences associated with the establishment of AWS networks have appeared in a number of publication (e.g., Titlow and Robbins 1988; Curry et al. 1988; Howell et al. 1984; Thompson et al. 1984; Hubbard et al. 1983).

Day-to-day operation of AWS networks mainly involves sensor calibration and maintenance and data dissemination. A regular program of maintenance and sensor calibration is essential. Maintenance is classified as either scheduled or emergency. Scheduled maintenance is performed at regular intervals on AWS equipment that is functioning properly. Emergency maintenance is performed after an AWS component is determined to have failed or drifted out of calibration. An effective scheduled maintenance program can significantly reduce costly emergency maintenance calls.

Environmental conditions, the type of datalogger, the sensors used at the AWS, and financial resources combine to determine the frequency of scheduled maintenance. Some networks schedule maintenance monthly (Ley 1988); others only twice annually (Curry et al. 1988). The general recommendation is that scheduled maintenance be performed at least four times per year to ensure high quality data. Local volunteers (e.g., experiment station personnel) are occasionally used to perform routine chores such as mowing grass and cleaning pyranometers, reducing the need for visits by network personnel (Curry et al. 1988).

The procedures performed during scheduled maintenance vary with the type of AWS and its location. Site maintenance performed during scheduled maintenance, includes the removal of trash and debris from the station enclosure, mowing of grass, fence repair, and removal of vegetation from the bare-soil area containing T_s (soil temperature) sensors. Maintenance of AWS equipment includes the cleaning of filter screens, radiation shelters, rain gauges and pyranometers; replacing disposable parts (e.g., batteries, humidity chips, bearings, etc.), leveling pyranometers and rain gauges, and inspection of all sensor cables. Where RF telemetry is used to retrieve data, technical personnel must also inspect and/or adjust transceiver equipment, antennae, and system clocks if needed.

Battery voltage should be recorded at the DPC and non-rechargeable batteries changed when needed. An auxiliary battery can be used with some dataloggers during battery changes in order to prevent loss of the program.

A brief, on-site calibration is often included as part of the scheduled maintenance program (Snyder et al. 1985c). During this process AWS sensor outputs are compared with field calibration standards. For rain gauges, a known quantity of water is poured into the gauge to ensure that PPT is recorded correctly. More rigorous on-site calibrations are performed on occasion by setting up a second weather station containing the field calibration standards for a period of a day or more. Sensors found to be in error during calibration are replaced.

Preventive maintenance procedures should be part of the scheduled maintenance program of all AWS networks in order to avoid serious measurement error or total sensor failure by replacing parts and/or whole sensors prior to the development of problems. Examples of regular preventive maintenance include the replacement of bearings in anemometers, potentiometers in wind vanes, humidity chips, and sensor cables. An inventory of spare sensors is used to rotate sensors. A set of sensors is first rehabilitated and recalibrated at a location where the necessary tools and calibration equipment are available. Sensors receiving preventive maintenance are replaced with the rehabilitated sensors and are in turn rehabilitated and recalibrated. The manufacturer's recommendations and personal experience are used to set the required frequency for sensor rotation. The cost of preventive maintenance must be compared with the consequences of lost or questionable data, and the generally higher costs of emergency maintenance.

Maintenance personnel should develop a checklist of actions that are to be taken routinely upon each site inspection. A maintenance frequency list based on the California, Arizona, and Nebraska networks is given in Appendix D. Maintenance and calibration frequency depends on the station environment. In California, stations near the ocean require more frequent servicing than inland stations, and desert stations require the least servicing.

An optimal calibration process would include comparing sensors against standards traceable to national or worldwide standards. In practice, most AWS networks calibrate network sensors against some form of local "standards" (Snyder et al. 1985c; Curry et al. 1988) having better accuracy than existing AWS sensors. Local standards should also be returned for approved factory calibration at regular intervals. A list of commonly used local standards is provided in Table 4.

As previously discussed, some networks do single-point or short-term calibrations at the AWS site. Others carry out more intensive field calibrations where a battery of AWS sensors are compared against local standards for a number of days (Aceves-Navarro et al. 1988) so that the sensors are evaluated over a wider range of environmental conditions.

Emergency maintenance must be performed on occasion due to adverse weather, animal damage, vandalism, and sudden instrument failures. The network must have available on short notice a trained technician to repair malfunctioning equipment, a vehicle, and the necessary diagnostic equipment. The network must maintain an adequate inventory of spare parts and calibrated sensors for use in emergency situations. Technical support needs depend on the number of AWSs and geographic coverage. Twenty stations would require one full-time technician. The number of spares for each type of sensor can be determined through practical experience with the sensors, but a ratio of one spare for every five installed sensors is a reasonable first estimate.

Major problems typically encountered while operating AWSs vary due to differences in climate, sensors, data retrieval techniques, and funding. Major climate-related problems include lightning, dust, and high humidity. In areas with significant thunderstorm activity, lightning is the major source of climate-induced problems. Direct or nearby lightning strikes can destroy or damage AWS equipment (Curry et al. 1988). Proper grounding and use of lightning rods can minimize, but never totally eliminate, problems associated with lightning. Most lightning damage results when current from an indirect strike travels through phone or power lines to the AWS. Spark gaps on incoming telephone lines are essential. Surge protectors are recommended and widely used to protect equipment connected to AC power.

Table 4. Standards commonly used for local calibration of AWS sensors

Parameter	Standard
Temperature	Platinum resistance thermometer
Solar Radiation	Precision thermopile pyranometer
Humidity	Ventilated psychrometer or chilled mirror hygrometer
Wind Speed	Anemometer (photochopper-type)
Wind Direction	Compass corrected for magnetic deviation from true north

Dust is impossible to eliminate and eventually penetrates all AWS equipment. Over time, anemometer bearings and wind-vane rheostats will fail due to accumulations of dust and dirt. Filter screens on temperature and humidity sensors eventually become clogged, producing non-representative sensor readings or poor response times. Rain gauges and pyranometers also suffer from dust problems. Dust can plug rain-gauge funnels or be deposited in sufficient quantities in the tipping buckets to affect the calibration. Dust accumulation on pyranometers is a chronic problem.

The solution to most dust-induced problems is a sound program of scheduled maintenance. The dust load in a particular environment should be a consideration when determining the required frequency of scheduled maintenance. It is also important to schedule maintenance visits shortly after severe dust conditions such as windy weather prior to spring planting or an actual dust storm.

Extended periods of high humidity can create serious problems for AWS dataloggers and sensors. Most dataloggers require protective enclosures and desiccants to control humidity related problems. Sensors, however, are subject to problems such as corrosion or short circuits (e.g., Snyder et al. 1985c) and should contain a minimum number of corrodible parts. External sensor connectors and cable splices must be sealed to minimize the possibility of water penetration.

Another source of problems can occur when using telephone lines to retrieve data. Telephone systems in the USA are reasonably reliable, but phone-service quality does vary with location and service company. Experience in Arizona and elsewhere suggests that most retrieval problems are caused by the telephone system itself and not the modem or datalogger. The response time of repair personnel varies, and it is not uncommon to be without telephone service for several days in areas serviced by small telephone companies.

Many problems associated with running AWSs can be eliminated or minimized if adequate funds are available to support labor, travel, and supplies necessary for a sound maintenance program. Also, facilities and personnel must be available for regular calibration of network sensors.

4.2 Data Quality Control, Archival, and Dissemination

Quality control (QC) routines are used by nearly all AWS networks to monitor network operation and data quality (Snyder et al. 1985a; Hubbard and Hines 1991). Most networks use both automated and manual QC routines. Automated QC routines are designed for unattended operation and utilize computers and special software to ensure that (1) all AWSs report, (2) data received are formatted correctly, and (3) all data values fall within acceptable limits. The first two QC checks are relatively easy to make while the third is somewhat more difficult. Most networks check incoming data to ensure that all parameters are within some acceptable range defined by global extremes – values that either cannot occur (e.g., negative solar radiation) or occur rarely. Examples of global extremes are provided by Snyder et al. (1985c) and Hubbard et al. (1983). Some networks employ additional checks as part of the automated QC process, such as comparing incoming data (1) with data collected from nearest neighbor stations, (2) against historical extremes, and (3) against "acceptable" ranges generated with standard statistical techniques (Titlow and Robbins 1988). The more sophisticated QC systems, e.g., Nebraska's AWDN, even provide interpolation routines to estimate missing or replace bad data (Hubbard 1988). Manual inspection and/or investigation of incoming data follows the automated QC process. Data values deemed questionable by automated QC routines are flagged and passed on to a system technician who decides the fate of questionable data. Computer-generated graphics have become an invaluable tool in this manual QC process. Graphical presentations of data can streamline the QC process and help detect malfunctioning sensors. In some cases, a series of plots and graphs are developed as part of the automated QC process (Titlow and Robbins 1988).

Measurements at a particular station can also be compared to the same measurements taken at nearby stations. A quality-control flag should be stored with each measurement to indicate the confidence in the data or the data status. This flag can be in the form of a letter, e.g., "E" for estimated, "M" for missing, or "Q" for questionable. In no case should valid data be replaced with estimated data in the archives. Bad data should be flagged and an estimate provided for dissemination purposes.

A well-organized system for archiving data is an essential element of network management. An effective archival system provides for the safe, long-term storage of data and records while providing access to those data and records as needed. Archival systems are, in fact, database management systems and make extensive use of computer technology. Developing archival software is expensive, therefore, some of the newer networks have chosen to use commercially available database management software for archival of data (e.g., Arizona and Louisiana). Commercial software provides report generators, programming languages, graphics, and flexible file import and export capabilities. CLICOM, a software package developed by the USA National Climatic Data Center (NOAA 1988) and designed specifically for QC and archival of weather data, is used by several networks (Titlow and Robbins 1988; Ley and Evans 1990).

Archival systems should be developed with an eye toward the future. Additional measurements or new instrumentation may be required at some future data and so flexibility should be built into the archival system (Hubbard 1988).

The archival process is integrated into the regular daily operation of many larger networks and typically is performed automatically upon completion of the QC process (Snyder et al. 1985c; Hubbard 1988). This ensures archival is performed regularly and minimizes the possibility of losing data. Smaller, less automated networks may archive data less frequently (e.g., weekly or monthly) using a combination of automated and manual procedures.

Improved computers and computer peripherals have greatly simplified the archival process. Personal computers (PCs) now have the power and storage capacity of the mini and mainframe computers used less than a decade ago and are widely used to assist with data archival (Hubbard 1988; Ley and Kroeger 1988). Vast improvements in data-storage technologies have also altered the archival process. Much of the early AWS data was archived on magnetic tape. However, new data-storage peripherals such as high-capacity fixed and removable cartridge disks are rapidly replacing tape as a long-term storage media. The removable cartridge disk, an example of improved storage technology, is accessed like a standard floppy or internal disk drive, but holds 44 Mbytes of data and costs about $100. Optical disk systems with even higher storage capacities appear to be the probable next advance in data storage.

The dissemination of information generated by AWS networks proceeds through three main pathways: (1) some form of dial-up computer system, (2) hardcopy publication of reports, and (3) electronic mass media. Nearly all AWS networks operate a public-access computer system. The sophistication of the system varies, ranging from interactive mainframe and minicomputer systems where a user can log-in and run programs and models (e.g., Wright and Ley 1989) to simple bulletin-board systems (BBS) offering reports and data in discrete files for viewing and/or downloading by remote users (e.g., AZMET; Brown 1987). BBSs have become popular in recent years because they are less expensive to develop and operate than interactive systems. Improved PCs and new farm-management software are also responsible for the shift to BBSs as a means of disseminating data. Many farmers and agribusinesses now own computers and farm-management software, and they need raw weather files as input for their software. BBSs are well suited to supply this need. A few products available on the High Plains Climate Center Bulletin Board are listed in Appendix E.

Another data dissemination option is the use of nationwide computer networks and the connection of these computers via gateways such as the NSF NET and other INTERNETS. In this type of system, file-transfer protocols are employed by the user (who may instigate a request on a mainframe computer at the opposite end of the INTERNET) to collect data from the dissemination system. Personal computers linked to a mainframe at the DPC can function as the file server for other mainframe computers on the INTERNET. Recently, MOSAIC, a retrieval tool for INTERNET has made it possible to display maps and graphs from AWS networks.

Dissemination by computer offers advantages and disadvantages. The major advantages are (1) much of the dissemination proceeds without direct human intervention saving on labor costs and (2) the computer systems are generally operational outside working hours (e.g., at nights and weekends) and therefore are more convenient for many users.

The major disadvantage of computer dissemination is the lack of computer experience and expertise among many potential consumers of weather information. Requests for assistance from these individuals can quickly overburden network staff and eliminate the reduction in labor that computer dissemination is supposed to provide. Regional workshops aimed at helping first-time users with computer access can be an effective means of dealing with this problem as can assistance from computer sales personnel and private consultants. The latter often will set up new users with automated access systems that make use of the programming capability that comes with most communication software.

Hard-copy dissemination remains a popular and viable means of disseminating AWS information. Direct mailing of newsletters and advisories, and publication in local newspapers are the two most common means of hard-copy transfer. Agricultural extension specialists and agents are responsible for much of the direct mailing in rural areas – in part because many networks are affiliated with universities. Use of facsimile (FAX) machines for hard-copy transfer of weather information is increasing as this reduces the lag time between measurement and receipt of the information.

The electronic media remains the major conduit by which rural consumers receive current weather information (Bartholic and Fear 1988). Several networks are directly involved in developing regular advisories for dissemination via the electronic media. In Nebraska, an agricultural situation committee consisting of university specialists from many agricultural disciplines meets weekly to discuss the impact of the current weather on agricultural production and to record a crop and weather advisory that is subsequently broadcast over many rural radio stations (Pride 1984). In Minnesota, a weekly weather advisory is transferred to the mass media via the NWS weather wire (Spoden and Seeley 1985). The NWS in Washington and Arizona use data from AWSs to develop frost advisories broadcast over both commercial and NOAA Weather Radio.

Widespread use of the mass media (both printed and electronic) does not come without cost. The mass media requires that advisories and information for dissemination be delivered on time and in a concise, ready-to-use format. Networks utilizing mass media for dissemination devote a considerable amount of labor to such efforts.

5 Computations

Several AWS variables are used extensively to improve management of agricultural systems. The most commonly used variables include air temperature (T_a) for crop growth and production modeling and pest models; potential evapotranspiration (ET_p) for irrigation scheduling; net radiation (R_n) for freeze forecasting and calculation of ET_p; relative humidity (RH) for calculation of vapor pressure (e) and the dew point temperature (T_d); wind speed (U) and direction (WD) for use in spray applications, wind damage assessment, and pest management; solar radiation (Rs) for use in growth models, ET_p

calculations, and pest models; precipitation for water budget computations and flood forecasting; and growing degree days (°D) or heat units for use in development models and pest management.

Air temperature during the day is perhaps the most important of the variables used in crop and pest models, and attempts to model diurnal temperature changes have been only partially successful; hourly data may be necessary to achieve the desired accuracy with some models (Reicosky et al. 1989). Similarly, use of hourly ET_p estimates may eliminate much of the inaccuracy and controversy surrounding the use of models based on mean daily data. Clearly, the use of hourly weather data improves our ability to characterize what has happened in the past and it may improve agricultural forecasts in the future.

5.1 Evapotranspiration

Potential evapotranspiration (ET_p) is defined by Rosenberg et al. (1983) as "the total process of water transfer into the atmosphere from vegetated land surfaces." They also quote the earlier definition of ET_p as "the evaporation from an extended surface of a short green crop which fully shades the ground, exerts little or negligible resistance to the flow of water, and is always well supplied with water. ET_p cannot exceed free-water evaporation under the same weather conditions." Although these definitions seem reasonable, it is also important that the short green crop used be considered when estimating ET_p.

Many equations for estimating ET_p using daily weather data are reported in the literature (Allen et al. 1989; Jensen et al. 1990). Differences in estimated ET_p often result from the empiricism used to develop the equations. The accuracies of equations based more on theory are often better than empirical equations, but even theoretical equations may require local calibration to improve accuracy (Pruitt and Doorenbos 1977).

Alfalfa (30–80 cm) and cool-season grass (10–15 cm) are typically used in the USA as the "short green crop" for estimating ET_p. However, alfalfa uses more water than the grass under arid conditions, so ET_p equations will give different results depending on which short green crop is used to derive calibration coefficients in the equations. The difference in itself does not present a problem until one attempts to estimate crop evapotranspiration (ET_c) by multiplying ET_p by a crop coefficient (K_c) to account for plant factors. Values of K_c developed by comparing with alfalfa should be used with equations predicting the ET_p of alfalfa and the same applies for K_c values developed by comparing with grass. An extensive listing of K_c values relative to a grass reference is given by Doorenbos and Pruitt (1977). To differentiate between the short green crops used for the development of ET_p equations, the term "reference evapotranspiration" (ET_o) is widely used for the ET_p of cool-season grass. Cool-season grass is the reference crop used for ET in the CIMIS and AZMET networks, and has also been used by the United Nations Food and Agriculture

Organization (Doorenbos and Pruitt 1977). Some western states within the USA use alfalfa to approximate ET. Since alfalfa has higher ET_c than cool-season grass, the use of the word "reference" and the symbol "ET_o" are typically used for a grass crop and the word "potential" and the symbol "ET_p" for an alfalfa crop is suggested to differentiate between ET equations.

Estimates of ET_p or ET_o are provided by nearly all agricultural AWS networks. Which equation to use depends on the frequency of data measurement, the variables measured, and the desired accuracy. AWDN uses the Kincaid and Heerman (1974) modified Penman equation using daily data. Pruitt and Doorenbos (1977), Slatyer and McIlroy (1961), Tanner and Pelton (1960), and van Bavel (1966) have recommended the use of hourly data to gain the most accurate ET_p estimates. Therefore, CIMIS and AZMET use Pruitt and Doorenbos's (1977) modified Penman equation requiring hourly weather data (Snyder and Pruitt 1992).

The Pruitt and Doorenbos (1977) equation requires hourly R_n data to compute ET_o. Although direct measurement of R_n is possible, past experience in California suggests that it is difficult to keep net radiometers functional for extended periods (Snyder et al. 1985c). Procedures for estimating daily R_n from simple linear relationships with R_s (Fritschen 1967; Jensen 1973) are available. However, more elaborate methods are required to estimate hourly R_n (Jensen 1973; Doorenbos and Pruitt 1977; Wright 1982; Dong et al. 1992). In areas with little cloudiness and relatively constant vapor pressure, R_n during daytime hours can be estimated from R_s alone with reasonable accuracy (Fritschen 1967) because the terrestrial radiation balance is small, relative to the solar radiation balance, and remains relatively constant. In cloudy regions, the more elaborate methods require estimates of surface albedo, values of R_s, e, and T_a, and some method of estimating cloud cover (Dong et al. 1992).

5.2 Humidity

Relative humidity (RH) is the main humidity variable measured by most AWS networks. Other humidity variables such as saturation vapor pressure (e_s), vapor pressure (e), wet-bulb temperature (T_w), dew-point temperature (T_d), and vapor pressure deficit (VPD), are typically computed. Most dataloggers can be programmed to compute several if not all of these variables on site. However, it is not uncommon to have some humidity variables computed at the DPC after the data have passed through the QC process.

Temperature and RH are the variables most likely to be measured with an AWS. It is important to note that RH is of limited use unless air temperature is recorded at the same time. Mean RH should never be calculated. It is best to calculate the dew-point temperature or vapor pressure from the T and RH at each sampling and then calculate the mean dew-point temperature or vapor pressure over the sampling interval to characterize humidity.

Saturation vapor pressure (e_s) is a function of T_a and can be computed using the formulas of Lowe (1976), Goff and Gratch (1946), or Tetens (1930). In turn, e is obtained by multiplying e_s by the percentage RH divided by 100. The vapor pressure deficit (VPD) is computed by subtracting e from e_s. Formulas to compute the dew-point (T_d) and wet-bulb (T_w) temperatures from RH, T_a, and barometric pressure (P) are provided by Snyder et al. (1987). In most operational situations, P is estimated from elevation (List 1963; Snyder et al. 1987).

5.3 Degree Days

Networks providing pest and crop management information generally offer several °D values (Brown 1989; Wright and Ley 1989). Threshold temperatures and calculation procedures for °D vary widely, reflecting the wide diversity of crop and pest management models currently in use. Procedures used to calculate °D from daily maximum and minimum temperatures range from simple mean temperature methods to more complicated sine and double sine curve procedures (Zalom et al. 1983). In California and Arizona, the single sine method is most commonly used. A listing of degree day thresholds compiled by the University of California Integrated Pest Management Project is included in Appendix F.

Although not widely used at this time, a more accurate measure for °D is possible using hourly data from AWSs. $°D_l$, the degree days above a lower threshold (T_l), are calculated as the sum over 24 h of T_a minus T_l divided by 24:

$$°D = \sum (T_i - T_l)/24 \text{ for } i = 1 \text{ to } 24 \text{ where } T_i - T_l = 0 \text{ whenever } T_i < T_l.$$

In some cases, an organism may also have an upper threshold. Above this threshold, temperature still contributes to development; however, there is no increase in the development rate relative to that at the upper threshold temperature (T_u). $°D_u$, the degree days above an upper threshold (T_u), are calculated as the sum over 24 h of T_a minus T_u divided by 24:

$$°D_u = \sum (T_i - T_u)/24 \text{ for } i = 1 \text{ to } 24 \text{ where } T_i - T_u = 0 \text{ whenever } T_i < T_u.$$

°D, the number of degree days between the two thresholds, determines the rate of development ($°D = °D_l - °D_u$). If there is no upper threshold temperature, then °D equals $°D_l$.

5.4 Wind Power Calculation

Potential wind power generation is directly related to wind speed cubed (Park 1981). The extractable power (P_e) in Watts from a wind power generator is expressed as:

$$P_e = 0.5 \rho u^3 AE,$$

where ρ is air density ($g\,m^{-3}$), u is instantaneous wind velocity)($m\,s^{-1}$), A is area swept out by the generator blade (m^2), and E is efficiency of the wind generator (expressed as a fraction). Inaccurate average extractable power over a specific time interval is obtained by cubing the mean wind velocity, and so, the wind speed cubed should be computed for each measurement and the mean of those cubed values recorded over the averaging period.

6 Survey of AWS Networks

The 1980s saw a tremendous expansion in the number of AWS networks across the USA and Canada. Fueled by the need for more specific meteorological data in real or near-real time, the AWS became established in many regions. In the late 1970s and early 1980s, California, Minnesota, Nebraska, North Carolina, and Ohio were among the first to establish AWS networks. By the end of 1983, networks were developed in Colorado, Florida, Idaho, Louisiana, New Mexico, New York, Oregon, and South Dakota. Since the mid-1980s, networks have become established in Arizona, Illinois, Iowa, Kansas, North Dakota, Utah, Washington, and Wisconsin. Most recently, Alabama, Georgia, Nevada, and Oklahoma have developed programs to install extensive AWS networks.

The functional use of data generated by AWSs has increased as rapidly as the number of stations and networks. Some of the uses for data generated from these stations in agricultural decision making include: crop-water use estimates (Meyer et al. 1989), irrigation scheduling (Heermann 1981), livestock management (Hahn 1981), integrated pest management (Jones et al. 1981), crop-canopy temperature estimates (Sagar et al. 1988), forestry management (Running 1981), crop and soil moisture modeling (Robinson and Hubbard 1990; Meyer et al. 1991), frost and freeze warnings and forecasts (Martsolf 1981; Ley and Kroeger 1988), crop-growth monitoring (Arkin and Dugas 1981), crop consulting and determination of crop insurance rates (Snyder 1991, pers. comm.), and drainage design and management (Curry et al. 1988).

6.1 Survey Objectives

The number of AWSs and AWS networks is rapidly increasing and the functions served are becoming more diverse. To quantify these factors, a questionnaire focusing on current uses and operations of AWSs was designed and a survey conducted (Meyer and Hubbard 1992). The objectives of the survey were to: (1) determine the number and location of AWSs, (2) determine the type of measurements taken, (3) learn more about operating procedures (i.e., maintenance and collection techniques), and (4) uses of the data (e.g., public dissemination, research analysis, model input, etc.).

This survey was aimed at a broad cross section of AWS managers. However, we attempted to survey only those individuals operating networks for the benefit of the public (e.g., research, service, etc.). No effort was made to identify private-sector operators (e.g., golf course greenskeepers, grain elevator operators, etc.), although their use of AWSs is increasing considerably.

6.2 Survey Results

A total of 214 surveys were mailed to AWS operators in the USA and Canada in December, 1990. A second mailing was sent to operators not responding by February, 1991. A total of 165 (77%) surveys were returned.

Of the 165 survey respondents, 100 (61%) are currently operating AWSs. However, the number of AWSs managed by AWS operators is highly skewed. The majority of operators (51%) manage five stations or fewer (26% manage just one AWS), and 35% manage between 6 and 20 stations (Fig. 1). However, a significant percentage of respondents (14%) manage more than 20 stations. Also, the majority of AWSs (55%) are managed by operators with more than 20 stations.

A total of 939 permanent and approximately 165 portable stations were identified. Locations of permanent stations are shown in Fig. 2. Dense areas of AWS networking are observed in the economically important fruit and vegetable producing regions of California, and respective regions in Arizona, New Mexico, and Florida. With economically important agronomic and livestock industries in the region, a high density of AWS stations is observed in the Corn and Wheat Belts of the Great Plains and Midwest. The Pacific Northwest and intermountain states, with economic interests in wheat, orchards, forests, and pastures boast a dense network of AWSs. A high density of

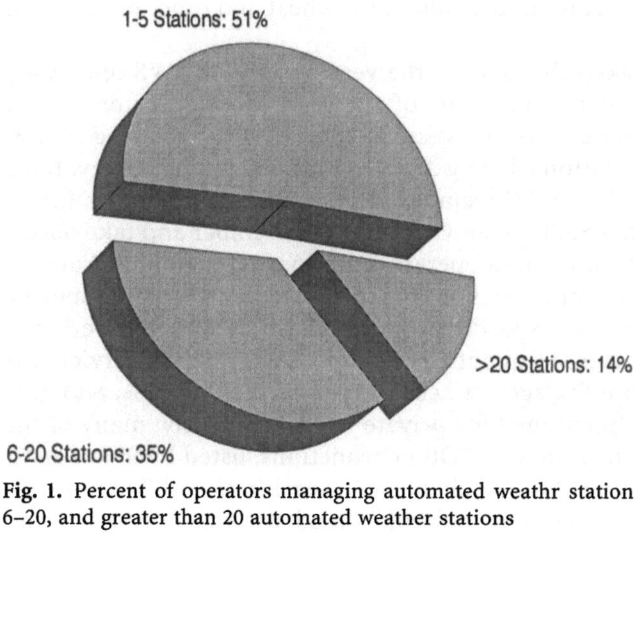

Fig. 1. Percent of operators managing automated weathr station networks composed of 1–5, 6–20, and greater than 20 automated weather stations

Fig. 2. Locations of permanent automated weather stations in the USA and Canada

AWSs is also observed in the Canadian Rocky Mountains for avalanche forecasting, and on the prairies and in the valleys for wheat and other small-grain production.

Measurements are taken throughout the year by 81% of AWS operators, while 19% take measurements only part of the year. Most operators (79%) taking measurements only part of the year do so during the growing season months – installing their stations during April or May and taking observations through September, October, or November. The remaining operators (21%) install their stations during September, October, or November and take observations through April or May; these operators are located mainly in Canada. These results suggest that operators taking measurements during specific months of the year do so for a specific purpose. Listed by 82% of operators, research is the function most frequently served by AWSs. Public service was listed by 39% of the operators, agency needs were specified by 28%, and 21% replied that the stations functioned for private needs; obviously, many of the AWSs serve more than one function. "Other" functions, listed by 9% of AWS operators, included "testing station reliability," "teaching" and "demonstration" purposes, "land use decision making," "land use management," and "avalanche forecasting."

6.2.1 Functions

Functions served by AWSs appear to be related to the number of stations managed. The smallest networks (1–5 AWSs) function to a greater extent for research (Table 5). As the size of the network increases, operators list a greater variety of functions served by AWSs. Although 86% of the networks are composed of 20 stations or less, these networks account for only 45% of the total number of AWSs in use (Table 5). The remaining 14% of the networks – those composed of more than 20 stations – actually operate the majority of AWSs in the USA and Canada (55%).

An examination of functions served by stations operated only part of the year showed that in all instances in which an operator manages a "small" network (1–5 AWSs), those stations function entirely for research. Also, 14 of 15 networks operated during summer months function exclusively or to a great extent for research. In contrast, the three networks operated during winter months function exclusively for public service.

6.2.2 Underlying Surface Cover

The majority of AWSs (60%) which comprise "smaller" AWS networks (networks of 1–10 AWSs) are maintained over a grass surface (Table 6). Conversely, the majority of AWSs (57%) which comprise "larger" networks (networks of 16 or more stations) are maintained over some "other" surface. A spatial distribution indicates that nearly all AWSs in the Midwest, Great Plains, and Canadian prairie provinces are maintained over a grass surface. Individual networks sited over grass tend to be smaller in size. AWSs maintained over some "other" surface are mainly sited in Rocky Mountain states and provinces and in forested areas of the Pacific Northwest. Networks sited over some "other" surface tend to be larger in size.

Table 5. Services provided by automated weather stations as a function of the AWS network size

Category			Percentage[a] of stations serving the following purposes:				
(Stations in network)	Networks in category	Stations in category	Research	Public service	Agency	Private	Other
1–5	50	104	84	35	7	14	7
6–10	16	121	77	41	38	17	13
11–15	9	122	100	56	34	22	0
16–20	8	149	50	49	50	37	0
>20	14	608	72	71	56	47	9

[a] The percentages within a category exceed 100 because many stations function for more than one purpose.

Table 6. Automated weather stations maintained over a grass surface and some "other" surface as a function of the size of the network

Category				
Stations in network	Networks in category	Stations in category	Stations over grass surface	Stations over "other" surface
1–5	50	104	62	42
6–10	16	121	72	49
11–15	9	122	62	60
16–20	8	149	55	94
>20	14	608	186	230

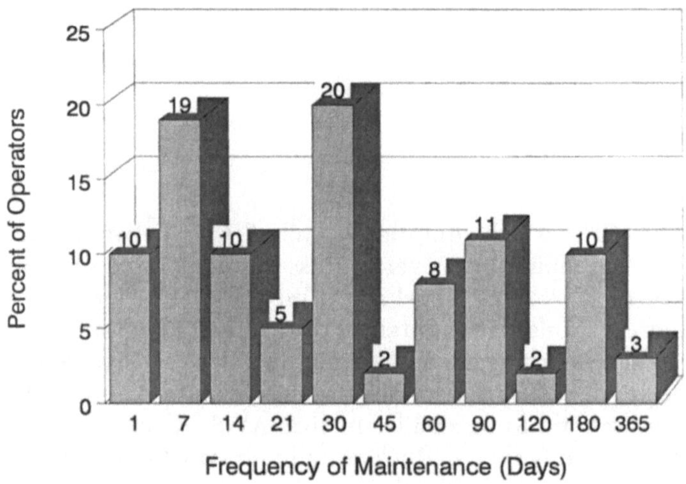

Fig. 3. Frequency of automated weather station maintenance

6.2.3 Maintenance

Most AWS operators regularly maintain their stations, the most common maintenance intervals being weekly or monthly (Fig. 3). Other common maintenance intervals include daily, biweekly, bimonthly, quarterly, and semiannually. There is no specific pattern with respect to the number of stations operated and frequency of maintenance. Generally, the fewer the number of stations operated, the more frequent the maintenance visits. Nevertheless, examples of operators managing three stations or less who maintain their stations only once or twice a year are common.

A significant percentage of operators (13%) do not recalibrate sensors. Although some sensors (e.g., thermistors) can operate effectively for extended periods without recalibration, other sensors (e.g., rain gauges) need relatively frequent calibrations to provide unbiased measurements.

6.2.4 Method of Data Retrieval

The majority of operators (52%) retrieve data via transmission through telephone communication. Transmission through satellite communication and radio-frequency devices were listed by 10 and 15% of the operators, respectively. Downloading data directly from the datalogger to some storage device is another method of retrieving data; 17% of those surveyed download directly to a laptop computer and 30% download directly to storage modules or digital data devices. Many operators (19%) use alternative methods of collecting data such as cassette tapes; direct connection to a personal computer; and dumping directly to a printer. Obviously, some operators use more than one method of collecting data.

6.2.5 Sensor Heights

Data collected from AWSs have a wide array of applications. Often, however, AWS operators have special applications for data (research or otherwise) requiring specific sensor placement with regard to both height and exposure. For this reason, no universally applicable standards for AWS sensor height and installation have been set. However, according to the American Association of State Climatologists (AASC 1985), where requirements do not dictate the height and exposure of sensors, a set of standards are recommended "in order to provide some measure of uniformity among the many networks being installed." The recommended standards are directed primarily to the needs of routine climatological monitoring, agricultural meteorology, and related applications.

The following sections discuss the placement of several sensors – solar radiation, soil temperature, air temperature, relative humidity, wind speed, wind direction, and precipitation – commonly found on AWSs.

6.2.5.1 Solar Radiation

According to the AASC, an R_s sensor may be mounted at any height provided that no object is greater than 10° above the horizon, with respect to the sensor. The majority (72%) of AWSs measuring R_s have sensors mounted at a height of 200 cm and 88% have sensors mounted between 200 and 300 cm (Fig. 4). Five percent of AWSs have R_s sensors mounted above 300 cm.

6.2.5.2 Soil Temperature

The 10 cm depth for measuring T_s recommended by the AASC is consistent with the USA National Weather Service standards. The AASC also recommends that sensors be placed below a 1 m^2 area maintained in a state consistent

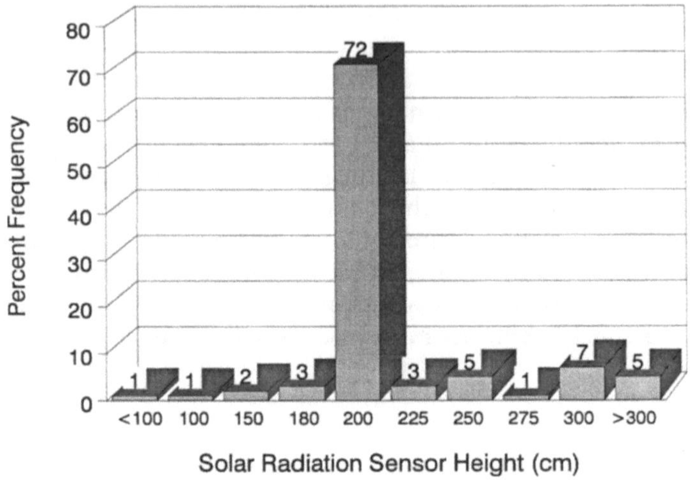

Fig. 4. Discrete frequency distribution of solar-radiation sensor heights at automated weather stations

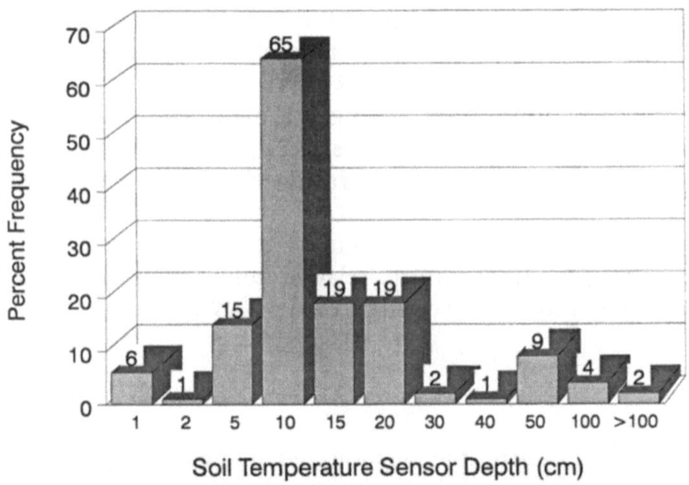

Fig. 5. Discrete frequency distribution of soil-temperature sensor depths at automated weather station sites

with the surface type of interest. Of all AWSs measuring T_s, 87% are buried at a depth between surface level and 10 cm, and 39% are buried at between 11 and 20 cm depth (Fig. 5). Two percent of AWSs have sensors buried below a depth of 100 cm (obviously, many AWSs measure T_s at more than one depth).

6.2.5.3 *Air Temperature and Relative Humidity*

RH sensors are co-located with T_a sensors because simultaneous measurement of RH and T_a is required to make conversions to other humidity, parameters e.g., actual vapor pressure and dew-point temperature. The AASC recommends that T_a and RH sensors be mounted at 150 cm. Of all AWSs measuring T_a and RH, 55 and 28% have sensors mounted at 150 and 200 cm, respectively (Fig. 6). Two percent of all AWSs have sensors mounted beyond 500 cm. Sensors were placed at these heights due to either project specifics or special terrain features (e.g., forested area).

6.2.5.4 *Wind Speed and Direction*

In a similar fashion to T_a and RH sensors, U and WD sensors are generally mounted side by side. The AASC recommends a mounting height of 300 cm for U and WD sensors as a reasonable compromise between the expense of setting up tall towers required to meet NWS and World Meteorological Organization (WMO) standards (1000 cm) and the problems of exposure that increase at lower heights. An optional height suggested by the AASC is 200 cm since U at this height is required for ET_p estimates using the Penman equation (of course, wind speed at 200 cm can be estimated from wind speed at other heights). Of all AWSs measuring U and WD, 25, 31, and 31% have sensors mounted at 200, 300, and 1000 cm, respectively (Fig. 7). Special projects or terrain features require wind sensors to be mounted at 1000 cm or higher (2% of AWSs).

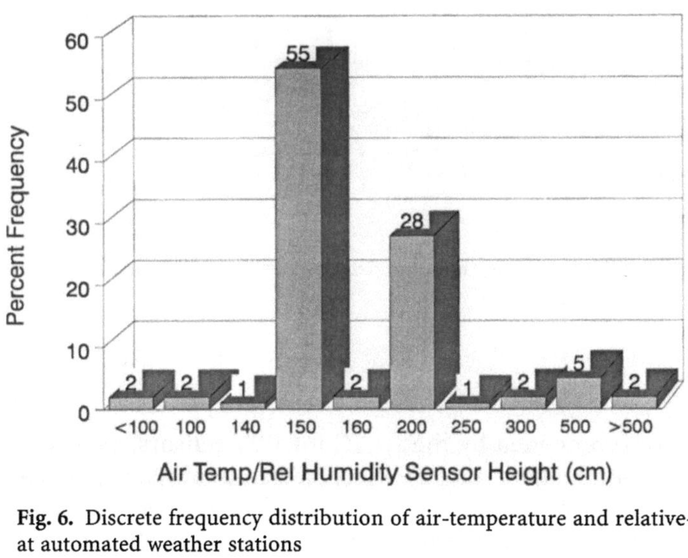

Fig. 6. Discrete frequency distribution of air-temperature and relative-humidity sensor heights at automated weather stations

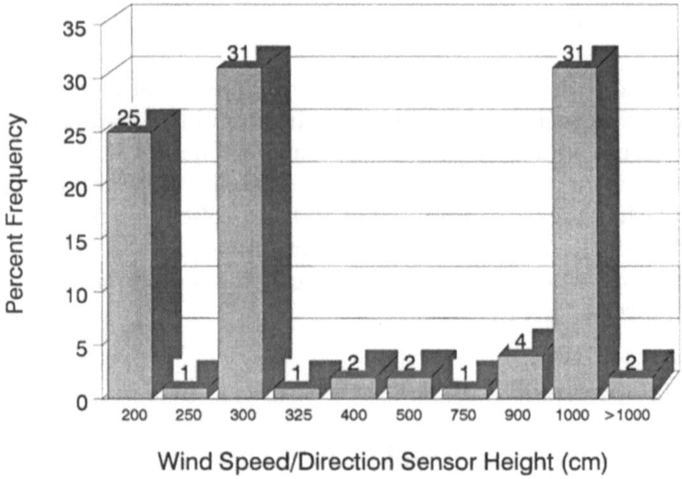

Fig. 7. Discrete frequency distribution of wind-speed and direction sensors heights at automated weather stations

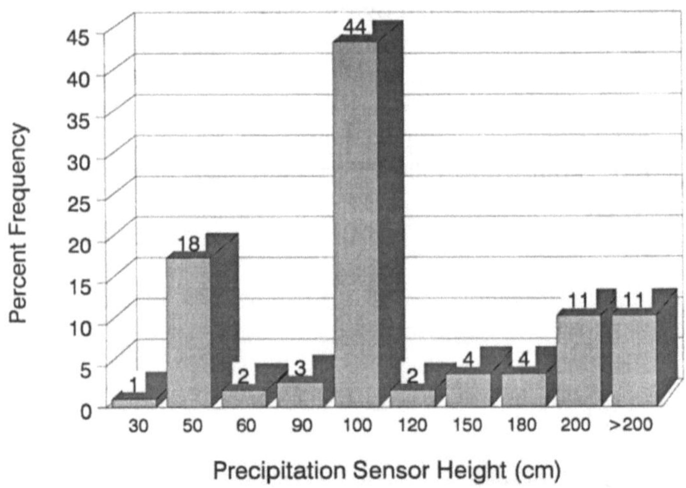

Fig. 8. Discrete frequency distribution of precipitation sensor heights at automated weather stations

6.2.5.5 Precipitation

A height of 100 cm is recommended by the AASC for PPT sensors, except in cold climates where sensor heights may be increased to avoid snow cover. Nearly all AWSs measuring precipitation have sensors placed between ground level and 200 cm (Fig. 8). The most common sensor heights are 50, 100, and 200 cm, representing 18, 44, and 11% of all AWSs measuring precipitation.

Sensors mounted higher than 200 cm were common, but are so mounted due to either special terrain features or cold climates.

6.2.5.6 Miscellaneous Sensors

Operators identified many other sensors that are monitored by AWSs. The more common "miscellaneous" sensors measure pan evaporation (sensor mounted at ground level, 30, and 40 cm), air pressure (sensor mounted at 50, 500, 1000 cm), leaf wetness (sensor mounted at leaf level, usually around 30 cm), and snow depth (sensor mounted at 300 cm). Less frequently identified sensors mounted on AWSs measure gamma radiation, sunshine duration, rainfall conductivity (sensor mounted at 500 and 1000 cm), diffuse radiation (sensor mounted at 200 cm), photosynthetically active radiation (sensor mounted at 250 cm), forest canopy precipitation throughfall (sensor mounted at 100 cm), and temperature inversion measurements from a tower (sensor mounted at 150, 9750, and 15 240 cm).

6.3 Survey Conclusions

Survey results reported here show that data collected by AWSs are used for a multitude of reasons, but most commonly for agriculturally related purposes. Station locations – e.g., fruit and vegetable growing areas of California, Arizona, and Florida; crop and livestock regions of the Corn and Wheat Belts of the Great Plains, Midwest, Pacific Northwest; and intermountain states and provinces – emphasize this relationship with agriculture.

Most AWS operators regularly maintain their stations. However, it is recommended that all stations be inspected no less than four times per year. In this way, data integrity can be maintained, since the data collected are only as good as the condition of the sensor that measured it.

The study also showed that similar sensors are mounted at a wide array of heights. Project specifics or special terrain features sometimes require unique sensor placement. However, often the instrument is placed at some arbitrary height, which could affect the magnitude of the measured variable. Therefore, it is strongly recommended that some standard be developed for AWS sensor placement. This is particularly true for those stations collecting data primarily to address the needs of routine climatological monitoring and agricultural meteorology.

Finally, the growth in AWS networking should be viewed as a positive step toward expanding the data available for meteorological research and service. As networks continue to grow in number and spatial coverage, it is recommended that an AWS climatic database be established. Such a data base could become a valuable source for future climatic and agronomic studies. With proper logistical coordination and the cooperation of network operators, development of such a database can become a reality.

7 Application – Case Studies

7.1 Irrigation Scheduling

Most large AWS networks are located in areas where water conservation is a major concern, and the main purpose used to justify installation of these networks is to provide ET information to farmers for irrigation scheduling. Although there are many examples where providing ET information has benefited irrigated agriculture, one of the best is reported by Fulton et al. (1991). They studied irrigation practices in a walnut orchard located near Hanford, California from 1983 through 1988. Starting in 1984, irrigations were scheduled for the farmer cooperator using CIMIS ET_0 data to estimate crop ET (ET_c) for the walnut orchard. In 1983, before the researchers started scheduling irrigations, the sum of the net application and precipitation equaled approximately 685 mm. The estimated ET_c was about 1140 mm. In 1984, the researchers convinced the farmer cooperator to apply more irrigation water. A net application of about 810 mm and precipitation equal to 125 mm resulted in a net total of about 935 mm. The estimated ET_c was 1320 mm in 1984. In subsequent years (1985–1988), the total of the net application and precipitation was nearly the same or slightly exceeded ET_c. Production from the scheduled orchard was compared with the regional average production in each of the 6 years.

In 1983, prior to the first year of scheduling, the test orchard produced the same in-shell yield as the regional average. In 1984, the first year of scheduling, the yield of the orchard fell to about 955 $kg\,ha^{-1}$ while the regional average was near 1090 $kg\,ha^{-1}$. In 1985, the first year when the net total water application was near ET_c, the production increased to 2045 $kg\,ha^{-1}$ for the scheduled orchard while the regional average was 1590 $kg\,ha^{-1}$ (29% higher). Production for the scheduled orchard exceeded the regional average by 455 $kg\,ha^{-1}$ (25%) in 1986 and 1987 and by more than 680 $kg\,ha^{-1}$ (about 40%) in 1988. Although this example did not demonstrate water savings, it clearly showed the benefits derived from using ET from an AWS network. The results of the trial have been used to encourage other walnut farmers to adopt the use of CIMIS ET_0 for irrigation scheduling.

Another example pertaining to the use of ET estimates occurred in an urban setting during the 1987–1992 drought in California. The city of Santa Barbara, California, was experiencing severe water shortages and the local water conservation specialists developed a strong program to promote the use of CIMIS data for scheduling irrigations in the urban landscape. As most landscapes contain a large variety of plant materials with differing water requirements the K_c values are difficult to generalize. In this case, homeowners and businesses were encouraged to apply between $0.6 \times ET_0$ and $0.7 \times ET_0$ to their landscapes. No attempt was made to evaluate irrigation application efficiency or uniformity, but the clientele were encouraged to design, install, and maintain their irrigation systems. The local water conservation specialists estimated water savings to vary between 10 to 35% with a mean of about 20%

in individual landscapes based on measured water applications (J Debra, personal communication). The employment of near-real time ET_0 information to reduce urban landscape water use was clearly demonstrated.

In the fall of 1993, a survey was conducted in California to determine if the ET_0 data from CIMIS are being used for irrigation management. Questionnaires were sent to 1578 registered CIMIS users and 531 (35%) returned the forms. Irrigators comprised 341 of the 531 respondents. The other 200 respondents were mainly researchers, government agencies, and consultants. CIMIS data are transferred to the public through several methods, but only those users receiving data directly from the main CIMIS computer received questionnaires.

The 341 irrigator respondents were asked which crops they grow, how much land they irrigate, and what irrigation systems they use. Percentages using ET_0 to schedule irrigations of the indicated crops were: tree crops (51%); turfgrass (37%); vineyards (29%); vegetables, melons, and strawberries (19%); alfalfa (12%); processing tomatoes (9%); cotton (6%); other field crops (10%); and other crops and landscapes (16%). The percentages do not add up to 100% because the respondents often grew more than one crop. Percentages of irrigators using ET_0 by land area were: 0–20 ha (27%), 20–80 ha (22%), 80–200 ha (23%), 200–400 ha (2%), >400 ha (19%). Percentages of irrigators using ET_0 by irrigation method were: drip and micro-sprinkler (53%); center pivot/linear move sprinkler (9%); impact sprinklers on ornamentals (12%); other sprinklers (39%); furrow (10%); border and flood (16%); and other methods (4%).

Clearly, most usage of ET_0 for irrigation scheduling occurs on tree crops, vine crops, and turfgrass. Irrigators using drip and micro-sprinkler methods and other sprinklers are most likely to use ET_0 for scheduling. The land area farmed does not seem to be a factor in determining usage except for moderately large California farms where usage is less. This is the group most likely to employ consultants for irrigation management, and perhaps the reduced usage occurs because the consultant rather than the farmer is accessing the ET_0 information. As the trend in California is to change to higher-value tree and vine crops and to drip and micro-sprinkler systems, the use of CIMIS ET_0 is expected to increase.

7.2 Integrated Pest Management

Data from CIMIS is used extensively for pest management. The Integrated Pest Management (IPM) group in California has a computer system (IMPACT) that is separate from CIMIS but uses daily CIMIS data for computation of degree days. The pest and growth models are based mainly on °D calculated with daily maxium and minimum temperature data using the single sine method (Zalom et al. 1983). Pest and crop growth and development information is disseminated to farmers and consultants.

Many of the field trials conducted by IPM researchers employ extensive sets of hourly data provided by CIMIS directly. The newly developed models often include the effects of hourly values of solar radiation, temperature, humidity, wind speed, wind direction, etc. on an hourly basis in addition to daily maximum and minimum temperature. At this time, the dissemination of pest and crop-growth model information based on AWS network data is limited because further testing is needed. Also, the CIMIS and IMPACT systems have not been fully integrated. Currently, the CIMIS data are available to IPM researchers but not to the public through the IMPACT system. However, models based on AWS data are likely to expand in the future.

One limitation to using network data for pest and crop mdels is that AWSs are typically located over a standard underlying surface (grass) rather than in the crop being studied. This is true because most of the existing large networks were developed to provide ET data rather than pest and crop information. However, if a pest problem is serious enough to make network installation cost effective, a separate network with AWSs located in a specific crop could improve pest and growth models.

7.3 Other Uses

Data from AWS networks are used in many ways in addition to providing ET for irrigation scheduling and °D for pest and crop modeling. Near-real time data from an AWS network improve the gathering and dissemination of extreme weather hazards related to livestock and poultry production. In addition, least-cost ration programs for livestock can use the AWS information. The hourly data from CIMIS are used extensively by freeze-forecast meteorologist with the National Weather Service (NWS) in California to update their forecasts during freeze nights. There are examples when freeze warnings have been greatly improved because wind speed and direction and temperature and humidity data were readily available from CIMIS to NWS forecasters. Wind speed and direction data are also used by aviators to determine when conditions are good for aerial application of pesticides and by firefighters during forest fires and urban firestorms to save lives and property. The data are used by highway maintenance crews to identify when icy roads need to be salted and to identify dust storm and fog hazards. The potential application of AWS data is only now being realized and it may take many years to fully appreciate the benefits.

Appendix A Weather Station Placement

Weather Station Placement in the AWDN Operated by the High Plains Regional Climate Center

The stations in the AWDN network are located to avoid measurement of microclimate features. For this reason, stations are placed over grass using the assumption that microclimate features of the underlying surface disappear at about four times the surface height (the sensors are five to six times the height). Microclimate effects of tall trees, buildings or nearby terrain features are also avoided. Stations are located a distance away from such features equivalent to six to ten times their height. A fetch of this length is assumed to be sufficient although a greater fetch is desirable.

The AWSs in the AWDN are placed at intervals of about 50 miles so as to represent the macroclimate of the region or, in other words, the climate of large areas of the region. Sites that represent mesoclimates (i.e., small valleys, frost hollows, frost clearings and tops of hills) are avoided.

Stations at this spacing can be expected to give relatively large differences in readings even when experiencing the same air mass because they represent areas that have a different topoclimate. The reader is referred to Mahrt (1986) for additional information on topoclimatology. Stations in a large valley would be expected to have lower average wind speed, higher humidity, etc., than stations outside the valley. Even so, stations at this spacing will be statistically correlated. This fact may allow interpolation between two stations as an adjustment even when the point of interest is known to have a different topoclimate.

AWSs are placed in rural areas to establish a reference away from city influences. Stations are generally located at some distance from isolated irrigated fields because the microclimate in such fields will vary considerably with the specific irrigation schedule. Hopefully, measurements from the AWSs reflect the average characteristics of a given vicinity and can be used as a reference point for both irrigators and dryland farmers alike.

Consideration should also be given to the reference station concept by researchers who may wish to apply their results to wide areas and who typically measure conditions above a specific field of interest. To make such an application, the total environment must be considered. The microclimate is important because it is the local environment of the crop, insect, etc. The microclimate inside a field responds to variations outside the field making the macroclimate important as well. Of primary importance, however, is the fact that on a large scale it is only practical to measure the macroclimate, therefore, the majority of climate networks measure macroscale features of the climate.

Appendix B Sources of Weather Sensors

Manufacturers and/or distributors of dataloggers and weather sensors. Instrumentation available is listed by code adjacent to the manufacturer/distributor. This list is only a partial list of manufacturers and distributors with a bias towards USA companies. Undoubtedly, many sensor vendors have been overlooked and we encourage readers to look for other sources in addition to those listed.

Instrumentation codes used in the manufacturer/distributor table

Code	Instrumentation
AT	Atmometers
CM	Chilled mirror hygrometers
DL	Dataloggers
EP	Evaporation pans
ER	Evaporation pan recorders
LC	LiCl hygrometers
NR	Net radiation meters
PR	Platinum resistance thermometer
PS	PAR sensors
PY	Pyranometers
RG	Rain gauges
RH	Relative humidity sensors
RS	Radiation shields
ST	Weather stations
TI	Thermistors
TO	Thermocouples
WD	Wet/dry bulb psychrometers
WS	Wind sensors

Manufacturer/distributor	*Instrumentation available*
Automata 10551 Bennett Street Grass Valley, California 95945, USA Phone: 916-273-0380 FAX: 916-273-0381	DL, PY, RG, RH, RS, ST, WS, AT, ER
Bay Technical Associates, Inc. Highway 603, P.O. Box 387 Bay St. Louis, Missouri 39520, USA Phone: 601-467-8231 Telex: 910-333-1618 (BAYTECH)	ST
BCP Electronics P.O. Box 314 Clovis, California 93613, USA Phone: 209-298-8068	EP, ER
Belfort Instrument Co. 727 S. Wolfe St. Baltimore, Maryland 21231, USA Phone: 301-342-2626	DL, LC, PR, PY, RG, RH, RS, ST, TI, WD, WS
Campbell Scientific, Inc. P.O. Box 551	CM, DL, NR, PS, PY, RG, RH,

Manufacturer/distributor	*Instrumentation available*
Logan, Utah 84321, USA Phone: 801-753-2342 (UK): 0509-672516 FAX: 801-752-3268 (UK): 0509-674928	RS, ST, TI, TO, WD, WS
Climatronics 140 Wilbur Place Bohemia, New York 11716, USA Phone: 516-587-7300 FAX: 516-567-7585	DL, LC, PR, PY, RG, RH, RS, ST, TI, WD, WS
Delta-T Devices Ltd. 128 Low Rd., Burwell, Cambridge, CB5 OEJ, UK Phone: 0638-742922 FAX: 0638-743155	DL, NR, PR, PS, PY, RG, RH, RS, ST, TI, TO, WD, WS
Didcot Instrument Co. Ltd. Station Rd., Abingdon Oxon, OX14 3LD, UK	DL, NR, PR, PS, PY, RG, RH, RS, ST, TI, TO, WD, WS
Eppley Laboratories P.O. Box 419 Newport, Rhode Island 02840, USA Phone: 401-847-1020 FAX: 401-847-1031	PY
Fenwal Electronics 450 Fortune Blvd. Milford, Massachusetts 01757, USA Phone: 508-478-6000 FAX: 508-473-6035	TI
General Eastern Instruments 50 Hunt Street Watertown, Massachusetts 02172, USA Phone: 800-255-3208 FAX: 617-923-0218	CM, LC, RH, WD
Handar, Inc. 1188 Bordeaux Drive Sunnyvale, California 94089, USA Phone: 408-734-9640 FAX: 408-734-0655	PR, PS, PY, RG, RH, RS, ST, TI, WS
Hygrometrix, Inc. 7800 MacArthur Blvd.	RH

Manufacturer/distributor	*Instrumentation available*
Oakland, California 94605, USA Phone: 415-639-7800 FAX: 415-639-0901	
Kipp & Zonen Delft B.V. Mercuriusweg 1 2624 BC Delft The Netherlands Phone: 015-561000 (USA): 516-589-2885 FAX: 015-620351 (USA): 516-589-2068	PY
Li-Cor, Inc. Box 4425 Lincoln, Nebraska 68504, USA Phone: 402-467-3576 FAX: 402-467-2819	DL, PS, PY, RG, RH, ST, TI, WS
Met One Instruments 479 California Avenue Grants Pass, Oregon 97526, USA Phone: 503-471-7111 FAX: 503-471-7116	PR, PY, RG, RH, RS, ST, TI, WS
Omega Engineering, Inc. Box 4047 Stamford, Connecticut 06907, USA Phone: 800-826-6343 (Foreign): 203-359-1660 FAX: 203-359-7700 (Foreign): 203-359-7807	PR, TI, TO
Omnidata International 750 W. 200 N. Logan, Utah 84321, USA Phone: 801-753-7760 FAX: 801-753-6756	DL, PR, PS, PY, RG, RH, RS, ST, TI, WS
Peek Electronic Products, Inc. 1316 Lymric Way Bakersfield, California 93309, USA Phone: 805-872-4600 FAX: 805-827-4589	ST
Pelican Wire Co., Inc. 6266 Taylor Road	TO

Manufacturer/distributor	*Instrumentation available*
Naples, Florida 33942, USA Phone: 813-597-8555	
Philipp Schenk GmbH Wien & Co. KG Jedleseer Strasse 59, A-1210 Wien, Austria Phone: 0222/385131 Telex: 112172 schen-a	NR, PY
Phys-Chemical Research Corp. 36 West 20th Street New York, New York 10011 Phone: 212-924-2070 TELEX: 620706 Phys Chem	RH
R.M. Young Company 2801 Aero Park Drive Traverse City, Michigan 49684, USA Phone: 616-946-3980 FAX: 616-946-4772	PR, RG, RS, TI, WS
REBS P.O. Box 15512 Seattle, Washington 98115, USA Phone: 206-624-7221	NR, PY
Rotronic Instrument Corp. 7 High Street Huntington, New York 11743, USA Phone: 516-427-3994 (Zurich): 01-492-3211	RH
Science Associates, Inc Box 230 Princeton, New Jersey 08540, USA Phone: 609-924-4470	EP, NR, PY, RG, RS, WS
Sierra-Misco, Inc. 1825 Eastshore Highway Berkeley, California 94710, USA Phone: 800-544-9229 FAX: 415-843-9381	PR, PS, PY, RG, RH, RS, ST, TI, WS
Swissteco Instruments Stegweg, Eichenwies CH-9463 Oberriet	NR, PY

Manufacturer/distributor	*Instrumentation available*
Switzerland Phone:07178-2717	
Skye Instruments Ltd. Ddole Industrial Estate Llandrindod Wells, Powys, LDI 6DF, UK Phone: 0597-4811 (USA): 215-453-9484 FAX: 0597-4812	PR, PS, PY, RG, RH, RS, WS, TI, WS
Sutron Corp. 2190 Fox Mill Road Herndon, Virginia 22071, USA Phone: 703-471-0810 FAX: 703-435-0232	PR, PY, RG, RH, RS, ST, TI, WS
Thermo Electric Co., Inc. Saddle Brook, New Jersey 07662, USA Phone: 201-843-5800 FAX: 201-843-7144	PR, TO
C.W. Thornthwaite Associates Route 1, Centerton Elmer, New Jersey 08318, USA Phone: 609-358-2350	NR, WS
Vaisala, Inc. 100 Commerce Way Woburn Massachusetts 01801, USA Phone: 617-933-4500 FAX: 617-933-8029	PR, PY, RG, RH, RS, ST, TI, WS
Vector Instruments 113 Marsh Rd., RHYL Clwyd, LL18 2AB, UK Phone: 0745-50700	PR, RS, TO, WD, WS
Weathermeasure/Weathertronics Division of Qualimetrics, Inc. P.O. Box 41039 Sacramento, California 95841, USA Phone: 916-923-0055 FAX: 916-923-5737	CM, DL, EP, LC, NR, PR, PY, RG, RH, RS, ST, TI, WD, WS

Manufacturer/distributor	*Instrumentation available*
YSI Yellow Springs Instrument Co. P.O. Box 465 Yellow Springs, Ohio 45387, USA Phone: 513-767-7241	LC, PR, TI

Appendix C Standards for Sensor Height and Exposure

A resolution passed by the American Association of State Climatologists (AASC) in 1985 suggesting standards for sensor height and exposure at AWS sites

There are many different applications for meteorological data. Any particular application may have unique requirements which dictate the height and/or exposure of sensors. Therefore, no set of standards is universally applicable. However, where requirements do not dictate sensor height and exposure, the following standards are recommended in order to provide some measure of uniformity among the many networks being installed. These standards are directed primarily to the needs of routine climatological monitoring, agricultural meteorology, and related applications. We encourage anyone planning to install an AWS to follow these guidelines if possible.

1. Wind

a) Height: 3 ± 0.1 m Optional: 2 ± 0.1 m; 10 ± 0.5 m

b) Exposure

The sensor shall be no closer to an object than ten times the height of the object.

c) Comments

The 3 m height is currently being used in many of the existing networks. It is a reasonable compromise between the expense of setting up the tall towers required for higher heights and the problems of exposure that increase at lower heights. The optional 10 m height conforms to the standards of the National Weather Service and the World Meteorological Organization. The optional 2 m height is included because the 2 m wind speed is required as input to the Penman ET_p equation. However, the Penman equation is rather insensitive to errors in wind speed, e.g., Merva and Fernandez (1985) and extrapolation of 3 m data down to 2 m using the logarithmic wind profile is acceptable in most cases.

2. Temperature

a) Height: 1.5 ± 0.1 m

b) Exposure

The sensor shall be shielded both from direct solar radiation and from long-wave radiation emitted by the Earth or man-made structure. If the shield is not aspirated, then the design of the shield shall provide for the free flow of air over the sensing elements.

c) Comments

Under calm conditions, non-aspirated shields may result in a more slowly varying signal than with an aspirated shield.

3. Precipitation

a) Height: 1.0 ± 0.1 m

b) Exposure

The sensor shall be no closer to an object than four times the height of the object. The open end of the gauge must be in a horizontal plane.

c) Comments

In cold climates, the height of the gauge may need to be higher than 1 m to avoid being covered by snowdrifts.

4. Solar Radiation on a horizontal surface

a) Height: Any height is acceptable.

b) Exposure

The sensor shall be mounted so that no object is at any angle with respect to the sensor of greater than 10° above horizontal.

5. Soil Temperature

a) Depth: 10 ± 1 cm

b) Exposure

A one meter square area of the surface above the sensor shall be maintained in a state consistent with the type of surface of interest. In most situations, this will be a bare surface although other surface types may be of interest. The ground surface should be level (neither depressed or elevated) with respect to the immediate (10 m radius) area.

c) Comments

The 10 cm depth with a bare surface is consistent with national Weather Service standards. However, there are many applications where different depths or exposures are useful. It is suggested that additional sensors be used if other depths are required.

6. Soil Moisture

a) Depth: 15 ± 0.1 m

b) Exposure

The ground surface should be level (neither depressed nor elevated) with respect to the immediate (10 m radius) area.

c) Comments

The depth may depend on the type of instrument being used, whether it integrates over some depth (neutron probe) or whether it is a point sensor (gypsum block).

Appendix D Weather Station Maintenance Checklist

Check sensor readings (daily)
Clipping of vegetation (as needed)
Onsite testing (4–6 months)
Cleaning of sensors (as needed)
Calibration of tipping bucket (annual)
Calibration of solar sensors (annual)
Test and maintenance of humidity sensor (annual or as needed)
Replace bearings in anemometer (2 years or as needed)
Replace potentiometer in wind vane (2–3 years)

Appendix E Selected Bulletin Board Products

NEO1.HR	Hourly weather data from Nebraska station #1
NEO1.14D	Last 2 weeks of daily data from Nebraska station #1

NEO1.ET	Potential evapotranspiration for crops at Nebraska station #1
KS79.NML	Daily normal data for Kansas station #79
GDDCO	Growing degree day assessment for Colorado stations
PRECAWDN.IA	Precipitation summary for Iowa automated weather data network
PHENOLWHT.SD	Wheat phenological growth stage estimates for the South Dakota sites
PHENOLCRN.IA	Corn phenological growth stage estimate for all Iowa sites
PHENOLSOY.NE	Soybean phenological growth stage estimates for all Nebraska sites
PHENOLSGM.KS	Sorghum phenological growth stage estimates for all Kansas sites
SOIL.ND	Soil temperature summary for all North Dakota sites
DECPREC.NE	Precipitation probabilities for Nebraska in the month of December
WYZONE.FCT	Local zone weather forecasts for Wyoming
US6–10DA.FCT	Six- to 10-day weather forecast for the USA

Appendix F Degree-Day Threshold Temperatures

Common pest name	Scientific pest name	Lower threshold (°C)	Upper threshold (°C)
Alfalfa weevil	*Hypera postica*	10.0	31.0
Apple maggot	*Rhagoletis pomonella*	6.4	
Armyworm	*Pseudaletia unipuncta*	10.0	29.0
Artichoke plume moth	*Platyptilia carduidactyla*	7.0	25.0
Asparagus beetle	*Crioceris asparagi*	10.0	30.0
Beet armyworm	*Spodoptera exigua*	12.2	
Beet leafhopper	*Circulifer tenellus*	14.4	35.0
Black cutworm	*Agrotis ipsilon*	10.4	
Blackberry leafhopper	*Dikrella californica*	8.1	28.0
Brown lacewing	*Micromus tasmaniae*	5.8	
Cabbage aphid	*Brevicoryne brassicae*	6.7	
Cabbage butterfly	*Pierus rapae*	9.4	
Cabbage looper	*Trichoplusia ni*	10.0	32.2
Cabbage maggot	*Delia (Hylemya) radicum*	6.1	
California red scale	*Aonidiella aurantii*	11.5	30.0
Carrot weevil	*Listronotus oregonensis*	7.0	
Cereal leaf beetle	*Oulema melanopus*	7.0	30.0
Citrus bloom		9.4	
Citrus red mite	*Panonychus citri*	7.9	
Codling moth	*Cydia pomonella*	10.0	31.1
Corn earworm	*Helicoverpa (Heliothis) zea*	12.6	33.3

Common pest name	Scientific pest name	Lower threshold (°C)	Upper threshold (°C)
Cotton bollworm	*Helicoverpa (Heliothis) zea*	12.6	33.3
Covergent lady beetle	*Hippodamia convergens*	9.0	
Crucifer flea beetle	*Phyllotreta cruciferrea*	11.0	
Diamondback moth	*Plutella xylostella*	7.3	
Egyptian alfalfa weevil	*Hypera brunneipennis*	7.2	
Elm leaf bettle	*Pyrrhalta (Xanthogaleruca) luteola*	11.0	
European red mite	*Panonychus ulmi*	10.6	
Grape leafhopper	*Erythroneura elegantula*	10.3	35.0
Green peach aphid	*Muzus persicae*	4.0	30.0
Greenhouse whitefly	*Trialeurodes vaporariorum*	8.3	
Imported cabbageworm	*Pieris rapae*	9.4	
Lygus bug	*Lygus hesperus*	12.2	
Meadow spittlebug	*Philaenus spumarius*	2.8	26.7
Mediterranean fruit fly	*Ceratitis capitata*	9.7	
Melon fly	*Dacus cucurbitae*	8.1	
Mexican bean beetle	*Epilachna varivestis*	11.5	
Nantucket pine tip moth	*Rhyacionia frustrana*	5.5	37.35
Navel orange bloom		9.4	
Navel orangeworm	*Amyelois transitella*	12.8	34.4
Northern corn rootworm	*Diabrotica barberi*	11.1	
Omnivorous leafroller	*Platynota sultana*	8.9	
Onion maggot	*Delia (Hylemya) antiqua*	4.4	
Orange tortrix	*Argyrotaenia citrana*	6.1	25.6
Oriental fruit fly	*Dacus Dorsalis*	12.8	36.4
Oriental fruit moth	*Grapholitha molesta*	7.2	32.2
Pacific spider mite	*Tetranychus pacificus*	11.1	
Pea aphid	*Acyrthosiphon pisum*	5.5	28.0
Peachtree borer	*Synanthedon exitiosa*	10.0	31.1
Peach twig borer	*Anarsia lineatella*	10.0	31.1
Pink bollworm	*Pectiniophera gossypiella*	15.6	
Plum fruit moth	*Grapholitha funebrana*	10.0	
Potato leafhopper	*Empoasca fabae*	11.4	30.0
Potato tuberworm	*Phthorimaea operculella*	14.6	
Saltmarsh caterpillar	*Estigminea acrea*	10.7	32.2
San Jose scale	*Quadraspidiotus perniciosus*	10.6	32.2
Seedcorn maggot	*Delia (Hylemya) platura*	3.9	29.0
Serpentine fruit fly	*Anastrepha serpentina*	8.8	30.0
Serpentine leafminer	*Liriomyza trifolii*	10.8	
Sevenspotted lady beetle	*Coccinella septempunctata*	12.1	30.0
Spotted tentiform leafminer	*Phyllonorycter blancordella*	6.7	
Spruce budworm	*Choristoneura fumiferana*	8.0	
Strawberry aphid	*Chaetosiphon fragaefolii*	14.2	
Strawberry mite	*Tetranychus turkestani*	12.0	
Sunflower moth	*Homoeosoma electellum*	13.3	
Tobacco budworm	*Helicoverpa (Heliothis) virescens*	12.8	30.0
Tomato fruitworm	*Helicoverpa (Heliothis) zea*	12.6	33.3
Tomato pinworm	*Keiferia lycopersicella*	10.2	35.0
Transverse lady beetle	*Coccinella transversoguttata richardsoni*	12.2	
Twospotted lady beetle		9.0	30.0

Common pest name	Scientific pest name	Lower threshold (°C)	Upper threshold (°C)
Twospotted spider mite	*Tetranychus urticae*	11.7	
Variegated cutworm	*Peridroma saucia*	7.2	26.7
Vegetable leafminer	*Liriomyza sativae*	10.0	
Western cherry fruit fly	*Rhagoletis indifferens*	5.0	
Western grape leafhopper	*Erythroneura elegantula*	10.3	

References

AASC (American Association of State Climatologists) (1985) Standards for instrumentation at automated stations. US Dep Commerce, National Climatic Data Center, Asheville, North Carolina. The State Climatologist 9 (4), pp 12–14

Aceves-Navarro LA, Hubbard KG, Schmidt J (1988) Group calibration of silicon cell pyranometers for use in an automated network. J Atmos Oceanic Technol 5:875–879

Allen RG, Jensen ME, Wright JL, Burman RD (1989) Operational estimates of evapotranspiration. Agron J 81:650–662

Arkin GF, Dugas WA (1981) Making weather and climate dependent crop management decisions. In: Weiss A (ed) Computer techniques and meteorological data applied to problems of agriculture and forestry: a workshop. Am Meteorol Soc, Boston, pp 223–237

Baldridge M, Byrn J, McElroy JH (1983) International meteorological satellite system: issues and opinions. Nov 1983, US Dep Comm, Washington, DC, pp 9–33

Bausch WC (1990) Sensor height effect on calculated reference evapotranspiration. Trans ASAE 33 (3):791–798

Bartholic J, Fear F (1988) Agricultural weather information systems evaluation project. Final Rep Mich St Univ and USDA Ext Ser

Baynton HW (1976) Errors in wind run estimates from rotational anemometers. Bull Am Meteorol Soc 57:1127–1130

Brock FV (ed) (1984) Instructor's handbook on meteorological instrumentation. NCAR/TN-237 + IA. Natl Center Atmos Res, Boulder

Brown PW (1987) Using a computer bulletin board as an agricultural weather information system. In: Preprint Vol 18th Conf Agric For Meteor, 8th Conf Biometeor Aerobiol, 14–18 Sept 1987, AMS, Boston, pp 67–69

Brown PW (1989) Accessing the Arizona meteorological network by computer. Ext Rep 8733, Univ of Arizona, Tucson

Brown PW, Owen-Joyce SJ, Daughtry CST, Kustas WP (1990) Effect of underlying surface on ground-based environmental data collected in an arid region. Poster: ASA Annu Meet 21–26 Oct 1990, San Antonio

Curry RB, Klinck JC, Holman JR, Elwell DL, Sciarini MJ (1988) Current Ohio experience with an automated weather station network. App Eng Agric 4 (2):150–155

Doorenbos J (1976) Agro-meteorological field stations. Irrig Drain Pap 27, FAO Rome

Doorenbos J, Pruitt WO (1977) Crop water requirements. FAO Irri Drain Pap No 24. UN FAO, Rome

Dong A, Grattan SR, Carroll JJ, Prashar CRK (1992) Estimation of daytime net radiation over well-watered grass. ASCE J Irrig Drain 118 (3):466–479

Dugas WA Jr, Arkin GF, Heuer ML (1984) A Texas weather advisory program. Texas Agric Exp Stn, MP-1562

Field RT, Fritschen LJ, Kanemasu ET, Smith EA, Stewart JB, Verma SB, Kustas WP (1992) Calibration, comparison, and correction of net radiation instruments used during FIFE. J Geophys Res 97:18681–18695

Fritschen LJ (1967) Net and solar radiation relations over irrigated fields. Agric Meteorol 4:55–62

Fritschen LJ, Gay LW (1979) Environmental instrumentation. Springer, Berlin Heidelberg New York

Fulton AE, Beede RH, Phene RC (1991) Implementing CIMIS at the farm level: a grower's experience in walnuts. California Agric 45 (5):38–40

Gandin LS (1970) The planning of meteorological station networks. WMO Tech Note 111, World Meteor Organ, Geneva

Gill GC (1983) Comparison testing of selected naturally ventilated solar radiation shields. Final Rep Contract #NA-82-0A-A-266 NOAA, St Louis

Goff J, Gratch S (1946) Low-pressure properties of water from −160 to 212F. Trans ASHVE 52:95–122

Hahn GL (1981) Use of weather data in the rational selection of livestock management practices. In: Weiss A (ed) Computer techniques and meteorological data applied to problems of agriculture and forestry: a workshop. Am Meteorol Soc, Boston, pp 362–382

Heerman DF (1981) Management of irrigation systems with water budgets. In: Weiss A (ed) Computer techniques and meteorological data applied to problems of agriculture and forestry: a workshop. Am Meteorol Soc, Boston, pp 247–256

Howell TA, Mack DW, Phene CJ, Davis KR, McCormick RL (1984) Automated weather data collection for research in irrigation scheduling. Trans ASAE 27 (2):386–391

Hubbard KG (1988) Collection, quality control and dissemination of weather data for irrigation and other operations in an automated setting. In: Hay D (ed) Planning now for irrigation and drainage in the 21st century. 18–21 July 1988 ASCE, New York, pp 732–739

Hubbard KG (1994) Spatial variability of daily weather variables in the high plains of the USA. Agric For Meteorol 68:29–41

Hubbard KG, Hines JR (1991) User guide for the automated weather data network system. High Plains Climate Center, Lincoln, Nebraska, 39 pp

Hubbard KG, Wilhite DA, Meyer SJ, Booysen J, Sagar R, Schmidt JJ, Hines JR (1987) A demonstration and evaluation of the use of climate information to support irrigation scheduling and other agricultural operations. Ctr Agric Meteorol Climatol Prog Rep 87–4, Univ of Nebraska, Lincoln

Hubbard KG, Rosenberg NJ, Nielsen DC (1983) Automated weather data network for agriculture. J Water Resour Plan Management ASCE 109:213–222

Jensen ME (1973) Consumptive use of water and irrigation requirements. ASCE, New York

Jensen ME, Burman RD, Allen RG (1990) Evapotranspiration and irrigation water requirements. ASCE Manuals Rep on Eng Pract No 70. Am Soc Civ Eng, New York

Jones AL, Fisher PD, Croft BA (1981) Implementing pest management models using real time weather data. In: Weiss A (ed) Computer techniques and meteorological data applied to problems of agriculture and forestry: a workshop. Am Meteorol Soc, Boston, pp 266–272

Kincaid DC, Heermann DF (1974) Scheduling irrigations using a programmable calculator. US Dep Agric, Agric Res Serv, ARS-NC-12, 55 pp

Ley TW (1988) Washington public agricultural weather system. Wash St Univ Ext and Wash Energy Ext Serv. Wash Irrig 9:1–8

Ley TW, Evans RG (1990) Washington public agricultural weather system. Pap OP-311 3rd Natl Irrig Symp, 28 Oct—1 Nov 1990, Phoenix

Ley TW, Kroeger MW (1988) Washington public agricultural weather system. Paper PNR880-106, ASAE Annu Meet, 28–30 Sep 1988, Pendleton, pp 12

List RJ (1963) Smithsonian meteorological tables. Smithsonian Inst, Washington, DC

Lowe PR (1976) An approximating polynomial for the computation of saturation vapor pressure. J Appl Meteorol 16:100–103

MacCallum DH, Nestlebush MJ (1983) The geostationary operational environmental satellite data collection system. NOAA Tech Mem NESDIS 2. US Dep Commerce, Washington, DC
Mahrt L (1986) Nocturnal topoclimatology. World Meteorological Organization, Geneva, WCD-117 WMO/TD No 132, 76 pp
Merva G, Fernandez A (1985) Simplified application of Penman's equation for humid regions. Trans ASAE 28:819–825
Martsolf JD (1981) Satellite frost forecasting system for Florida. In: Weiss A (ed) Computer techniques and meteorological data applied to problems of agriculture and forestry: a workshop. Am Meteorol Soc Boston, pp 143–163
Meyer SJ, Hubbard KG (1992) Nonfederal automated weather stations and networks in the United States and Canada: a preliminary survey. Bull Am Meteorol Soc 73 (4):449–457
Meyer SJ, Hubbard KG, Wilhite DA (1991) A crop specific drought index for corn. I. Model development and validation. Agron J 85:388–395
Meyer SJ, Hubbard KG, Wilhite DA (1989) Estimating potential evapotranspiration: the effect of random and systematic error. Agric For Meteorol 46:285–296
Newman JE, Shaw RH, Suomi VE (1959) The agricultural weather station: its instruments, observations, and site requirements. Bull 537, Univ of Wisconsin Agric Exp Stn, Madison
NOAA (1988) US CLICOM system program manual. NOAA NCDC, Asheville
NOAA/NESDIS (1990) User interface manual for the data collection system automatic processing system. NOAA/NESDIS Contract No 50-DDNE-7-00037, US Dep Commerce, Washington, DC
NESDIS (1985) Users guide for random reporting: an introduction to GOES random reporting services. Contract No 0-07-10-S0104, US Dep Commerce, Washington DC
Park J (1981) The wind power book. Chelshire, Palo Alto
Penman HL (1948) Natural evaporation from open water, bare soil and grass. Proc R Soc Lond A193:120–145
Phene CJ, Campbell RB (1975) Automating pan evaporation measurements for irrigation control. Agric Meteorol 15:181–191
Powers AR (1988) AgriMet – "ET" phones home – (adaptation of satellite telemetry for crop water use modeling in the Pacific Northwest). In: Hay D (ed) Conf proc planning now for irrigation and drainage in the 21st century. 18–21 July 1988, ASCE, New York, pp 359–367
Powers AR, Lute D, Rigby R (1988) Regional water management and conservation in the Pacific Northwest – technical and institutional approaches. In: 11th Conf on Irrigation and drainage, 14–17 Sept 1988, US Comm Irrig Drain, San Diego
Pride (1984) The changing direction in weather information. Farming with Pride 4 (1):4–7; 10–11
Pruitt WO, Doorenbos J (1977) Empirical calibration a requisite for formulae based on daily or longer mean climatic data? Proc Int Round Table Conf on Evapotranspiration, Budapest, Int Comm Irrig and Drain, 20 pp
Reicosky LJ, Windelman LJ, Baker JM, Baker DG (1989) Accuracy of hourly air temperatures calculated from daily minima and maxima. Agric For Meteorol 46:193–209
Robbins KO, Titlow JK (1989) Louisiana agricclimatic information system annual climatic summary. Bull 801, Louisiana Agric Exp Stn
Robinson JM, Hubbard KG (1990) Soil water assessment model for several crops in the High Plains. Agron J 82:1141–1148
Running SW (1981) The influence of microclimate on forest productivity: a system to predict the biophysical site quality of forest land. In: Weiss A (ed) Computer techniques and meteorological data applied to problems of agriculture and forestry: a workshop. Am Meteorol Soc, Boston, pp 266–272
Rosenberg NJ, Blad BL, Verma SB (1983) Microclimate – the biological environment, 2nd edn. Wiley, New York, pp 210–211
Sagar RM, Hubbard KG, Norman JM, Holtzer TO (1988) Estimation of corn canopy temperature and water budget using automated weather station data. Thesis, Univ of Nebraska-Lincoln, 143 pp

Schwerdtfeger P (1976) Physical principles of micro-meteorological measurements. Elsevier, New York

Skaar J, Hegg K, Moe T, Smedsrud K (1989) WMO international hygrometer intercomparison (Norway, 1987–1989). World Meteorol Org Geneva

Slatyer RO, McIlroy IC (1961) Practical microclimatology, CSIRO, Melbourne (UNESCO)

Snyder RL, Pruitt WO (1992) Evapotranspiration data management in California. J Irrig Drain (ASCE) Proc Water forum '92, Aug 2–6, 1992, Baltimore

Snyder RL, Pruitt WO, Dong A (1985a) An automatic weather station network for estimation of evapotranspiration. In: Perrier A, Riou C (eds) Crop water requirements. Int Comm Irrig Drain, Versailles, pp 133–142

Snyder R, Pruitt WO, Henderson DW, Dong A (1985b) California irrigation management information system final report, vol 1. Land air and water resources papers 10013A. Univ California, Davis

Snyder R, Pruit WO, Henderson DW, Dong A (1985c) California irrigation management information system final report, vol 3. Land air and water resources papers 10013C. Univ California, Davis

Snyder R, Shaw R, Paw-U KT (1987) Humidity conversions using a computer program. Appl Agric Res 2:183–192

Spoden GJ, Seeley MW (1985) The Minnesota cooperative agriculture weather advisory program. Preprint Vol 17th Conf Agric For Meteorol, 7th Conf Biometeorol Aerobiol, 21–24 May 1985, AMS Boston, pp 109–111

Tanner BD (1990) Automated weather stations. In: Goel NS, Norman JM (eds) Remote Sensing Reviews, vol 5(1). Gordon and Breach, New York, pp 73–98

Tanner CB, Pelton WL (1960) Potential ET estimates by the approximate energy balance method of Penman. J Geophys Res 65:3391–3413

Tetens VO (1930) Über einige meteorologische Begriffe. Z Geophys 6:297–309

Thompson RC, Costello TA, McLaughlin JD (1984) Design of a permanent automated agroclimatic data acquisition station. ASAE Pap No 84-4540

Titlow JK, Robbins KD (1988) Climate data: collection, quality control and management. Pap: 1988 Winter Meet ASAE, 13–16 Dec 1988, Chicago

Van Bavel CHM (1966) Potential evaporation: the combination concept and its experimental verification. Water Resources Res 2:455–467

Wang JY, Felton CMM (1983) Instruments for physical environmental measurements with special emphasis on atmospheric instruments, vol I. Kendall/Hunt, Dubuque, Iowa

Warren JR, Vance DL (1981) Remote automatic weather station for resource and fire management agencies. Gen Tech Rep INT-116. Intermountain For Range Exp Stn, USDA, Intermountain Res Stn, Ogden, Utah

WMO (1983) Guide to meteorological instruments and methods of observation, 5th edn. World Meteorol Org No 8, Geneva

Wright JL (1982) New evapotranspiration crop coefficients. J Irrig Drain 108 (IR2):57–74

Wright MA, Ley TW (1989) PAWS: Washington public agricultural weather system user manual. EB 1547 Wash State Univ, Pullman

Zalom FG, Goodell PB, Wilson LT, Bantley WJ (1983) Degree-days: the calculation and use of heat units in pest management. Univ California, Div Agric and Nat Resour, Leaflet 21373, Oakland

Satellite Mapping of the Surface Radiation Budget

MANUEL NUNEZ[1] and JETSE D. KALMA[2]

1 General Introduction

Bioclimatology is defined in the preface to *Advances in Bioclimatology* as "the study of the relations between the physical environment and the form and function of living organisms". Such a definition largely confines the discipline to the proximity of the land surface, the interface between the atmospheric boundary layer, vegetation and the soil surface.

Physical climatology is concerned with the processes by which energy and mass are transferred, converted and stored in the Earth–atmosphere system (Oke 1988). The transfer of energy occurs in three modes: conduction, convection and radiation. Radiation is the transfer of energy by the rapid oscillations of electromagnetic fields. These oscillations may be considered as travelling waves, characterized by wavelength λ. In environmental physics and bioclimatology we are concerned with wavelengths in the approximate range of 0.001–100 µm (1 µm = 10^{-6} m = 1000 nm = 10^4 Å).

Solar radiation is the sole source of energy to living organisms, directly in the case of plants and indirectly to animals – including man. All physical and biological processes at and near the land surface are driven by energy from the sun. Shortwave radiation falls in the 0.15–3 µm range and longwave radiation lies in the range 3–100 µm. Ultraviolet (UV) radiation has wavelengths between 0.001 and 0.4 µm, i.e., 1–400 nm. UV-C radiation (λ < 280 nm) never reaches the Earth's surface because of absorption by the atmosphere; the 280–320 nm range is called UV-B; and UV-A comprises the 320–400 nm part of the spectrum. UV-A and UV-B radiation is of major significance for human health and plays a role in biological productivity.

All processes, including the biological, involve energy transformation or transfer. The energy balance describes the energy available to land-surface processes and how this energy is used. Sellers (1965) observes that "the physical state of any entity, from a small leaf to a whole continent is determined by the way it utilizes the available energy". The radiation and energy balances are

[1] Department of Geography and Environmental Studies, The University of Tasmania, Hobart Tas. 7001, Australia

[2] Department of Civil Engineering and Surveying, The University of Newcastle, Callaghan NSW 2308, Australia

therefore important in describing the state of the physical environment to which living organisms (plants, animals, human beings) are exposed.

If an infinitesimally thin layer of zero thickness without energy storage is considered, the surface heat budget or energy balance is given by

$$Q_* = Q_h + Q_e + Q_g, \tag{1}$$

where Q_* is the net gain of shortwave (0.15–3 μm) and longwave (3–100 μm) radiation at the surface, Q_h is the sensible heat exchange between the Earth's surface and the atmosphere (convection), Q_e is the latent heat exchange (evaporation, condensation) between the ground surface and the atmosphere, and Q_g is the sensible heat flux into the soil. Equation (1) expresses conservation of energy both instantaneously and integrated over time. Stull (1991) points out that the neglect of a storage layer is acceptable for quasi-steady state situations without appreciable change in mean temperature. He notes that it works well for flat bare land surfaces.

When considering a layer of finite thickness, above the soil surface, comprising crops, trees, buildings or animals, various processes can occur within the layer including the storage of sensible heat in plant canopies or buildings; the storage of chemical energy by photosynthesis; and flux divergence in sensible and latent heat fluxes associated with evaporation, transpiration and condensation. These processes are all included in a heat storage term ΔQ_S so that

$$Q_* = Q_h + Q_e + \Delta Q_s + Q_g. \tag{2}$$

In Eq. (2) the soil-heat flux Q_g leaves/enters through the bottom of the surface layer. The fluxes of net radiation, sensible heat and latent heat leave/enter through the top of the layer. Energy balance Eqs. (1) and (2) are of crucial importance in providing a rational basis for many physical and biological processes considered in climatology, meteorology, agronomy and hydrology. Separate energy balances may be drawn up for each organism (leaf, plant, animal etc.) within the surface layer.

Conservation of energy requires that gains and losses of energy at the land surface must balance both instantaneously and integrated over time. All energy for physical and biological land-surface processes comes from the sun, and environmental physics "is concerned with the ways in which this energy is dispersed or stored in thermal, mechanical or chemical form" (Monteith and Unsworth 1990). As shown earlier, a key component of the surface energy budget is the net amount of radiation Q_* which is defined by the radiation balance equation

$$Q_* = K_d - K_u + L_d - L_u = (1 - \alpha_g) K_d + L_d - L_u \tag{3}$$

where K_d is the total (direct + diffuse) downward solar (shortwave) radiation incident on a horizontal surface, K_u is the upward solar radiation reflected by the ground (with a reflection coefficient α_g), L_d is the downward (incoming) longwave radiation and L_u is the upward (outgoing) longwave radiation.

It should be noted that biology plays a key part in the energy balance through the release of latent heat in transpiration. The other direct biological components of the energy balance, i.e., the release of heat by respiration, the storage of heat by photosynthesis and the storage of heat in biomass, are usually very small. On all but the largest spatial scale the downwelling radiation fluxes reaching the land surface, i.e., K_d and L_d, are independent of the properties of the surface. This is not the case for the upwelling fluxes of reflected shortwave (K_u) and emitted longwave radiation (L_u). The energy balance Q_* depends on the reflectivity, emissivity and surface temperature of plant associations, thus providing an important bio-feedback control system.

Information on the temporal and spatial distribution of the components of Eq. (3) are essential in estimating the surface energy budget. Various measuring and estimating techniques have been developed for short- and longwave radiation components.

Short-wave radiation (K_d, K_u) may be measured directly with pyranometers, longwave radiation (L_d, L_u) is measured with pyrgeometers and the net radiation Q_* may be measured with net radiometers or pyradiometers. Such instrumentation is costly and difficult to maintain. Routine radiation measurements are usually restricted to sparse networks of pyranometers. In the absence of radiation measurements, the components of Eq. (3) are usually estimated from various relationships involving cloudcover observations or observations of the duration of bright sunshine (sunshine hours) in the case of K_d; measurements of surface reflectivity (albedo) in the case of K_u; and various (semi-) empirical relationships for L_d and L_u employing temperature and humidity data obtained with a network of climate stations and atmospheric soundings.

Present capital costs of measuring and recording even the simplest and most basic elements of the radiation balance run into several thousands of US dollars per station. Regular maintenance and calibration costs must be added to the capital costs, so that the purchase and maintenance of a solar radiation network is an expensive investment. As a result, most solar radiation networks, usually maintained by national meteorological organizations, consist of only a limited number of stations.

The spacing between pyranometer stations is usually not sufficient to resolve the finer details in the regional solar radiation field. Measurements taken with pyranometer arrays in various regions (Wilson and Petzold 1972; Hay and Suckling 1979; Hay 1984) provide estimates of distance-correlation functions developed for daily global radiation. The results are highly dependent on the degree and type of local cloudiness which in turn may vary with season. Typically, a station separation of less than 20 km is needed in cloudy conditions and 80 km in cloudless conditions if errors in the interpolated global irradiance are to be kept within 10% of the mean monthly value (Hay 1981).

Cloud and sunshine duration data may also be used in models to supplement instrumental measurements (Hounam 1963; Davies et al. 1975; Paltridge and Proctor 1976; Suckling and Hay 1976, 1977; Davies and McKay 1982). However, there are difficulties in applying this technique on the regional scale. The number of meteorological stations recording cloud-cover information is usually sparse, so that only the broadest features of the global radiation field may be observed. More importantly, cloud type and coverage, as well as total cloud coverage, are needed for estimating the global radiation field at frequencies of a day or shorter (Davies et al. 1975; Suckling and Hay 1977). Observations of cloud type are made less frequently than those of total cover, so that most mapping applications are restricted to averages over monthly or longer periods (Hounam 1963; Nunez 1978).

It has become apparent since the early 1970s that satellite remote sensing provides a range of methods for determining the Earth's radiation and energy budgets which are globally consistent and economically feasible (Sellers et al. 1990; Carleton 1991). It may be argued that many remote sensing techniques can now provide information at substantial cost savings compared to more traditional approaches.

Clearly, in the absence of closely spaced instrumental measurements on the ground, a modelling approach using atmospheric opacity data is the only alternative to mapping the regional global radiation field. The high-frequency and high-density cloud information (or cloud-related brightness information) is needed as an index of atmospheric opacity for use in radiation models. This requirement may be largely satisfied by the use of satellite data. Satellites can provide high temporal and spatial coverage and regional-scale raster-based information. However, the resolution of the satellite sensors limits small-scale applications.

In this review we provide a survey of methods for estimating the components of the surface radiation budget with remote sensing data. In separate sections we discuss techniques for estimating global radiation, ultraviolet radiation, reflected shortwave radiation and net all-wave radiation. The format adopted is different from that of more traditional reviews: we survey recent advances in satellite mapping of the surface radiation budget and, by way of illustration, report on relevant case studies carried out in Australia and the tropical western Pacific region. Particular emphasis is placed on the use of operational (meteorological) satellites because of the frequency and accessibility of the data as well as their relatively low cost. These features are particularly important for applications in agriculture and hydrology.

Throughout the review we emphasize the use of satellite-based methods for regional rather than global scale mapping of the radiation budget components. We discuss a range of approaches developed for the estimation of solar irradiance K_d from satellite-based remote sensing. These fall into three broad groups: methods based on physical principles which account for scattering, absorption and reflection processes; statistical/empirical techniques which relate brightness to insolation; and methods which combine physical and statistical approaches.

There is also an urgent need for regional and global climatologies to be developed for ultraviolet irradiance at the Earth's surface. Such climatologies are not yet available because of several obstacles. These include a lack of universally acceptable instrumentation and of a global measurement network of systematic measurements; uncertainties as to which spectral bands to cover; and a lack of process studies which address the depletion of UV radiation by atmospheric constituents such as cloud and aerosols. This review will also discuss the use of satellite data for regional-scale estimation of UV-A and UV-B irradiance. UV-A estimates may be based on satellite-based maps of incoming broad-band solar irradiance (K_d) and an empirical relationship between (narrow-band) UV-A irradiance and (broad-band) global solar irradiance. UV-B is obtained with a relationship between measured daily cloudy sky transmissivity for UV-B and a satellite-derived cloudiness index.

Reflected radiances sensed by the satellite consist of atmospheric and land surface contributions. The review describes the operations necessary to obtain regional surface albedo from these reflected radiances: estimating the contribution from atmospheric absorption and scattering; accounting for non-Lambertian behaivour; and converting narrow-band radiances to broad-band radiances.

Section 5 reviews estimation methods of net (all-wave) radiation, $Q_* = K_* + L_*$. The net shortwave component K_* may be derived from satellite-derived estimates of K_d and K_u. For cloudless conditions net longwave radiation $L_* = L_d - L_u$ may be obtained by estimating L_d and L_u separately. Satellite-derived surface temperatures may be used to estimate the upwelling longwave radiation L_u and the downwelling longwave radiation L_d may be obtained from semi-empirical techniques using near-surface air temperature. Empirical relations between L_* and K_* over various surfaces for use in cloudy and clear conditions are discussed. Results obtained with a convective boundary layer model are presented to show the important role of surface roughness (texture) in determining the L_*/K_* ratio.

Finally, a comment on the use of satellite data used for estimating components of the surface radiation budget. The main satellite systems with visible and infrared radiometers providing data for use in the development of radiation climatologies are satellites in the TIROS and NOAA polar orbiters series and geostationary platforms, notably the ATS/SMS/GOES series (USA), Meteosat (Europe) and GMS (Japan).

Polar-orbiting satellites cover the entire globe with an increasing density of coverage approaching the poles. TIROS and NOAA polar-orbiting satellites are equiped with the Advanced Very High Resolution Radiometer (AVHRR). The AVHRR scanning radiometer is sensitive in the visible, near-infrared and infrared window regions. Its footprint is 1.1 km. The TIROS Operational Vertical Sounder (TOVS) sensors include the High Resolution Infrared Sounder (HIRS). Its footprint is 17.4 km.

Geostationary satellites do not provide coverage at latitudes greater than 60° and there are gaps in the coverage of the 40–60° zone in both hemispheres. Early geostationary satellites were equiped with the imaging Visible (0.5–0.7 μm) and Infrared (10–12 μm) Spin-Scan Radiometer (VISSR) with a spatial resolution of between 1 and 8 km. The VISSR Atmospheric Sounder (VAS) is a combined imaging and sounding device on current GOES satellites. It has a visible imager with a 1-km footprint, an infrared imager with a 7-km footprint and a 12-channel infrared sounder with a footprint of 14 km. Further details on current satellite systems and sensors can be found in many recent publications, including Barrett and Curtis (1992) and Carleton (1991).

2 Global radiation

2.1 Introduction

On its path through the atmosphere the beam of direct solar radiation is subjected to absorption by water vapour, ozone and carbon dioxide, and to Rayleigh and Mie (aerosol) scattering. Conservation of radiative energy implies that the extra-terrestrial irradiance, K_{extra} (calculated from latitude, date, time and solar declination) must equal the sum of the fluxes of radiation absorbed in the atmosphere (K_{abs}), the radiation reflected back to space $(K_u)TOA$ and the net shortwave irradiance absorbed by the Earth's surface (K_*). Thus

$$K_{extra} = K_{abs} + (K_u)TOA + K_* . \tag{4}$$

The net shortwave irradiance at the surface (K_*) is given by

$$K_* = (1 - \alpha_g)K_d = K_d - K_u , \tag{5}$$

where α_g is the surface albedo, K_d is the downwelling shortwave irradiance incident on the ground and K_u is the upwelling (reflected) shortwave irradiance. K_d may be divided into direct and diffuse components.

Solar irradiance K_d can now be estimated from satellite brightness observations (which depend on surface albedo and cloud conditions). Brightness data as an index of atmospheric opacity are obtained from the low-resolution visible channels of geostationary or polar-orbiting satellites. Clear-sky insolation values are calculated from existing simple relationships. Clouds reduce these clear sky values by an amount which depends on their albedo as determined from observations of reflected radiation.

Sellers et al. (1990) have summarized algorithms for determining insolation at the land surface from remote sensing. They note that there are three groups: algorithms based on (1) physical techniques which account for scattering, absorption and reflection processes (see Sect. 2.3); (2) statistical/ empirical techniques which relate brightness to insolation (see Sect. 2.4); and (3) approaches which combine physical and statistical approaches.

It should also be noted that complementary to the use of satellite brightness values for estimation of K_d is its use for determining the surface albedo α_g as discussed in Section 4.2.

2.2 The Satellite Approach

The usefulness of the satellite approach can best be illustrated with the following simple example. Consider the transmission of daily solar radiation by an atmosphere composed of an absorbing layer lying above a scattering layer. Assume that no further absorption occurs in the scattering layer. Some of the backscattering is lost to space and this is considered to be atmospheric reflectance. On a daily basis the global radiation at the surface (K_d) can then be written in terms of the extraterrestrial global radiation (K_{extra}), calculated from latitude, date, time and solar declination, and daily values of atmospheric reflectance (α_a) and absorptance (ϕ_a):

$$K_d = K_{extra}(1 - \alpha_a)(1 - \phi_a). \tag{6}$$

The direct beam absorption of solar radiation by water vapour and ozone may be estimated with the technique of Lacis and Hansen (1974). Figure 1 shows the daily absorptance of K_{extra} for the two Australian cities of Hobart and Darwin as a function of precipitable water vapour. These sites were chosen as examples of temperate and tropical climates. Superimposed on the curves are the absorptances corresponding to the 10 (point A), 50 (point B) and 90 (point C) percentile values of the precipitable water vapour for January at each city (Pierrehumbert 1972).

The results illustrate the conservative nature of ϕ_a when compared to α_a. Typically, daily atmospheric absorptance values may depart by one or two percent (in absolute terms) from their mean value. However, larger differences in ϕ_a may exist between climatic regions as illustrated in Fig. 1 which shows the absorptances for tropical Darwin during the wet season of January with equivalent figures for mid-latitude Hobart. This conservative nature of ϕ_a implies that a single daily value based on precipitable water vapour may suffice for radiation estimates over regional areas.

By contrast, atmospheric reflectance (α_a) varies strongly with cloud cover and type since clouds are good reflectors but poor absorbers of solar radiation (Paltridge and Platt 1976; Welch et al. 1980; Rawlins 1989). Essentially the satellite technique supplies information related to α_a in a timely, repetitive and comprehensive fashion from which K_d may be calculated.

In the real world the above assumptions do not fully describe all the processes that act on the irradiance on its passage through the atmosphere and

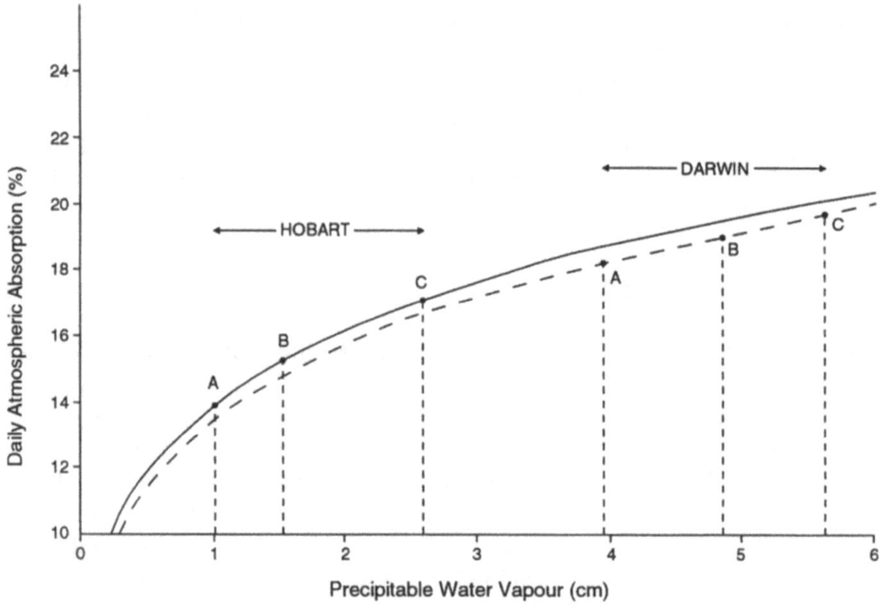

Fig. 1. Daily atmospheric absorption of extraterrestrial solar radiation using the relationship of Lacis and Hansen (1974)

it may be necessary to incorporate such effects as narrow-band/broad-band differences, cloud absorption, non-Lambertian reflection from clouds, etc. Variability in aerosol load also presents a potential problem as discussed by Nunez (1993). Previous modelling attempts at estimating global radiation from satellites may include some of these processes. These approaches range from purely empirical and site-specific to a detailed description of the radiation transfer involved (Sellers et al. 1990).

2.3 Physical Models

Physical models attempt to reproduce the physical processes in such a way so as to minimize the degree of empiricism and to make the approaches potentially applicable to a wide variety of environments. Since the role of the satellite is essentially to provide cloud-related atmospheric opacity information, it is convenient to classify the various satellite models according to their treatment of cloud cover.

The physical approach uses measurements of surface albedo to differentiate between cloudy and cloudless regions. Radiation theory is then used to estimate global radiation in clear and cloudy conditions and to correct for atmospheric depletion between cloud top and satellite sensor (Dedieu et al. 1987).

In the general case, the narrow-band radiance (N′) received by the satellite from a pixel which is partly covered by clouds is given by

$$N' = (1-C)N'_o + CN'_c, \tag{7}$$

where N'_o and N'_c denote the radiance from the cloudfree and cloudy portions of the pixel respectively as incident on the sensor, and C is the fractional cloud cover. N′ may increase from pixel to pixel if either C increases with N'_c constant, or N′ increases with C constant. There is not enough information at the sub-pixel scale to resolve which of the two processes will dominate. Some approaches (Dedieu et al. 1987) treat only the combined radiance N′. An Earth–atmosphere albedo may be obtained as

$$\alpha'_{ea} = (\pi N')/\left[I'\cos(z)r^{-2}\right], \tag{8}$$

where I' is the solar constant in the satellite bandpass, z is the local solar zenith angle, and r is the radius vector of the sun–Earth distance. Note that Eq. (8) assumes that the reflection from the Earth–atmosphere system is Lambertian so that the upward irradiance is given by πN′. Dedieu et al. (1987) compute the surface global irradiance as

$$K_d = I\cos(z)r^{-2}Tr_o(z)(1-\alpha'_a)/(1-\alpha_g), \tag{9}$$

where I is the broad-band solar constant, $Tr_o(z)$ is a clear sky transmissivity obtained from the parameterization of Lacis and Hansen (1974) for Rayleigh scattering, Mie scattering and gaseous absorption, and the formulation of Tanré et al. (1979) for aerosol depletion. The calculation of the broad-band surface albedo (α_g) is described in the next section. The narrow-band atmospheric albedo (α'_a) is derived from the satellite observation. The assumption is made that no further atmospheric depletion takes place on the beam reflected from the cloud tops (or the Earth's surface) and incident on the satellite sensor.

Darnell et al. (1988) assume that the surface irradiance is a product of a "clear sky" transmissivity ($Tr_o(z)$) and a satellite-derived cloudy transmissivity (Tr_c), both acting on the extraterrestrial irradiance K_{extra}:

$$K_d = K_{extra}Tr_o(z)Tr_c. \tag{10}$$

The novel feature of this approach is that $Tr_o(z)$ is obtained by examining five pyranometer sites in the United States under cloudless conditions. After accounting for attenuation by Rayleigh scattering and gaseous absorption, the aerosol depletion is given as a residual term provided that K_d is known. In turn, the aerosol depletion term is parameterized as a linear function of precipitable water vapour which restricts the generalization of the results.

The cloudy sky transmissivity Tr_c was obtained from the 2.5° × 2.5° heat budget product (HBP) of AVHRR. Tr_c is a normalized transmission function relying on the top of the atmosphere (TOA) albedo of HBP. It relied on the development of a normalized albedo function as given by Eq. (11a)

$$\alpha_n = (\alpha'_{ea} - \alpha'_1)/(\alpha'_0 - \alpha'_1), \tag{11a}$$

where α'_{ea} is the monthly average TOA albedo and values of α'_0 and α'_1 are obtained by extrapolating the linear regression between the monthly transmissivity (Tr) based on pyranometer measurements (K_d/K_{extra}) and α'_{ea} to values of Tr = 0 and Tr = 1, respectively. Hence, the transmissivity due to clouds (Tr_c) may be obtained from

$$Tr_c = 1 - \alpha_n, \tag{11b}$$

where α_n may be considered as the normalized albedo of a pseudo cloud cover.

A more theoretical approach has been described by Pinker and Ewing (1985) who used a three-layer model with a homogeneous cloud layer extending from 2 to 5.5 km. A Delta-Eddington technique was developed with parameterization for water vapour, ozone absorption and Rayleigh scattering. Single-scattering albedos for various optical depths and solar-zenith angles were obtained. In the operational framework the satellite-derived albedo is used to derive a cloud optical depth. Having determined this value the model then calculates the surface irradiance. This approach has been refined (Pinker and Lazlo 1992) for use with satellite global data sets. It involves incorporating greater flexibility in algorithms which estimate surface albedo as well as radiation extinction due to aerosols and clouds.

Another normalized cloud-cover approach has been developed by Raschke and Preuss (1979), Moser and Raschke (1983, 1984), Tuzet et al. (1984) and Diekmann et al. (1988). A normalized albedo is derived from each image as:

$$\alpha_n = \left[\alpha'_{ea} - (\alpha'_{ea})_{min}\right] / \left[(\alpha'_{ea})_{max} - (\alpha'_{ea})_{min}\right], \tag{12}$$

where $(\alpha'_{ea})_{min}$ and $(\alpha'_{ea})_{max}$ are the minimum and maximum TOA albedo values as observed by the satellite in a given scene. The concept is very similar to the normalized transmission function of Darnell et al. (1988) described in Eqs. (11a) and (11b). Raschke and Preuss (1979) used NIMBUS-3 data with a 250 by 250 km resolution to derive the surface insolation

$$K_d = (1 - \alpha_n)K_1 + \alpha_n K_2, \tag{13}$$

where K_1 and K_2 are the cloudless and overcast global solar radiation using a two-stream approximation model (Kerschgens et al. 1978). Equation (13) is analogous to the classical insolation formula if α_n is substituted for cloud cover (C).

In a later paper, Moser and Raschke (1984) used visible band Meteosat data to derive an insolation relationship of the form

$$K_d = K_1\left[1 - f(Z, \alpha_N)\right], \tag{14}$$

where the function f was derived from the theoretical relationship of Kerschgens et al. (1978). Stuhlman et al. (1990) developed this model further by considering the atmosphere to be composed of separate layers influenced by ozone absorption, Rayleigh scattering, cloud absorption/scattering and water vapour/aerosol scattering and absorption. The satellite-based technique of Moser and Raschke (1983) was used to obtain the cloud transmitance. Various relationships published in the literature were used to obtain the albedo and transmitance of the other layers. The method resulted in estimates of direct, diffuse and global irradiance at the surface.

A different physical approach was developed by Gautier et al. (1980). The technique assumes that the pixel seen by the satellite views a completely cloudless or cloudy scene. The model traces the extra-terrestrial global radiation in its path to the Earth's surface where it is reflected back to space and to the satellite sensor. In its passage through the atmosphere the solar beam is absorbed by water vapour; Rayleigh scattered; absorbed, reflected and transmitted by clouds; and reflected by the Earth's surface. The formulations of Lacis and Hansen (1974) and Coulson (1959) are used to estimate water-vapour and ozone absorption, and Rayleigh scattering respectively. The model for cloudy conditions is essentially the clear-sky model with the added effect of clouds. The clouds are assumed to occur in a discrete layer. Scattering and absorption coefficients based on other studies are assigned to the airmass above, within, and below the cloudy layer. Thresholds are set for pixel reflectivity to distinguish cloudless from cloudy pixels. Surface and cloud albedos are obtained from the cloudless and cloudy albedos respectively. The cloudy and cloudless irradiance at the Earth's surface may be calculated once the albedos are obtained.

When the model of Gautier et al. (1980) was used for hourly estimation with GOES data, it was found that for daily total insolation values its estimates were within 5% of mean daily measured insolation on clear days. This figure increases to 14 and 15% for partially cloudy and completely overcast conditions, respectively. When all daily values were combined, the standard errors of the estimates were within 9% of the measured daily values of incoming solar radiation.

In a later paper, Diak and Gautier (1983) modified the radiation transfer algorithms, omitting water-vapour absorption in the GOES spectral band but including improved ozone absorption and Rayleigh scattering parameterization. Their method operates on individual pixels. Insolation for clear pixels was computed with an atmospheric transmission model. In the case of cloudy pixels, the atmospheric transmission model assumed a certain cloud-absorption value and used observed cloud reflection. Gautier and Frouin (1985) included an aerosol-depletion term which was obtained from visibility measurements and the parameterization of Tanré et al. (1979).

Frouin and Chertok (1992) used monthly Nimbus-7 Earth radiation budget (ERB) planetary albedo data to generate a long-term climatology of net solar irradiance at the ocean surface. Their model considered a clear-sky atmosphere positioned above a cloud/surface layer. The incoming solar irradiance for cloudy conditions at the ocean surface (K_{dc}) would then be

$$K_{dc} = K_{do}\left(1 - \alpha_{cg} - \phi_c\right)\left(1 - \alpha_a\alpha_{cg}\right)^{-1}, \tag{15}$$

where K_{do} is the clear sky irradiance reaching the ocean surface, further depleted by the albedo of the cloud/surface system (α_{cg}) and the absorption by clouds (ϕ_c). The term ($1 - \alpha_a\alpha_{cg}$) accounts for multiple reflections between the cloud/surface system and the overlying atmosphere characterized by an atmospheric albedo (α_a). The authors used the radiative transfer model of Tanré et al. (1985) to relate α_{cg} to the satellite reflectivity. The above model relies on climatological values of water vapour and ozone amounts with an aerosol load corresponding to a 23-km visibility marine atmosphere. The term α_{cg} may be related explicitly to satellite derived surface albedo, cloud albedo and cloud cover. Validation (Chertok et al. 1992) was done with high-resolution satellite data sets as well as surface-based climatologies. A similar approach has been taken by Breon et al. (1994) using high-resolution ERB data in the S4 format.

2.4 Empirical Models

Empirical techniques generally are based on comparing satellite brightness data with simultaneous pyranometer measurements within the same pixel. Transmission functions are then developed which may be used to map solar radiation over the region of interest.

In some applications an empirical approach offers distinct advantages. These could occur, for example, when the satellite calibration is not well known, or if some of the atmospheric depletion terms are unknown or difficult to estimate. These models consider implicitly aerosol effects and non-Lambertian reflection from the cloud top. In most cases these relationships are easy to use and not demanding in terms of computing time. However, they lack generality and cannot be used in other environments without surface validation.

Some of the earlier relationships have used analogue hard-copy satellite data, such as photo-transparencies of GOES images (Hiser and Senn 1980) which are scanned and converted into grey levels representing a range of conditions from cloudless to overcast. For a given pixel these digitized cloud cover data can then be related to concurrent pyranometer measurements.

Rimoczi-Paal (1983) also used analogue images from Meteosat to derive a relation for Hungary. A circle 50 km in diameter was drawn around various pyranometer stations and the cloud cover was determined visually. The following relationship was derived for 50 days in September, October and November 1979:

$$K_d/K_{extra} = c_1C^2 + c_2V^2 + c_3VC + c_4C + c_5V + c_6, \tag{16}$$

where V is visibility (km), C is cloud cover in tenths (0–10), K_d/K_{extra} is expressed as a percentage (0–100) and c_1–c_6 are regression constants.

One advantage of empirical models is their ability to represent complex radiation interactions without the need to develop a rigorous theoretical framework. Tarpley (1979) implemented a modelling procedure which required the development of a clear (minimum) brightness parameterization; the determination of cloud amounts; specification of mean target-area brightness and mean cloud brightness; determination of atmospheric transmittance; and formulation of final regression equations based on simultaneous satellite reflectance measurements and ground-based insolation measurements.

The clear day minimum brightness (B) is the radiance from the Earth's surface in the absence of clouds. Tarpley (1979) has obtained B as visible reflectance counts by the United States GOES satellite over a region in the continental USA, spanning 10° longitude and 20° latitude. B is represented as a function of the local solar zenith angle (z), and the azimuth angle between sun and satellite ($\Delta\phi$). This relationship takes into account the non-Lambertian properties of the surface. It has the form

$$B = a + b\cos z + c\sin z\cos\Delta\phi + d\sin z\cos^2\Delta\phi, \tag{17}$$

where a, b, c and d are regression constants.

Using two threshold values for observed brightness, each pixel in the target area is then labelled as being cloud-free (N_1), partly cloudy (N_2) or totally cloudy (N_3). The total cloud fraction (n) of the entire target area may then be represented by

$$n = \left(0.5N_2 + N_3\right)/\left(N_1 + N_2 + N_3\right), \tag{18}$$

where the denominator represents the total number of pixels in the target area.

Total atmospheric transmittance is taken as the product of the partial transmittances due to water-vapour scattering, water-vapour absorption and Rayleigh scattering. These three terms are expressed as functions of precipitable water and optical air mass.

Tarpley (1979) then developed regression equations for the three cloud categories which express hourly values of irradiance at the ground surface as functions of local solar-zenith angle (z), azimuth angle between sun and satellite ($\Delta\phi$), minimum (clear) brightness, mean target brightness and mean cloud brightness.

Brakke and Kanemasu (1981) obtained the following prediction relationship for hourly fluxes:

$$K_d = c_0 + c_1\text{DIFF} + c_2\cos z, \tag{19}$$

where DIFF is the difference in GOES digital counts between the clear sky and cloudy radiance, z is the zenith angle (radians), and c_0, c_1 and c_2 are regression constants.

Davis and Chatters (1981) used NOAA-6 and TIROS-N data in the eastern United States to derive a predictive relationship. The final form uses precipitable water vapour and seven types of cloud transmittance. Each cloud type is also characterized by a particular value of cloud cover.

A simple relationship relating solar-radiation transmissivity ($Tr = K_d/K_{extra}$) to satellite-observed reflectivity (α'_{ea}) has been derived by Hay (1981), Nunez et al. (1984) and Nunez (1987):

$$Tr = c_1 + c_2\alpha'_{ea}, \tag{20}$$

where c_1 and c_2 are regression constants.

Nunez et al. (1984) use a modelling framework to show that c_1 and c_2 are relatively insensitive with respect to the various atmospheric scatterers and absorbers. Typically, a 50% change in precipitable water vapour will cause a change in atmospheric absorptance of only 12% in the value of Tr (see Fig. 1).

A similar approach has been used by Cano et al. (1986) using Meteosat data. A transmission relationship similar to Eq. (20) was developed using pyranometer data from the French Meteorological Office. Their relationship differs in that in place of an Earth–atmosphere reflectivity, a cloud index (α_n) is used as in Eq. (12):

$$Tr = k_1 + k_2\alpha_n, \tag{21}$$

where k_1 and k_2 are regression constants.

The advantage of this relationship is that α_n takes into account any variability due to changes in surface albedo.

2.5 Comparison and Evaluation of Various Models

The various approaches discussed in the previous section have been summarized in Tables 1 and 2. The majority favour the use of geostationary satellites since these allow a high frequency of daily scans, and therefore the error arising from diurnal cloud variability will be reduced (Nunez et al. 1984). An exception is the work of Darnell et al. (1988) who used the Heat Budget Product (HBP) of NOAA-AVHRR to obtain their Earth–atmosphere reflectivity. Those authors were interested in a global data set which would estimate monthly averages of daily global irradiance. It was therefore possible to relax the constraints of a high frequency of daily scans, although the effect of temporal bias cannot be discounted.

Treatment of cloud depletion is evenly divided between estimates of cloud cover and estimates of cloud-layer transmission (Table 1). All physically based models surveyed assume that solar-radiation reflection from the cloud top is Lambertian. This is undoubtedly an approximation which theoretical models have shown not to be valid for large solar-zenith and/or satellite-viewing angles. It is likely that the problem is important for short averaging periods, hourly or daily, or when the spatial variability of cloud cover is large. Model

Table 1. Characteristics of models which use satellite data to derive surface solar irradiance

	Satellite used in most publications	Cloud-cover estimation	Cloud-layer transmission	Lambertian reflection from clouds	Narrow band–broad band conversion for cloud albedo	Aerosol depletion explicitly included	Aerosol depletion obtained from satellite		
Gautier et al. (1980)	GOES	No	Yes	Yes	No	No	No	Physically based models	
Diak and Gautier (1983)	GOES	No	Yes	Yes	No	No	No		
Gautier and Frouin (1985)	GOES	No	Yes	Yes	No	Yes	No		
Frouin and Chertok (1992)	Nimbus-7 ERB	Yes	Yes	Yes	No	Yes	No		
Breon et al. (1994)	Multi-Sat ERBE	Yes	Yes	Yes	No	Yes	No		
Pinker and Ewing (1985)	GOES	No	Yes	Yes	No	Yes	No		
Pinker and Lazlo (1992)	ISCCP	No	Yes	Yes	No	Yes	No		
Raschke and Preuss (1979)	Nimbus-3	Yes	Yes	Yes	No	Yes	No		
Moser and Raschke (1983, 1984)	Meteosat	Yes	Yes	Yes	No	Yes	No		
Tuzet et al. (1984)	Meteosat	Yes	Yes	Yes	No	Yes	No		
Diekmann et al. (1988)	Meteosat	Yes	Yes	Yes	No	Yes	No		
Stuhlmann et al. (1990)	Meteosat	Yes	Yes	Yes	No	Yes	No		
Dedieu et al. (1987)	Meteosat	No	Yes	Yes	No	Yes	No		
Darnell et al. (1988)	NOAA/AVHRR	Yes	No	Yes	No	Yes	Yes		
Hiser and Senn (1980)	GOES	Yes	No	Yes	No	No	No	Empirical models	Digital data
Rimoczi-Paal (1983)	Meteosat	Yes	No	Yes	No	Yes	No		
Tarpley (1979)	GOES	Yes	No	No	No	No	No		
Brakke and Kanemasu (1981)	GOES	Yes	No	No	No	No	No		
Hay (1981)	GOES	No	Yes	Yes	No	No	No		Analog data
Nunez et al. (1984)	GMS	No	Yes	Yes	No	No	No		
Nunez (1987)	GMS	No	Yes	Yes	No	No	No		
Nunez (1993)	GMS	No	Yes	Yes	No	Yes	No		
Cano et al. (1986)	Meteosat	Yes	No	Yes	No	No	No		

Table 2. Model validation with pyranometers

	Hourly SE ($W m^{-2}$)	Hourly SE (mean %)	Daily SE ($W m^{-2}$)	Daily SE (mean %)	Monthly SE ($W m^{-2}$)	Monthly SE (mean %)
Pinker and Ewing (1985)[a]	150	36	29	12		
Dedieu et al. (1987)[a]	109	20			37	6
Darnell et al. (1988)[a]			30	19	4	3
Moser and Raschke (1983)[a]	77					
Moser and Raschke (1984)[a]			24	11		
Tuzet et al. (1984)[a]			21	9		
Diekmann et al. (1988)[a]				9–21		1–5
Stuhlmann et al. (1990)[b]					19	11
Gautier et al. (1980)[b]				9		
Gautier and Frouin (1985)[b]			16	12		
Tarpley (1979)[a]	64–132		24–48	9–17		
Brakke and Kanemasu (1981)[b]	91–105	18–33	31	10		
Hay (1981)[a]	126	31	26	15		
Rimoczi-Paal (1983)[a]				10		
Nunez (1987)[a]			17	13		
Nunez (1993)[a]					14	7
Cano et al. (1986)	50–116					

[a] Error calculated as an RMS residual from the regression.
[b] Error calculated as an RMS of the difference (measured-calculated).

performance will be poor in certain subsets of large data sets which encompass a large range of solar and satellite-viewing angles. We suggest that this is one likely explanation for the poor wintertime performance in some models (Diekmann et al. 1988; Nunez 1990a,b). By contrast, some of the empirical models (Tarpley 1979; Brakke and Kanemasu 1981) explicitly take this effect into account by including a solar-zenith angle term in the empirical regressions.

Another source of error may be due to the failure to correct for the spectral band of the satellite sensor, which in the case of GOES and GMS is fairly narrow (0.55–0.75 µm). Spectral dependence is not considered in the cloud albedo determination, although experimental measurements (Paltridge 1988) fail to provide conclusive support for this omission. The problem is largely alleviated in the normalization process which is necessary for the estimation of cloud cover. However, models which estimate a cloud-layer transmission do not use a normalization process and thus significant errors could arise.

Aerosol depletion also merits closer scrutiny. Most physically based models incorporate an aerosol-depletion term in the clear-sky formulation. However these relationships do not take into account the substantial spatial and temporal variability in the tropospheric aerosol load. Only the relationship of Darnell et al. (1988) attempts to model the aerosol-depletion term from satellite observations. This is accomplished by relating the precipitable water vapour, obtained from a TOVS sounding, to the aerosol depletion obtained as a residual term in surface pyranometer observations. Further studies are needed to verify the general applicability of this relationship. Nunez (1993) has shown that, in tropical regions at least, other factors beside precipitable water vapour influence the magnitude of the aerosol depletion.

Table 2 presents the performance of the various models when compared with pyranometer data. Reported errors are calculated either as a mean error of estimate from a regression, or as a root mean square error of differences. These two terms approximate each other if there is no mean bias.

Table 2 confirms that increasing the period of averaging will decrease the error, and that this decrease overwhelms any difference in model performance for a given period of averaging. At hourly time scales errors range from 50 to 150 W m^{-2}, and typically represent between 18 and 36% of the mean measured global radiation. At daily time scales the errors drop from 17 to 48 W m^{-2} (9 to 21% of measured global radiation); and finally at monthly time scales the errors are 6% or less.

The above trend points to the importance of random effects in affecting the model performance, and their reduction after temporal averaging. Effects such as differences between pyranometer point sampling and the satellite spatial sampling are bound to be important at hourly time scales (Hay 1981). Inadequate accounting for cloud reflection will also contribute to the error. The frequency of satellite scans will contribute at daily time scales. Finally, systematic errors due aerosol depletion, temporal changes in surface albedo, or seasonal changes in solar-zenith angle will dominate at the monthly time scale.

The relatively high standard errors obtained at the monthly time scale by Stuhlman et al. (1990) and Nunez (1993) are interesting and deserve further comment. These two studies, conducted in Europe/Africa and in the western tropical Pacific region respectively, used a broad range of pyranometer data obtained in their respective study areas for validation. Two likely causes for the high errors are localised aerosol loads and instrument calibration errors. These results suggest that, pending the establishment of a global network of stations with well-calibrated pyranometers, monthly global solar irradiance from satellites might be accurate only to within 10% of the measured value.

2.6 Regional Mapping of Incoming Global Radiation

Various physical and empirical model relationships were discussed in the previous sections. The advantage of physical models lies in their ability to replicate the radiation field in a wide variety of environments. Various physical processes must be incorporated into these models to obtain a level of generality and usefulness. In many cases these requirements are fairly stringent and further research is needed to model all the processes involved.

Fortunately, some of these constraints may be relaxed at the regional time scale. The assumption might be made that at this spatial scale many of the physical processes (aerosol depletion, non-Lambertian cloud reflection, etc.) are constant in space and therefore do not need to be explicitly modelled. This was the approach taken by Nunez (1990a,b) in a project designed to provide statistics on global radiation in Australian capital city regions for use in feasibility studies of photovoltaics and other solar energy applications.

Reception of GMS in Australia is via a 5-m dish located at the Bureau of Meteorology headquarters in Melbourne. A subset of the full Earth disc covering the Australian region is stored in archive tapes three times daily (0, 3 and 6 GMT) in the visible band (0.55–0.75 μm). At monthly intervals a program was run which sectorized an area encompassing each of the eight Australian capital cities. It consisted of an area slightly larger than 2° longitude by 2° latitude (approximately 200 km × 200 km at mid-latitudes) with the city airport in the geometric centre (Table 3). Daily satellite data were collected in this fashion for a 3-year period; 1 March 1985 to 28 February 1988. Concurrent with these data, daily global radiation and precipitable water vapour data were obtained from the local airport.

Processing involved the following steps. Upon receipt of the monthly tapes the images were unpacked, registered to known geographic coordinates, the artificial coastline was removed, and the images were re-sectorized into a 2° by 2° region with the pyranometer station (the city airport) at the centre. While performing this step, the program also extracted the visible radiance from the pixel above the pyranometer station. This last data set was used in the model development.

The approach followed was to tune the model by comparison with the pyranometer measurements, and then use the relationship to map global solar

Table 3. Details of the study regions encompassing the capital cities

	Pyranometer station		Size of image	
City	Latitude	Longitude	N–S (km)	E–W (km)
Adelaide	34° 57′	138° 31′	219	177
Brisbane	27° 24′	153° 04′	219	201
Canberra	35° 18′	149° 11′	219	177
Darwin	12° 24′	130° 52′	219	213
Hobart	42° 50′	147° 32′	219	159
Melbourne	37° 51′	144° 45′	219	177
Perth	31° 54′	115° 57′	219	183
Sydney	33° 56′	151° 10′	219	183

radiation for the entire region. Equation (6) is used to calculate the daily irradiance (K_d) on the surface for every pixel on a daily basis:

$$K_d = K_{extra}(1-\alpha_a)(1-\phi_a), \tag{22}$$

where K_{extra} is the extraterrestrial daily radiation flux (irradiance) at the top of the atmosphere which may be readily calculated with the relationship of Paltridge and Platt (1976). α_a is the daily (broad-band) value of the reflectance of radiation by the atmosphere. ϕ_a is the daily value of absorptance by water vapour and ozone in the atmosphere and is calculated with the formula of Lacis and Hansen (1974) using precipitable water data from each airport site. The spatial variability in K_d is provided by K_{extra}, a slowly varying function of latitude and by α_a which is derived from the satellite data and varies with every pixel. This expression ignores the relatively small effect of surface reflection and subsequent backscattering from the atmosphere which has been addressed recently by Nunez (1993).

The following procedure is used to relate α_a to the satellite signal. Equation (22) may be solved for α_a with computed values of K_{extra}, daily pyranometer measurements of K_d and observations of precipitable water to estimate ϕ_a. Furthermore, this value of α_a is a broad-band estimate and accounts for all Lambertian and non-Lambertian reflection processes as well as depletion due to gaseous and particulate pollution. This value of α_a is next transformed into a broad-band Earth–atmosphere albedo α_{ea},

$$\alpha_{ea} = \alpha_a + \left[(1-\alpha_a)^2(1-\phi_a)\alpha_g\right]. \tag{23}$$

Equation (23) has been derived from a simple relationship of Paltridge and Platt (1976) in which the Earth–atmosphere albedo is given as the sum of two terms. The first term on the RHS of Eq. (23) is due to reflection of the incoming extraterrestrial radiation by the atmosphere; the second term refers to the extraterrestrial radiance that is transmitted to the surface after depletion by absorption and atmospheric scattering $(1 - \alpha_a)$ $(1 - \phi_a)$, reflection by the surface (α_g), and transmitted back to space after scattering by the atmosphere $(1 - \alpha_a)$. No further absorption has been assumed in the irradiance that has been reflected by the ground. For each pixel the broad-band albedo α_g is given by averaging the four lowest satellite reflectivities in a given month and by applying a broad-band atmospheric correction. This procedure will be described in Section 4.2 dealing with surface albedo.

The link to the narow-band satellite reflectivity (α'_{ea}) is now given as a linear multiple regression between α_{ea} and α'_{ea}:

$$\alpha_{ea} = c_0 + c_1[\alpha'_{ea}]_0 + c_2[\alpha'_{ea}]_3 + c_3[\alpha'_{ea}]_6, \tag{24}$$

where the numbers 0, 3 and 6 refer to the GMT times of the satellite observations and c_0, c_1, c_2, and c_3 are regression constants. This relationship links the narrow-band (0.55–0.75 μm) satellite observation to the (0.2–2.9 μm) broad-

Table 4. Results of multiple linear regression described by Eq. (24) between the daily surface-derived Earth–atmosphere reflectance (α_{ea}) and the three satellite-derived narrow-band reflectances $[\alpha'_{ea}]_0$, $[\alpha'_{ea}]_3$, $[\alpha'_{ea}]_6$. The results refer to Hobart Airport, using all data obtained between 1 March 1985 and 28 February 1988

Month	c_0	c_1	c_2	c_3	r^2
Jan	0.149	0.368	0.277	0.342	0.82
Feb	0.149	0.502	0.233	0.281	0.86
Mar	0.122	0.486	0.402	0.17	0.86
Apr	0.063	0.528	0.648	0.16	0.90
May	0.124	0.521	0.492	0.171	0.78
Jun	0.115	0.535	0.631	0.017	0.59
Jul	0.11	0.36	0.66	0.155	0.61
Aug	0.079	0.371	0.534	0.334	0.85
Sep	0.124	0.407	0.37	0.307	0.73
Oct	0.114	0.52	0.462	0.209	0.85
Nov	0.154	0.35	0.41	0.266	0.81
Dec	0.14	0.489	0.393	0.182	0.86

band pyranometer measurement, and also accounts for any non-Lambertian behaviour in cloud reflection and the inability of the satellite to monitor this effect.

The results of the regression for one of the regions, Hobart, are presented in Table 4 as a monthly statistic. Note that the percent variance accounted for by the regression drops during winter. This effect is likely to be due to a number of causes including anomalous non-Lambertian reflection by clouds, anomalous reflection by clouds and aerosols, narrow-band/broad-band differences, etc. These results illustrate the need to assess model relationships in a wide variety of environmental conditions (see also Diekmann et al. 1988).

In the operational mode Eq. (24) is solved for α_{ea} for every pixel in the image. Knowing α_{ea}, Eq. (23) is solved for α_a as a solution to a quadratic equation, also for every pixel in the image. The daily atmospheric absorptivity ϕ_a is assumed constant throughout, and α_g is obtained for every pixel using the four lowest radiances in the month as discussed above. Finally, the solar irradiance is calculated for every pixel using Eq. (22).

In addition to monthly averages, the standard deviation of daily global radiation and the number of sunny and cloudy days were also mapped. Cloudless days were obtained by searching through the pyranometer data and obtaining the highest value for the month. Substituting in Eq. (22) with the appropriate value of ϕ_a and K_{extra} will yield a value of α_a. Equation (22) is then re-calculated for the 15th of the month keeping α_a and ϕ_a constant, but varying the value of K_{extra} to the mid-month value. A day was considered cloudless if K_d was greater than 80% of the clear sky value K_1; and cloudy if K_d was less than 60% of K_1.

Table 5. Statistics which compare the satellite-derived estimates (x_s) with pyranometer measurements (y_p). Monthly data for the eight capital city regions have been grouped together. Each statistic is therefore based on 96 data pairs

	$y_p = ax_s$		D	$D/(x_s)_{av}$
	a	r^2	RMS of $(y_s - x_p)$	%
Mean monthly global radiation	1.00	0.99	0.41 MJ m^{-2} day^{-1}	2
SD of global radiation	1.03	0.95	0.37 MJ m^{-2} day^{-1}	8
Number of sunny days	0.96	0.89	1.46	9
Number of cloudy days	1.05	0.85	1.29	15

Table 5 presents the validation statistics for the four radiation flux fields that were mapped. Monthly data for all the capital regions have been grouped together to produce 96 data pairs (8 regions × 12 months) for each statistic. The approach followed used the above multiple linear regression and assumed that three daily satellite reflectivities were available on a daily basis. In fact, due to operational problems, the percentage of days in a month with three satellite observations varied from a minimum of less than 20% to a maximum of over 90%. Therefore, a linear interpolation scheme had to be devised to replace the missing data in one or two of the three possible daily satellite reflectivities (Nunez 1990a). No attempt was made to provide an interpolated value if all three daily satellite reflectivities were missing.

Despite these limitations, the validation with respect to the monthly airport pyranometer data, calculated for all available days of record, is very satisfactory. Table 5 shows that root mean square differences were 0.41 MJ m^{-2} day^{-1} for the monthly global radiation values (2% of the mean estimate); 0.37 MJ m^{-2} day^{-1} for the standard deviation of global radiation (8% of the mean estimate); 1.46 for the number of sunny days (9% of the mean estimate); and 1.29 for the number of cloudy days (19% of the mean estimate).

Based on the above results, all image data have been presented at intervals of 1.0 MJ m^{-2} day^{-1}, 0.5 MJ m^{-2} day^{-1}, 2, and 2 for K_d, the standard deviation of K_d and number of sunny and cloudy days, respectively. These intervals correspond to at least one standard deviation in the relationship between the pyranometer measurement and the satellite calculation.

A typical set of statistics for the Sydney region during the month of December is shown is Fig. 2. The four maps represent a laser-printed product after the image is transformed using a Lambert conformal projection, density sliced and smoothed using a 3 × 3 smoothing function. Also shown are the coastline, the outline of the city boundaries and larger towns, and at the bottom right, a 50-km scale showing 10-km intervals. True north is parallel to the y-axis, and the city airport is at the geometric centre of the image.

All four images show considerable spatial variability. Mean global radiation (Fig. 2A) varies between 26.9 and 21.0 MJ m^{-2} day^{-1}. It is lower inland

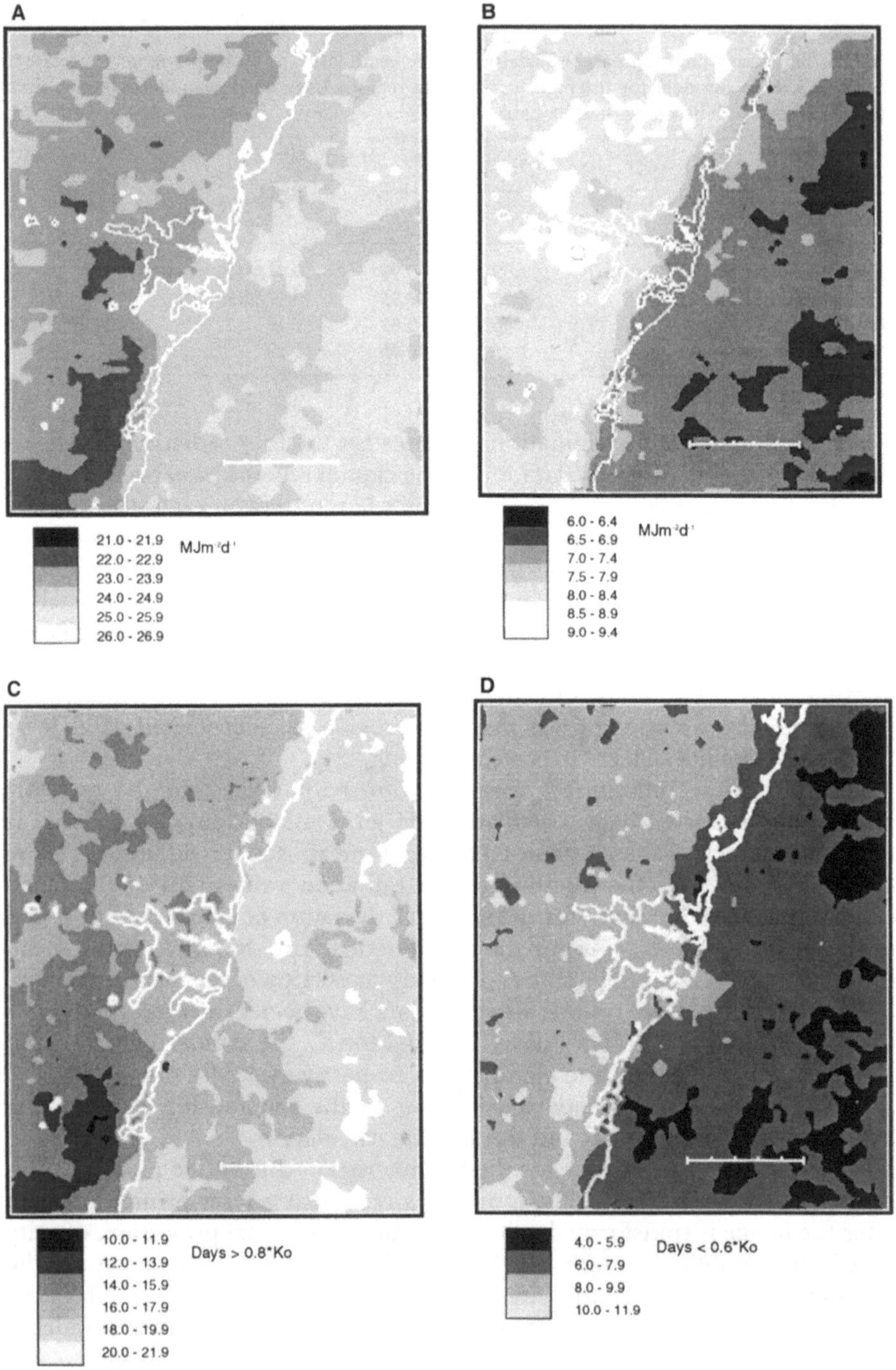

Fig. 2A–C. Maps of solar radiation statistics for Sydney, Australia, in December showing **A** mean solar irradiance (MJ m^{-2} day^{-1}); **B** standard deviation of solar irradiance (MJ m^{-2} day^{-1}); **C** number of cloudless days in the month; and **D** number of cloudy days in the month. A distance scale with 10-km intervals is shown at the *bottom right*

due to the greater convection and cloud cover being experienced in inland areas during summer. Minimum solar radiation in the range 21.0 to 22.9 $MJ\,m^{-2}\,day^{-1}$ is estimated along the mountain region located just inland from the coast. Over the ocean, the cooler water temperatures enhance atmospheric stability and, as a result, lower cloud cover and higher solar radiation values up to 26.0–26.9 $MJ\,m^{-2}\,day^{-1}$ are experienced.

The pattern of standard deviation (Fig. 2B) indicates lower values, corresponding to lower mean values of solar radiation. A higher standard deviation is estimated over inland regions since these may experience abrupt changes as their daily cloud regime changes from overcast to clear. Figure 2C,D are largely complementary. Lowest numbers of cloudless days are estimated southwest of the city boundaries, with typical values between 10 and 14 days (Fig. 2C). This area also receives the highest number of totally cloudy days, which range between 10 and 12 days (Fig. 2D). This complementarity also applies to the oceanic regions.

These Sydney results are typical of the results obtained for all capital city regions. They demonstrate the importance of obtaining detailed information on the spatial and temporal variability of incoming solar radiation for use in a wide range of domestic and industrial applications of solar energy.

3 Ultraviolet Radiation

3.1 Introduction

In recent years there has been a growing level of awareness, both in the scientific and lay community, of the potential dangers of ultraviolet (UV) radiation to human health. Results from a variety of studies document the increasing incidence of several forms of skin cancer in Caucasian populations and its relationship to long-term exposure to UV radiation in the wavelength range from 280 to 400 nm (Elwood et al. 1974, 1989; Hinds and Kolonel 1980; Holman et al. 1980; Fisher and Kripke 1982; Kraemer et al. 1984). Ultraviolet radiation is also a significant factor affecting biological productivity (e.g., Worrest 1982; Scotto et al. 1988).

A major research activity in meteorology has focused on the effect of man-made chlorofluorocarbons on stratospheric ozone, leading to its partial destruction in Antarctica during springtime (Stolarski et al. 1986; Farmer et al. 1987). Recent data from the Total Ozone Mapping Spectrometer (TOMS) aboard NIMBUS-7 reveal a long-term depletion of a few percent per decade at mid-latitudes, but with much larger values in south polar regions. The importance of stratospheric ozone changes on surface UV radiation has been widely documented in theoretical/empirical models (Green et al. 1974; Shettle and Green 1974; Dave and Halpern 1976; Green et al. 1980; Nagaraja et al. 1984; Frederick and Lubin 1988; amongst others) and in surface measurements as well (Barton and Robertson 1975; Roy et al. 1989; Bittar and McKenzie 1990), although other environmental effects are also important in determining surface levels (Scotto et al. 1988).

The above considerations support the urgent need to develop climatologies of surface UV radiation for use in epidemiological studies and for assessing the impact on biological productivity. Attempts to map the UV radiation field at the regional scale suffer from the same limitations already listed in Section 1, but these problems are exacerbated by the lack of measurement standards (Roberts 1991), the lack of standard instrumentation and uncertainties as to which spectral bands to cover. For the purposes of mapping the radiation field, it is important to distinguish between UV-A, UV-B and UV-C radiation, depending on the wavelength chosen (Fig. 3). At wavelengths of less than 280 nm, in the UV-C range, the ozone absorption coefficient is so large that practically no UV irradiance reaches the Earth's surface. At the other extreme, there is low ozone absorption at wavelengths above 320 nm in the

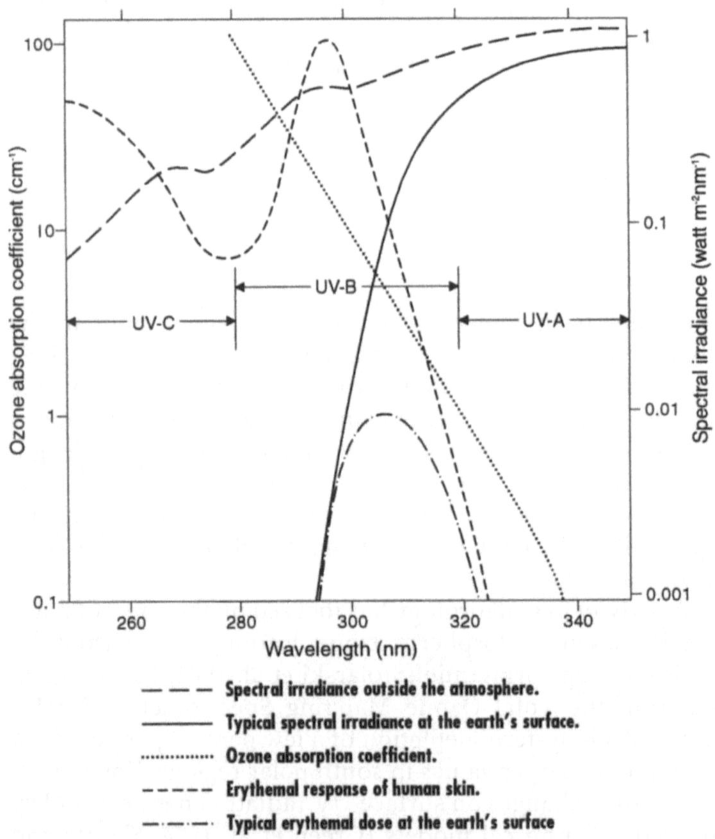

Fig. 3. Ozone absorption coefficient, erythemal response of human skin and solar irradiances for the middle ultraviolet. (After Barton and Paltridge 1979)

UV-A range. It is only in the UV-B range, between 280 and 320 nm, that sensitivity to stratospheric ozone levels is high. This is also the region of greatest biological activity and therefore importance.

As will be discussed below, Nunez et al. (1994) have shown that regional mapping of surface UV radiation may also be accomplished with GMS data. In a similar fashion to broad-band solar radiation, visible-reflectance data from GMS may be used to provide the spatial cloud pattern. However, it is useful to adopt differing mapping approaches for various regions of the UV radiation spectrum.

3.2 Mapping UV-A Radiation

The various processes which affect the receipt of UV radiation flux at the Earth's surface are illustrated in Fig. 4. In the UV-A range, Rayleigh scattering is the dominant depletion mechanism along with a small contribution from ozone absorption. No absorption by water vapour occurs in this wavelength range, although aerosols may scatter and absorb the radiation further. Cloud reflection and absorption represent important depletion mechanisms which may be monitored by GMS.

One approach that can be followed consists of developing an empirical relationship between UV-A and global solar irradiance. The rationale is as follows: not being strongly dependent on ozone absorption, it is expected that UV-A radiation should vary systematically with broad-band solar irradiance. The advantage of this technique is that full use can be made of historical solar-radiation data, both measured and modelled, for various regions.

As an illustration of the technique, UV-A global-radiation data monitored with an Eppley UV pyranometer were compared with simultaneous measurements of solar irradiance using a Middleton CN 27 pyranometer measuring in the 300 to 3000 nm range. Two years (1985, 1986) of daily data taken at the Cape Grim baseline station in Tasmania were used in the analysis. Equations (25) and (26) present the results of least square regressions between measured UV-A [$(K_d)_{uva}$] and solar irradiance (K_d), and between the ratio of $(K_d)_{uva}/K_d$ and daily sunshine duration(n), which is expressed as a fraction of the maximum possible hours of bright sunshine (N).

$$(K_d)_{uva} = 0.027 + 0.0374 K_d (\mathrm{MJ\,m^{-2}\,day^{-1}}) \quad (25)$$

with $r^2 = 0.988$ and SEE = 0.035

$$(K_d)_{uva}/K_d = 0.0458 - 0.0113(n/N) \quad (26)$$

with $r^2 = 0.670$ and SEE = 0.002.

Expressed in $\mathrm{MJ\,m^{-2}\,day^{-1}}$, Eq. (25) reveals that $(K_d)_{uva}$ and K_d are highly correlated, with about 99% of the variance explained in the relationship and a standard error of estimate of 0.0347 $\mathrm{MJ\,m^{-2}\,day^{-1}}$. The relationship was consid-

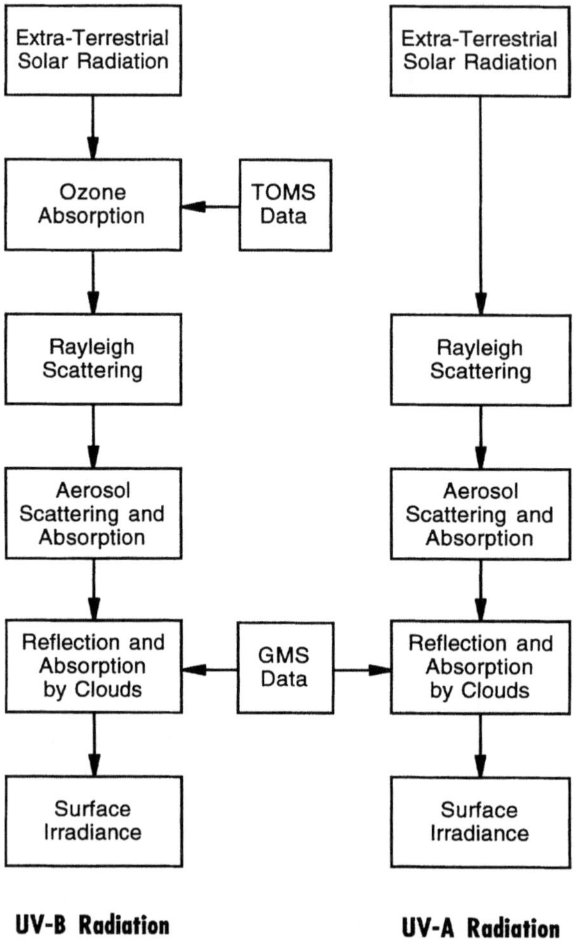

Fig. . Overview of the atmospheric processes affecting UV-A and UV-B irradiance at the surface. Ozone absorption in the stratosphere is monitored with the Total Ozone Mapping Spectrometer (TOMS) of Nimbus-7. Cloud reflection and absorption may be monitored with the GMS geostationary satellite

ered applicable for the entire Tasmanian region and is used below in the analysis of satellite data.

With an r^2 of 0.67 Eq. (26) is also acceptable. It provides some insight into the effect of cloud cover on the relationship. For totally cloudless conditions ($n/N = 1$), approximately 3.5% of the solar irradiance is in the UV-A spectral band. This percentage increases with cloud cover, and reaches a maximum of 4.6% when there is no bright sunshine ($n/N = 0$). This trend is very likely to be a result of the increased water-vapour and liquid-water absorption by clouds.

Equation (25) may be used to transform the regional global radiation field derived from GMS, as described in Section 2.6, into a map of $(K_d)_{uva}$. An image was selected of dimensions 219 km (north–south) by 159 km (east–west) which has Hobart Airport in its centre. The daily average solar irradiance for January was used in the tranformation. Figure 5 presents the results. The top half of the image is in the maximum range of 0.8 to 0.84 MJ m^{-2} day^{-1}. Lower values, reaching a minimum range of 0.60 to 0.64 MJ m^{-2} day^{-1} are obtained in south-western Tasmania. This is a wet region with relatively high elevations which is exposed to the prevailing moist westerly wind systems. The east/west zonation in UV-A irradiance reflects the presence of cloud cover and rainfall.

3.3 Mapping UV-B Radiation

Daily changes in stratospheric ozone load will affect the magnitude of the surface UV-B irradiance (see Fig. 4). Two differing spatial scales are involved in the depletion of UV-B radiation. Satellite soundings of ozone distribution, as seen by the TOVS system in NOAA-9 and NOAA-10 and the Total Ozone Mapping Spectrometer (TOMS) of NIMBUS-7, confirm a high degree of temporal and spatial variability (Rao et al. 1990; Bureau of Meteorology 1990). Ozone distribution is influenced by dynamic processes in the atmosphere and has spatial and temporal scales similar to synoptic weather systems. By contrast, cloud cover exhibits spatial variability in a wide range of scales, from the local to the synoptic.

Mapping of UV-B radiation may be accomplished by using data from TOMS to derive surface UV-B irradiance under essentially clear-sky conditions. The spatial resolution of TOMS is 50 × 50 km at nadir (Madrid 1978), and the assumption is made that, on a daily basis at least, the ozone load is constant over the region of interest. Having determined a clear-sky value, an empirical fit is developed between satellite reflectivity and the transmission of UV-B radiation by clouds (Nunez 1990c; Nunez et al. 1994).

The analysis used measurements of daily UV-B irradiance (280–315 nm) made at the Australian Radiation Laboratories in Yallambie, a suburb of Melbourne, Australia (38 °S, 145 °E). The concurrent satellite data collected for the Melbourne region comprised daily TOMS data from NIMBUS-7 and the three daily GMS images, as described in Section 2.6. The data set covered the period 1 September 1986 31 December 1987.

As an initial step a clear-sky model was tested for the Melbourne region. The pixel directly above Yallambie was selected and conditions were considered cloudless if all three daily satellite reflectivities were less than 16%. This procedure gave 36 days in the data set which met this criterion. The daily total UV-B irradiance for these 36 days formed the measurements against which various clear-sky models were evaluated.

The formulation described in the models of Green et al. (1974) and Barker et al. (1980) was found to best fit the experimental data. The model is semi-empirical in nature, and incorporates variations in wavelength, solar-zenith angle, ozone optical depth, aerosol optical depth and surface albedo. It has the form:

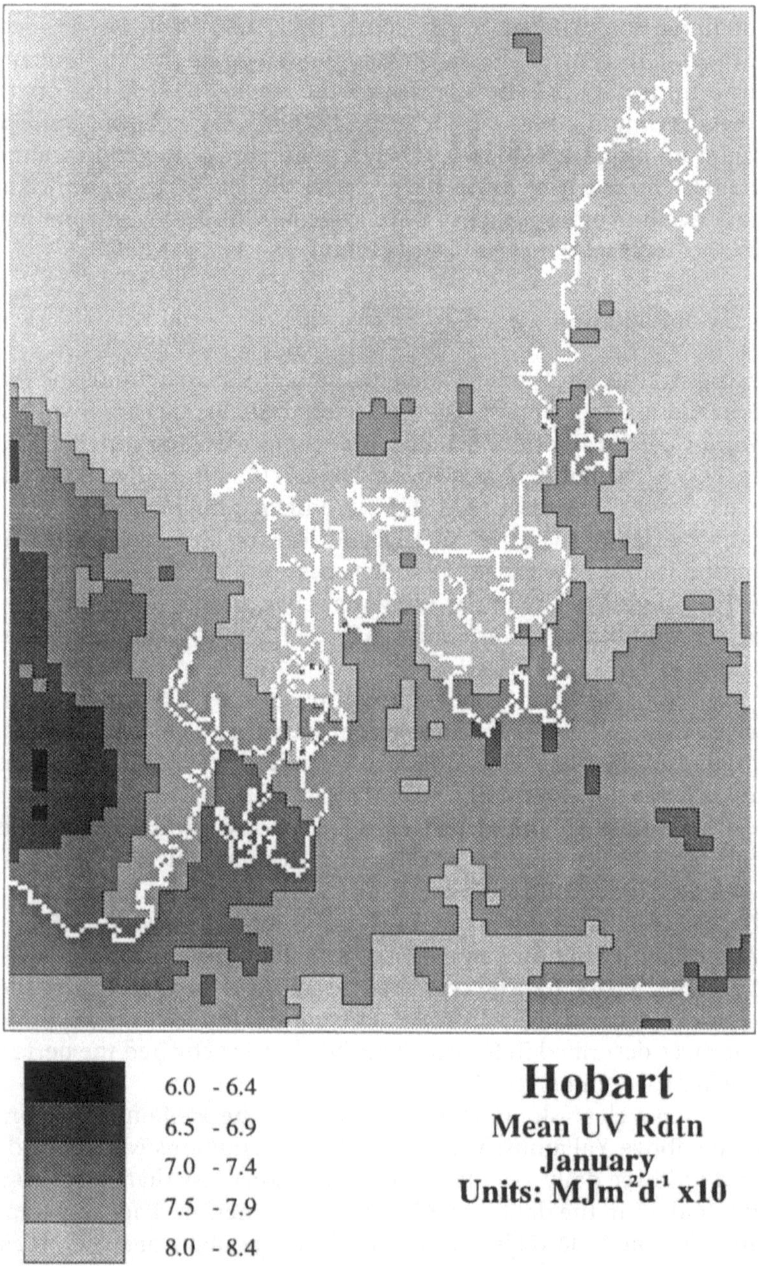

Fig. 5. Mean daily UV-A irradiance (MJ m^{-2} day^{-1} × 10) for the Hobart, Tasmania, region for January. The image represents a 2° × 2° region centred on Hobart Airport (42°50′ S; 147°32′ E). A distance scale with 10-km intervals is shown at the *bottom right*

$$K_{d\lambda} = I_\lambda \cos(z) \exp.\left(-\sum_{i=1}^{3} d_i k_i m_i\right) + I_\lambda \exp[-f(O_3)], \tag{27}$$

where $K_{d\lambda}$ is the UV-B radiation for wavelength λ, I_λ is the direct extraterrestrial spectral radiation, and d_i, k_i, and m_i are the optical thickness, extinction coefficient and air mass respectively for ozone ($i = 1$), water vapour ($i = 2$) and aerosols ($i = 3$). The first term on the RHS of Eq. (27) represents the direct irradiance. The second term relates to the diffuse flux which is adjusted using various empirical fits to the ozone function $f(O_3)$. This function describes the diffuse radiation components which can vary depending on the particular parameters used, which in turn depend on atmospheric turbidity and other variables.

The mean daily measured irradiance (296–315 nm) for the 36 days chosen as representing clear-sky conditions was $37.4\,\mathrm{KJ\,m^{-2}\,day^{-1}}$ which compares favourably with the mean of $38.2\,\mathrm{KJ\,m^{-2}\,day^{-1}}$ for the three forms of the empirical fit to $f(O_3)$ as listed in Barker et al. (1980). The mean bias error of $0.8\,\mathrm{KJ\,m^{-2}\,day^{-1}}$ (model measurement) is 2.1% of the mean measured value.

Daily RMS differences are higher, at $5.8\,\mathrm{KJ\,m^{-2}\,day^{-1}}$ and represent 15.5% of the mean measured value. Inspection of the data revealed that there were four days with substantially lower irradiance values, greater than twice the standard deviation of the differences. It is likely that the criterion used for choosing clear days did not remove all days containing some cloud. Removing these four values reduces the RMS differences to 9.5% of the mean measured value. Based on these results, the average of models 2 and 3 was taken as indicative of clear-sky conditions.

Using all satellite reflectivity data for the pixel above Yallambie, an empirical relationship was developed between the ratio of daily measured to cloudless UV-B irradiance $(K_d)_{uvb}/(K_{do})_{uvb}$ and satellite reflectivity. It had the form

$$\begin{aligned}(K_d)_{uvb}/(K_{do})_{uvb} &= 1.03 - 0.19C_0 - 0.18C_0^2 - 0.14C_3 + 0.053C_3^2 \\ &\quad -0.59C_6 + 0.31C_6^2,\end{aligned} \tag{28a}$$

where C_i is given as

$$C_i = [\alpha_i - (\alpha_{min})_i] / [(\alpha_{max})_i - (\alpha_{min})_i]. \tag{28b}$$

In the above relationship C_i is a normalized cloud index, obtained at time $i = 0$, 3 and 6 h GMT, α_i is the satellite reflectivity at time i, and α_{min_i}, α_{max_i} are the minimum and maximum reflectivities for time i, respectively. A minimum reflectivity α_{min_i} was obtained for every pixel, therefore incorporating localised effects such as surface-cover type and texture. By contrast, α_{max_i} was obtained for the entire image using all records in the study period.

Two data sets were used to test the above relationship. Concurrent measurements of UV-B irradiance and satellite reflectivity for a period 1 November

1988–28 February 1987 gave the coefficients of the multiple linear regression. The results show that there is much scatter in the relationship ($r^2 = 0.55$, SEE = 0.16). Equation (28a) was then used to calculate $(K_d)_{uvb}$ on the second data set encompassing the period 1 November 1987 to 31 December 1987. Daily RMS differences are 7.2 KJ m^{-2} day^{-1}, which correspond to 20.7% of the mean measured value. Clearly, the above relationship is not suitable for mapping on a daily basis. Alternatively, the relationship between the radiometer point measurements and the spatially averaged cloud information is subject to large errors at a daily time scale. Averaging calculated values of $(K_d)_{uvb}$ over weekly periods reduces the RMS difference to 2.9 KJ m^{-2} day^{-1}, or 8.4% of the mean value. Figure 6 presents a scatter plot of the weekly data.

The relationship derived in Eq. (28) was used to map the UV-B radiation field in the Melbourne region from 1 November 1986 to 20 November 1986. Daily values from TOMS were used and incorporated in Eq. (27). Daily clear-sky values from two empirical relationships were averaged to obtain a representative value for Melbourne. This clear-sky value $(K_{do})_{uvb}$, although derived for Yallambie, was assumed to be representative of the entire Melbourne region. Finally, Eq. (28), with individual cloud information, was used to calculate $(K_d)_{uvb}$ for every pixel.

The average pattern for the Melbourne region for the period listed above is shown in Fig. 7. Of interest are the lower range across the centre of the image, and the higher range to the north and south. The considerable variability that is observed is undoubtedly a result of orographic clouds to the north of Melbourne. Drier conditions prevail at the northern end of the image because moisture-bearing frontal systems are not common in this region. These results

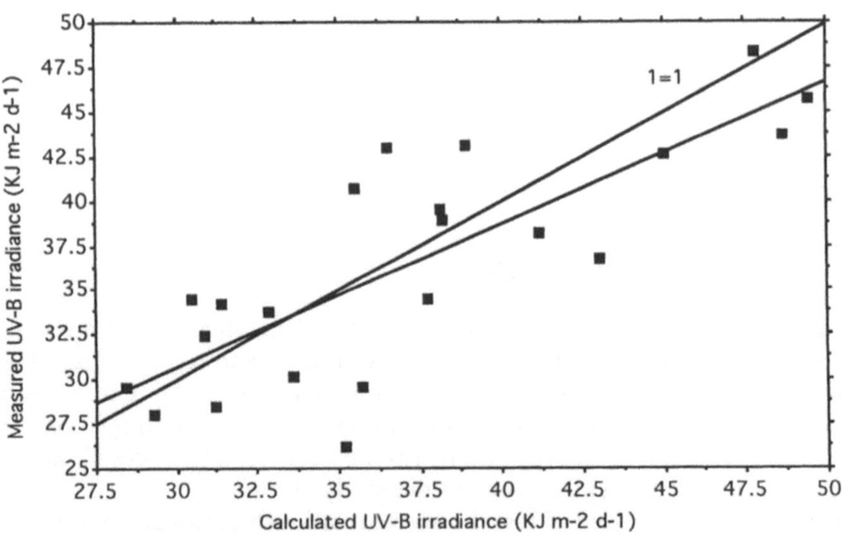

Fig. 6. Comparison of weekly values of measured and calculated UV-B irradiance (kJ m^{-2} day^{-1}) at Yallambie, Victoria

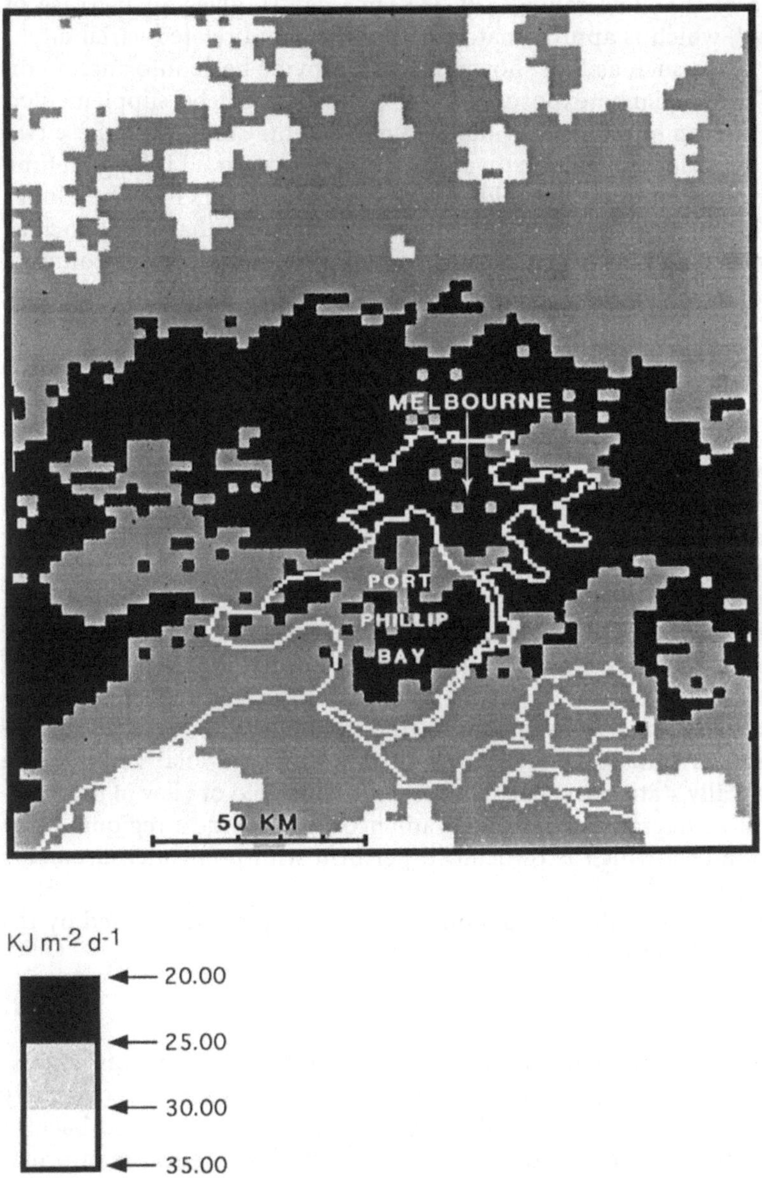

Fig. 7. Mean daily UV-B irradiance (kJ m^{-2} day^{-1}) for the Melbourne region for the period 1 November 1986 to 20 November 1986. The image represents a 2° × 2° region centred on Tullamarine Airport (37°40′ S; 144°50′ E). A distance scale with 10-km intervals is shown at the *bottom right*

illustrate the importance of cloud cover in relation to other processes such as changes in stratospheric ozone load. Radiometric measurements in clear-sky conditions (Barton and Robertson, 1975) reveal a 20% change in surface erythemal UV-B for a 10% change in ozone content. Applying a change of 20%

to the middle of the "low range" (20–25 KJ m^{-2} day^{-1}), gives an increase of 6.4 KJ m^{-2} day^{-1}, which is approximately half of the cloud related variability.

Regional maps such as that shown in Fig. 7 provide basic information on potential UV-B irradiance exposure. These patterns must be supplemented with information on human behaviour and movement. Combining these two data sets will provide information on actual human exposure. Linking lifetime exposures of UV radiation to the epidemiology of skin cancers must be a long-term goal of biomedical research. It is suggested therefore that maps such as the one in Fig. 7 could play a crucial role in defining the radiation environment of human beings.

4 Reflected Shortwave Radiation

4.1 Introduction

Surface albedo exerts an important control on net radiation. It is a highly variable term which depends on vegetation type and coverage, land-use features, soil type, soil moisture and colour, and even structural and textural features (Oke 1988). It is likely that, at a regional scale, this term will exhibit marked changes, thus affecting the overall pattern of net radiation.

A satellite sensor offers a practical means of observing reflected (cloud-free) solar radiation and thus obtaining the variability of regional surface albedo. The sensor can scan a given area quickly while the solar-zenith angle remains practically stationary. Furthermore, the wide field of view of the satellite will spatially integrate the reflected radiance and provide a regional value of the albedo, a task which is difficult to perform with point measurements taken at the surface.

For the purposes of this discussion, the reflected radiance sensed by the satellite (N′) may be written as

$$N' = N'_a + N'_g, \tag{29}$$

where N'_a represents the contribution to the satellite radiance from atmospheric sources exclusively, and N'_g contains the radiance which was originally transmitted to the surface, reflected by the surface and transmitted back to space to reach the satellite sensor. It is this term which contains the information on surface albedo. Various schemes to obtain N'_g will be discussed in the next section.

To obtain α_g using Eq. (29), a series of operations must be performed on the data. Firstly, atmospheric effects must be eliminated. This involves estimating N'_a and extracting from N'_g the contribution by atmospheric absorption and scattering. Once the upwelling radiance from the surface is determined, the effect of non-Lambertian behaviour must be accounted for. Finally, a correction factor needs to be determined to modify the radiance from a narrow-band observation, limited by the spectral filter of the satellite sensor, to a broad-band radiance encompassing all solar wavelengths. The albedo α_g would then be given as the ratio of the upwelling to downwelling broad-band irradiances.

4.2 Analysis of Satellite Data to Determine Surface Albedo

Some of the physical models listed in Section 2.3 need, as a first step, an estimate of the surface albedo. This is usually done by eliminating atmospheric effects from the satellite signal. Gautier et al. (1980) and Diak and Gautier (1983) define a threshold in the satellite radiance below which the pixel is considered cloudless. The pixel albedo is then solved as the only unknown in the radiative transfer equation. No further broad-band or bi-directional correction is performed. Sellers et al. (1990) note the difficulties associated with the narrow-band to broad-band conversion caused by the variation of surface reflectance with wavelength which is a function of soil and vegetation properties.

Dedieu et al. (1987) also derived a surface albedo as a parameterization of the incoming global irradiance [see Eq. (8)]. In their study, they follow the method of Tanré et al. (1979). They obtain one clear-sky (upwelling) radiance at the top of the atmosphere using 30 data points obtained over a month at a given hour. This clear-sky radiance is corrected for multiple scattering, absorption by water vapour and ozone and aerosol effects. It is also related to the ability of the surface to reflect radiation in a diffuse manner, which depends on surface structure and texture.

Dedieu et al. (1987) express the minimum reflectivity α'_{ea} for a given time series as

$$\alpha'_{ea} = \alpha'_a + \alpha'_g \, \mathrm{Tr}(z, \theta_s) / \left(1 - \rho_a \alpha'_g\right), \tag{30}$$

where α'_a is the atmospheric reflectance in the Meteosat band, α'_g is the narrow-band surface albedo corresponding to the Meteosat window, ρ_a is an atmospheric albedo for upwelling diffuse radiation from the surface and $\mathrm{Tr}(z, \theta_s)$ is the atmospheric transmissivity which depends on the zenith angles for the sun (z) and satellite radiometer (θ_s). The term $1/(1 - \rho_a\alpha'_g)$ accounts for multiple reflections between the surface and the atmosphere. The term $\mathrm{Tr}(z, \theta_s)$ may be partitioned into the individual transmissions for downwelling solar irradiance and upwelling reflected radiance, i.e.. $\mathrm{Tr}(z, \theta_s) \sim \mathrm{Tr}(z).\ \mathrm{Tr}(\theta_s)$. However, secondary interactions are also taken into account in this expression for $\mathrm{Tr}(z,\theta_s)$ by Tanré et al. (1979) who used this model to derive an albedo map for France for July 1979. No broad-band or bidirectional correction was applied.

Pinty et al. (1985) developed a different form of Eq. (30) using the extra-terrestrial irradiance K_{extra} and by expanding $1/(1 - \rho\alpha'_g)$:

$$K_{extra}\alpha'_{ea} = K_{extra}\alpha'_{a} + \alpha'_{g}\left(K_d^{\,2}/K_{extra}\right) + \alpha'_{g}\left(K_d^{\,2}/K_{extra}\right)\rho_a. \tag{31}$$

Therefore α'_g was obtained as a solution to a quadratic equation. The terms α'_a and ρ were obtained from the parameterizations of Deschamps et al. (1981). The authors assume that for small solar and satellite viewing angles $Tr(z) \approx Tr(\theta) \approx K_{dc}/K_{extra}$ where K_{dc} is the measured incoming irradiance at the surface. Using Meteosat data, surface albedos were mapped for the Sahel. No broadband or bi-directional corrections were applied.

Other corrections have also been used. Chen and Ohring (1984) applied the transmission parameterization of Lacis and Hansen (1974) to derive a simple model of clear-sky planetary albedo in terms of surface albedo, Rayleigh scattering and water-vapour and ozone absorption. They found that Earth-atmosphere and surface albedos were related linearly. They checked their results using zonally averaged surface albedos and minimum satellite reflectivities. Using their linear model as a predictor, the RMS error of estimates was approximately 0.02 in α_g, provided the data is grouped according to zenith angle.

Various empirical relationships have also been applied. Mekler and Joseph (1983) developed an expression for the upwelling radiance at nadir as seen by a satellite. It consisted of a quadratic expression in α'_g with three empirical coefficients. In turn, these coefficients could be determined with three concurrent measurements of satellite radiance and α'_g in one image. Knowing the coefficients, the entire α'_g field could be solved. The authors used this technique with Landsat data to calculate spectral albedos in the Sahel region.

Pinty and Szejwach (1985) developed a linear relationship between satellite radiances and surface albedos (α_i) for any two pixels (i,j) in a satellite image:

$$N'_j = \underset{\text{(A)}}{\left(\alpha_j/\alpha_i\right)}N'_i + \underset{\text{(B)}}{\left(N'_{aj} - \left(\alpha_j/\alpha_i\right)N'_{ai}\right)} + \varepsilon, \tag{32}$$

where N'_{ai}, N'_{aj} are the atmospheric radiances above the pixels i,j respectively and ε is an error term which is small and omitted in the analysis. In the case of Meteosat data used by these authors, all the above terms are spectrally averaged values in the satellite window (0.40–1.1 μm). Using successive images during the day, N'_j is linearly regressed against N'_i. The required data is then given as the slope of the relationship. All terms in (B) are assumed invariant during the diurnal cycle. This approach can then be used for all pixels in the image. The entire albedo field will be obtained provided that at least one pixel albedo is known.

Structural and textural effects on the satellite-determined albedo may be explored using the concept of bi-directional reflectance (Minnis and Harrison 1984; Taylor and Stowe 1984; Briegleb et al. 1986). It involves obtaining the radiant exitance M(z) for a particular target by successive satellite radiance observations at different viewing angles:

$$M(z) = \int_0^{2\pi}\int_0^{\pi/2} N'(z, \theta_s, \phi_s)\cos(\theta_s)\sin(\theta_s)d\theta_s d\phi_s, \tag{33}$$

where θ_s and ϕ_s are the satellite viewing zenith and azimuth angles, and M(z) is defined for a solar-zenith angle z. The departure from Lambertian reflection may be described in terms of a bi-directional radiance function:

$$F(z, \theta_s, \phi_s) = \pi N'(z, \theta_s, \phi_s)/M(z). \tag{34}$$

A perfectly diffuse reflector (i.e., a Lambertian surface) will reflect radiation independently of satellite viewing angles (θ_s, ϕ_s) and solar zenith angle (z). In addition, the radiant exitance may be given simply as $\pi N'(z, \theta_s, \phi_s)$. Therefore F should become unity under those conditions.

Comprehensive studies by Minnis and Harrison (1984) and Taylor and Stowe (1984) have shown that most cloud and Earth surfaces are non-Lambertian and that F is a complex function of z, θ_s and ϕ_s. Minnis and Harrison (1984) and Briegleb et al. (1986) converted the narrow-band data from GOES into a broad-band estimate by calibrating GOES counts against simultaneous broad-band radiances from NIMBUS-7. Using Eq. (34), characteristic F values were derived for land and ocean surfaces. This step allowed the calculation of M(z) for the entire GOES scene under cloudless conditions. Finally, the top of the atmosphere albedo is estimated by dividing M(z) by the extraterrestrial incoming irradiance.

Staylor and Shuttles (1986) used NIMBUS-7 ERB scanner measurements to observe the bi-directional behaviour of desert surfaces. Power-law relationships were derived between the satellite radiances, the solar-zenith angle and the satellite viewing angle. These relationships allow the estimation of hemispheric albedos, the azimuth-independent reflectance, and the azimuthal phase function. No correction is performed for atmospheric effects. Barker and Davies (1989) also use NIMBUS-7 data to derive surface albedo over four desert surfaces. A two-stream radiative transfer model was used to correct for atmospheric depletion. Results showed that seasonal values of precipitable water vapour and ozone loads were adequate model inputs. Aerosol optical depths less than 0.1 could be neglected in the calculations provided that the solar-zenith angle is less than 70°. Moran et al. (1990) compared SPOT-1 HRV data with concurrent surface and aircraft radiometric surveys. The surface measurements taken over bare soil gave view-angle correction factors which they applied to the SPOT data. The radiative transfer model of Tanré et al. (1985) was used to correct for atmospheric effects.

Some physical models of surface albedo have also been reported in the literature (Otterman 1984; Otterman and Tucker 1985). Otterman (1984) developed a model of a soil surface plane with protruding vegetation elements which were treated as cylinders. Lambertian reflection was assumed for soil and vegetation surfaces and the spatially integrated albedo was found to depend on both solar-zenith angle and cylinder spacing and width. Model results are used to interpret albedos from a NOAA-6 image containing a desert enclosure (Otterman and Tucker 1985).

Stum et al. (1985) and Pinty and Ramond (1987) used a simple radiation transmittance model to correct for the narrow-band radiance from Meteosat. The correction factor was found to depend on total surface albedo, solar-zenith angle and the ratio of the reflected radiation in the visible to the near-infrared band. Brest and Goward (1987) used Landsat MSS bands 4 and 7 to map broad-band surface albedo. The approach empirically related digital counts to surface albedo in each channel. Weighting factors (for the narrow-band to broad-band transformation) were given as the proportion of the total incoming solar irradiance in each of the two bands.

Lazlo et al. (1988) used NOAA–AVHRR channels 1 and 2 to map surface albedo. A radiative transfer model was used along with published spectral albedos for five typical surfaces. Their results showed that a linear combination of the two NOAA channels gave better results than when each channel was used individually. Valiente et al. (1995) used a radiative model and field measurements of spectral albedo to model TOA Meteosat albedo in terms of TOA albedo from channels 1 and 2 of NOAA–AVHRR. Both the TOA Meteosat albedo and broad-band surface albedo performed best with a linear combination of the two channels. Wydick et al. (1987) compared concurrent ERBE Nimbus 7 (broad-band) albedos with albedos from NOAA 1 and 2. As with the above two results, best agreement was found with a linear combination of the two NOAA channels.

Other analyses have concentrated on all three corrections: atmospheric, bi-directional and spectral (Gutman et al. 1989; Vulis and Cess 1989). Gutman et al. (1989) use a broad-band correction consisting of a linear combination of visible and near-infrared albedos. This correction was used with NOAA-9 data to map the albedo of the US Great Plains. Vulis and Cess (1989) used spectral albedo values in a delta-Eddington radiative transfer model to calculate broad-band and narrow-band top-of-the-atmosphere albedos. Significant differences were found in these two quantities with vegetation type.

Saunders (1990) used a linear combination of NOAA–AVHRR channels 1 and 2 to derive broad-band albedo maps for Great Britain. A radiative transfer scheme was used along with weighted corrections to transfer NOAA channel albedo to a broad-band value. View-angle corrections were applied using data presented by Taylor and Stowe (1984).

4.3 A Regional Estimate of Broad-Band Surface Albedo

Various relationships were described in the previous section for use in calculating surface albedo from satellites. Nearly all expressions use satelllite data to provide regional or global coverage, and then correct the signals for atmospheric, bi-directional or spectral effects. These correction relationships are fairly general, relying on theoretical or empirical relationships to correct for the effect of atmospheric depletion or narrow-band errors.

Regional studies have the advantage of being able to investigate these processes at a greater depth. In this section a technique will be described (Skirving 1984; Nunez et al. 1987) which focuses on direct albedo determinations at the surface, but which are then linked to GMS data to provide a regional estimate. This technique is feasible since the study area is, by definition, small enough to allow regional field observations in all the different environments that are encountered.

For cloudless conditions and any solar-zenith and satellite-nadir angle, the radiance detected by the satellite (N′) may be approximated by

$$N' = N'_a + \left(K'_d \alpha'_g \exp(-\tau' m)\right)/\pi, \tag{35}$$

where N'_a is the atmospheric radiance incident on the sensor in the wavelength band of the GMS satellite (0.55–0.75 μm), K'_d is the downwelling solar irradiance also in the band of the satellite, α'_g is the narrow-band surface albedo, τ' is the direct beam turbidity coefficient at 0.65 μm, and m is the optical air mass from the ground surface to the satellite sensor. Thus the second term on the right-hand side of Eq. (35) refers to the radiance from the target which is incident on the sensor after transmission through the atmosphere.

Equation (35) may be rewritten to given an expression for α'_g

$$\alpha'_g = \left[\alpha'_{ea} K'_{extra} - \pi N'_a\right] / \left[K'_d \exp(-\tau' m)\right], \tag{36}$$

where α'_{ea} is the satellite-derived reflectivity value (i.e. Earth-atmosphere albedo) and K'_{extra} is the incoming irradiance at the top of the atmosphere. Empirical formulae were developed for K'_d and τ' based on field measurements. In addition, simultaneous surface measurements of narrow-band albedo (α'_g) and broad-band albedo (α_g) allow a correction factor to be determined.

An Exotech (100 AX) four-channel spectral radiometer was used for the field measurements. This instrument is designed to measure the irradiance in the four Landsat MSS bands (0.5–0.6, 0.6–0.7, 0.7–0.8, 0.8–1.1 μm). The incoming spectral irradiance in the GMS wavelength band (K'_d) may be approximated by a linear combination of the incoming irradiance in the three channels:

$$K'_d \approx 0.5K'_{d(0.5-0.6)} + K'_{d(0.6-0.7)} + 0.5K'_{d(0.7-0.8)}. \tag{37}$$

A total of 84 spot measurements of K'_d were taken with this instrument in Hobart, Australia, during cloudless conditions in the autumn, winter and spring of 1983. They covered zenith angles extending from 80° to just under 45°. A linear regression was developed between solar-zenith angle z and K'_d:

$$K'_d = 383.5 - 4.5z\left(\mathrm{W\,m^{-2}}\right), \tag{38}$$

with $r^2 = 0.988$ and SEE = 3.3 W m^{-2}

This relationship was considered appropriate for substitution into Eq. (36).

A relation was then developed between α_g and α'_g using field data. A narrow-band spectral albedo may be estimated from the spectral radiometer:

$$\alpha'_g = \frac{\left[0.5K'_{u(0.5-0.6)} + K'_{u(0.6-0.7)} + 0.5K'_{u(0.7-0.8)}\right]}{\left[0.5K'_{d(0.5-0.6)} + K'_{d(0.6-0.7)} + 0.5K'_{d(0.7-0.8)}\right]}, \tag{39}$$

where $K'_{u(i)}$ refers to the upwelling spectral irradiance in the ith band. The experimental arrangement consisted of attaching the spectral radiometer to a boom at a height of 2 m above the ground and alternately pointing the radiometer at the surface and at the sky at 5-min intervals. Two Kipp and Zonen CM-11 pyranometers were also used in the measurement program. They were also placed on the boom, one facing downward and the other facing upward and next to the Exotech radiometer. This configuration ensured that all three radiometers were essentially observing the same portion of the ground surface and sky.

Aircraft albedo measurements were taken in the case of forested surfaces. The Exotech radiometer had a 15° field of view (FOV) to detect upwelling irradiance from specific surface types. Two Kipp and Zonen CM-11 pyranometers were used to determine the broad-band upwelling and downwelling irradiance. To determine the broad-band upwelling irradiance a stainless steel tube, blackened on the inside and providing a 57.5° FOV, was fitted to the downward facing pyranometer. This particular aperture accepted enough light to meet the sensitivity requirements of the pyranometer while enabling small regions to be measured. No measurements of incoming spectral irradiance were available with the aircraft measurements. These estimates were obtained using the relationships in Eq. (38).

The results for all concurrent broad-band (0.4–3.0 µm) and narrow-band (0.55–0.75 µm) measurements taken in 1984 under cloudless conditions are presented in Fig. 8. Although they are spot measurements, the relationship between the two data sets is well defined but non-linear. Forest vegetation prevails at low albedo values, with rainforests at the bottom of the scale. Paddock grass occupies the middle to high end of the scale, and dry aquatic herbland registered the highest albedo values which were between 35 and 40%.

Well-irrigated short grass failed to conform to the pattern in Fig. 8. We attribute this to be due to the surface not being a "natural" surface and behaving much like an individual leaf. Spectral albedo values from some leaf surfaces

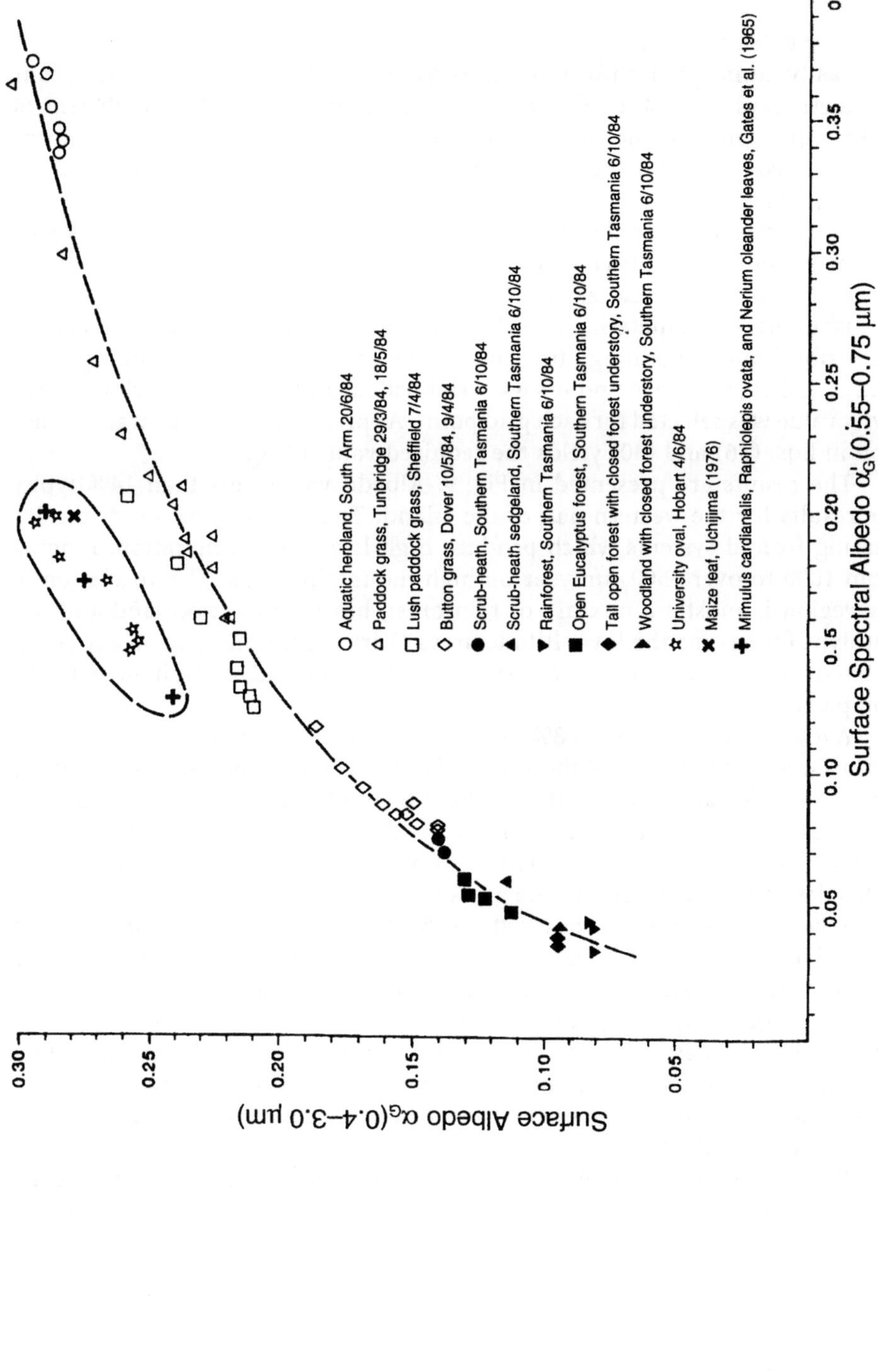

Fig. 8. Spectral albedo (α'_g, 0.55–0.75 µm) vs. total broad-band albedo (α_g, 0.4–3.0 µm) for various vegetated surfaces in Tasmania. Aircraft measurements were taken on 6 October 1984. The rest of the data represent surface observations

are included in Fig. 8 and are shown circled. As a result of these considerations, data from the university oval were not used in the analysis. A regression of the rest of the data takes the form

$$\alpha_g = 0.3904 + 0.0936 \ln\left(\alpha'_g\right), \tag{40}$$

with $r^2 = 0.984$ and SEE = 0.0086.

Lastly, atmospheric turbidity measurements (τ') were taken during four cloudless days in 1984 at Hobart, Tasmania. A mean turbidity coefficient of 0.197, represented as the sum of a Rayleigh and aerosol scattering term, was obtained using an Eko sun photometer. This value was then substituted for τ' in Eq. (36).

The atmospheric reflectance N'_a [Eq. (36)] was solved using the single scattering model of Bullrich (1964) which includes a parameterisation for both Rayleigh and aerosol scattering (Nunez et al. 1987).

With these parameterizations, Eq. (36) was applied to eight GMS images in the visible band containing little cloud and covering the Tasmanian region. A grid of 154 evenly spaced points was established, and a median satellite reflectivity value was selected for each grid point. Applying the required transformation in Eqs. (36) and (40) yields the required value of α_g.

The results are presented in Fig. 9. Albedo values less than 14% typify the results for the western half of the Island. The region is exposed to rain-bearing frontal systems which produce high levels of precipitation, ranging from 1000 to over 3500 mm/year on high mountain peaks. The vegetation in the region is mostly a mixture of rainforest, buttongrass moorland and wet eucalypt forests for the low-altitude areas (Kirkpatrick and Dickinson 1984). High-altitudinal areas are characterized by an alpine and subalpine vegetation complex.

A low albedo of less than 8% represents the contribution by water surfaces in the Great Lakes region of the Tasmanian Central Plateau. Cleared paddocks occupy albedo values greater than 18%, at the high end of the range, and these are located in a corridor extending from the north to the southeast of the island. These regions are expected to show a marked seasonal change as the fields dry in late summer and the grass starts to cure.

Vegetation on the east coast reflects the drier conditions, with dry eucalypt and sclerophyll forests being typical of the vegetation in the interior, and scrub, heathland and other vegetation dominating the coastal fringes. The albedo range occupied by these communities appears to extend from 14 to 18%. Albedo values less than 14% are also characteristic of the north-eastern highland areas which exhibit a high precipitation and rainforest/wet eucalypt vegetation. No data was available for the southern part of the region due to cloud contamination.

Broadband transformation plays a crucial role in the analysis. The assumption has been made that Eq. (37) typifies natural vegetated regions as

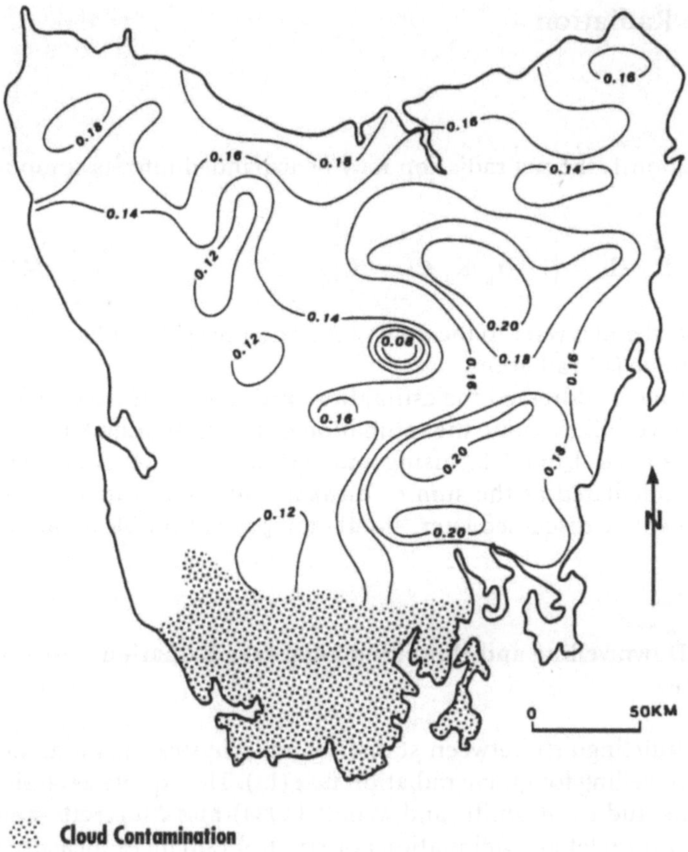

Fig. 9. Regional surface albedo (α_g, 0.4–3.0 μm) for Tasmania. Data used were eight images collected between 26 January 1985 and 9 February 1985. An isoline interval of 0.02 was used which exceeds the resolution of the model used to estimate the surface albedo

seen within the satellite field of view. Although other non-natural surfaces may be contained in the pixel (well-maintained football grounds, parks, etc.), these are not expected to significantly change the overall transformation, simply because of the size of the satellite pixel. Bare surfaces (sand, rock, etc.) also yield different transformations (Nunez et al. 1987). However, in Tasmania the vegetated environment dominates at the scale of the GMS pixel.

The above arguments illustrate the importance of local field measurements to determine a regional albedo. Single-band expressions such as Eq. (40) or double-band transformation (using, say the visible and near-infrared band of NOAA–AVHRR could be developed for a range of vegetated environments. In addition to providing accurate regional values, they would also serve to validate global data sets.

5 Net All-Wave Radiation

5.1 Introduction

As discussed in Section 1, the net radiation may be expanded into its component fluxes as

$$Q_* = K_d - K_u + L_d - L_u = (1-\alpha_g)K_d + L_d - L_u, \qquad (41)$$

where the subscripts d and u refer to the downward and upward irradiance for solar (K) and longwave (L) radiation.

Earlier sections have addressed the estimation of K_d and K_u from satellite data. In this section we will examine the estimation of the remaining terms of the net radiation balance, L_d and L_u, using satellite data. In theory the net radiation could be calculated as the sum of estimates of each of its components. However, as will be discussed later, cloud data present problems in the estimation of L_u.

5.2 Estimation of Downwelling and Upwelling Longwave Radiation at the Earth's Surface

Sellers et al. (1990) distinguish between statistical and physical methods for calculating the downwelling longwave radiation flux (L_d). They quote as statistical techniques the studies of Smith and Woolf (1984) and Morcrette and Deschamps (1986) who related combinations of spectral radiances measured by infrared sounders on GOES and NOAA satellites to on-ground measurements with pyrgeometers. Physical methods include the techniques of Darnell et al. (1983), Gautier and Frouin (1985) and Frouin et al. (1988). These methods use radiative transfer models with parameters determined from satellite radiance data.

The expression for the incoming longwave spectral radiance may be written as Eqs. (42) and (43) for clear skies $(L'_{do})_\lambda$ and cloudy skies $(L'_{dc})_\lambda$ respectively (Darnell et al. 1983).

For clear skies

$$\left(L'_{do}\right)_\lambda = \int_{Tr(\lambda,\theta,h)}^{1} B_\lambda(T_h)\,dTr(\lambda,\theta,h), \qquad (42)$$

and for cloudy skies

$$\left(L'_{dc}\right)_\lambda = B\lambda(T_H)Tr(\lambda,\theta_c,H) + \int_{Tr(\lambda,\theta,h)}^{1} B\lambda(T_h)\,dTr(\lambda,\theta,h). \qquad (43)$$

In these expressions variables T and θ refer to actual and potential temperature respectively. B_λ (T_h) is the Planck function defined at temperature T_h for an arbitrary height h. Tr (λ,θ, h) is the transmittance of longwave radiation between the surface and height h, which depends on wavelength λ and potential temperature θ. The cloudbase is at height h = H. Longwave radiation from the cloudbase, i.e., at temperature T_H is emitted as blackbody radiation [Bλ (T_H)]. The transmittance between the cloudbase and the ground is Tr (λ, θ_c, h). The second term on the RHS of Eq. (43) refers to the contribution by the atmosphere between the cloudbase and the ground. Both the radiative and transmittance terms have the same zenith and azimuth angle dependence, but, for the sake of clarity, they have not been included in Eqs. (42) and (43).

Incoming longwave radiation fluxes at the Earth's surface have been calculated using Tiros Operational Vertical Sounder (TOVS) data. It provides information on cloud cover as well as profiles of air temperature and relative humidity to solve Eqs. (42) and (43) using various radiative transfer codes (Darnell et al. 1983; Michael and Nunez 1991). Spatial-interpolation procedures are required to estimate temperature and humidity profiles in overcast conditions. Darnell et al. (1983) used HIRS-2–MSU data to obtain information on cloud cover and cloud-top altitude. Schmetz et al. (1986) used screen-level air temperatures to derive an empirical estimate of the clear-sky longwave radiation flux. These data were combined with cloud information derived from Meteosat data to obtain the total (cloudless + cloudy) longwave radiation flux.

Gupta (1989) derived an empirical expression for the net longwave radiation flux at the Earth's surface using simple parameters obtained from TOVS data. This empirical relation was obtained by comparing the TOVS parameter with simultaneous results from a radiative-transfer model. In a later publication Gupta et al. (1992) modified the regression constants to take into account the more appropriate use of ISCCP (International Satellite Cloud Climatology project) data. An empirical term estimating L_d for overcast conditions was also corrected so as to reduce systematic errors.

Sellers et al. (1990) have noted that the estimation of land-surface temperature and upwelling longwave radiation are complicated because of the complex structure of the land surface and the departure from unity emissivity. These factors cause angular dependence in the radiances measured by the satellite. They point out that the satellite radiometer measures thermal radiation emitted by the surface and by the atmosphere; thermal radiation reflected by the surface and solar radiation reflected by the surface and scattered by the atmosphere. Therefore the radiative transfer equation for upwelling longwave radiation incorporates surface temperature, atmospheric transmittance, surface emissivity and bidirectional reflectance, as well as the atmospheric emittance.

Applications include the use of GOES-1 thermal imagery to assess nocturnal farmland temperatures in Florida (Chen et al. 1979); of HCMM thermal imagery in frost mapping in Victoria, Australia (Kalma et al. 1983); of TOVS data from NOAA-6 to calculate near-surface air temperatures; and of HCMM thermal data to estimate surface temperatures in surface energy balance modelling (Carlson et al. 1981). Assuming no atmospheric depletion, upwelling longwave radiation (L_u) may be given as

$$\begin{aligned} L_u &= \sigma\left(T'_g\right)^4 + (1-\varepsilon)L_d \\ &= \varepsilon\sigma\left(T_g\right)^4 + (1-\varepsilon)L_d, \end{aligned} \tag{44}$$

where T'_g is the satellite derived radiative temperature, T_g is the true or kinetic temperature of the surface and ε is the surface emissivity which for vegetative surfaces may exceed 0.97 (Kalma et al. 1983, 1986).

Lagouarde et al. (1991) have used NOAA-AVHRR data, obtained in the early afternoon, to map L_u throughout the day. Their analysis is based on a simple model of diurnal surface temperature which uses both an afternoon satellite image and the minimum air temperature obtained from the meteorological network. Other components of the radiation balance are solved for essentially clear-sky conditions. This involves pyranometer measurements and use of the Ångström formula (for estimating K_d), NOAA-AVHRR data to estimate surface albedo and air temperature to estimate L_d with the Brutsaert (1982) formula. The relationship was used to map net radiation on cloudless days in France and in Niger.

In summary, it appears that data from polar-orbiting meteorological satellites such as NOAA-AVHRR may readily measure T'_g and therefore may be used to obtain L_u, provided that conditions remain cloudless. Absorption by clouds presents problems since they block the outgoing longwave radiation from the surface. In addition, the energy balance of the surface changes in cloudly conditions so that a simple clear-sky interpolation will not suffice to determine T'_g. And yet it is precisely in cloudy conditions that the satellite technique can provide useful data.

5.3 Regional Mapping of Net Radiation

The above sections have discussed how each of the components of the net radiation given by Eq. (3) may be obtained from satellite data. The visible channels of geostationary or polar-orbiting satellites may be used for estimating K_d and K_u, whilst data from the thermal channels of NOAA-AVHRR and TOVS may be used for L_u and L_d respectively. In principle, it is possible to apply these techniques to estimate each component of Q_* separately. Alternatively, Q_* may be obtained by a combination of satellite data and surface measurements (Lagouarde et al. 1991). However, as mentioned earlier, the technique is only applicable to cloudless conditions since the upwelling longwave radiation is absorbed by clouds. Therefore it seems that other approaches might be warranted in estimating Q_*.

One useful approach is to seek simple relationships between the TOA net solar irradiance and the net surface irradiance (see also the reference to Chen and Ohring 1984 in Sect. 4.2). It may be argued that these two variables are highly correlated given the conservative nature of atmospheric absorptance already discussed in Section 2.2. That is, the radiation fluxes entering the atmospheric "slab" from the top are highly correlated to the fluxes leaving the slab at the bottom. The results look encouraging and should be further explored as predictive tools for mapping regional surface net radiation.

Using a modelling framework, Vulis and Cess (1989) investigated the relationship between planetary and surface albedo. In a later paper, Cess et al. (1991) used clear-sky ERBE satellite data to derive TOA net solar irradiance. It was found to be highly correlated to measurements of net solar irradiance at the surface. The relationship agreed well with model parameterisation even when monthly precipitable water vapour and aerosol loads are used for the model. The authors conclude that net shortwave irradiance is climatically more important than incoming irradiance, and when considering these two variables, net solar irradiance may be estimated more accurately using satellite data.

In a later paper, Cess et al. (1993) examined the above relationships using days with clouds and found a mean bias (measured-estimated) of −4.4%. Li et al. (1993a) using a modelling approach, derived a linear relationship between surface and TOA net solar irradiance. A mean bias error of less than 1% (Li et al. 1993b) was obtained when tested against the data from Cess et al. (1993). The improved linear relationship used precipitable water vapour and solar-zenith angle, but no cloud information was explicitly needed.

A similar approach has been tried with net radiation. Initial studies by Pinker and Tarpley (1988) and Pinker et al. (1985) have shown that satellite derived estimates of net radiation at the top of the atmosphere are highly correlated to concurrent surface net radiation measurements. Empirical regressions between these two variables give typical correlation coefficients of over 0.90 and daily standard error of $20\,W\,m^{-2}$. Recent studies by Calvet and Viswanadham (1993) also find a good correlation over the Amazon Forest, although the slope of the relationship has a different sign to that reported by Pinker and Tarpley (1988). The authors argue that radiative properties of clouds could account for this difference.

Net radiation contains upwelling and downwelling longwave irradiance, which in turn are linked to the energy and water balance via the surface and air temperature. Furthermore, longwave radiation is highly absorbed in the atmosphere, so that the outgoing longwave irradiance from the surface is not readily related to its TOA value under cloudy conditions. It would appear that, unlike the case for solar irradiance, simple relationships between TOA and surface all-wave irradiance may be more difficult to obtain. An alternative might lie in exploring the relationships between various components of net radiation as determined by measurements.

There is large seasonal, daily and diurnal variation in the fraction Q_*/K_d between different land-surface and climatic environments. This has been documented and discussed by Linacre (1992).

Stanhill et al. (1966) studied the effect of surface characteristics associated with various plant associations for 11 surfaces in Israel on annual values of Q_*/K_d. The maximum value found was 0.58 for a fish pond and the minimum 0.25 for a desert wadi vegetation. The mean of the 11 surfaces was 0.45 with a standard deviation of 0.10. A similar mean value (0.49) and standard deviation (0.09) in the ratio was found in the literature for a global range of 11 plant associations (Stanhill et al. 1973). The maximum ratio of annual values was 0.67 for a sugar cane plantation in Hawai and the minimum ratio was 0.38 for a short grass surface in southern England.

Kalma and Stanhill (in preparation) have recently analysed the temporal variation in the fraction of global radiation retained at the surface of an intensively grazed catchment in SE Australia. Measurements extended over a five-year period and they observed the following relationhip between daily totals of net radiation (Q_*) and daily totals of incoming global radiation (K_d) for 1255 days

$$Q_* = 0.482K_d - 0.457\left(\text{MJ m}^{-2}\ \text{day}^{-1}\right), \tag{45}$$

with $r^2 = 0.851$ and SEE = 1.741.

When the regression is forced through the origin the results are

$$Q_* = 0.462K_d\left(\text{MJ m}^{-2}\ \text{day}^{-1}\right), \tag{46}$$

with $r^2 = 0.849$ and SEE = 1.751.

The slope of 0.46 is in good agreement with the values of 0.45 and 0.49 quoted above. However, there is very significant temporal variability in the Q_*/K_d ratio as evidenced by the significant standard error of the Q_* estimate.

An important reason for this variability is the dependence of net radiation Q_* on the surface properties of reflectivity, emissivity and temperature which are related to plant factors (e.g. quantity and composition of the available biomass), soil factors (soil-moisture status) and climate factors (especially vapour-pressure gradient and wind speed).

The concept of the heating coefficient, proposed by Monteith and Szeicz (1961), is explored as a means of obtaining all the components of the net radiation balance. The physical basis of this relationship is that as solar radiation decreases due to cloud cover, the net longwave radiation increases; this is because the presence of clouds increases L_d but decreases T_g and L_u. Therefore, net solar irradiance (K_*) and net longwave irradiance (L_*) should be negatively correlated.

Considerable hysteresis occurs in the surface temperatures and soil heat flux in response to net radiation during the day, so that development of simple one-to-one relationships is difficult at this time scale. However, the approach is much more tractable at the daily time scale or longer, which is of concern

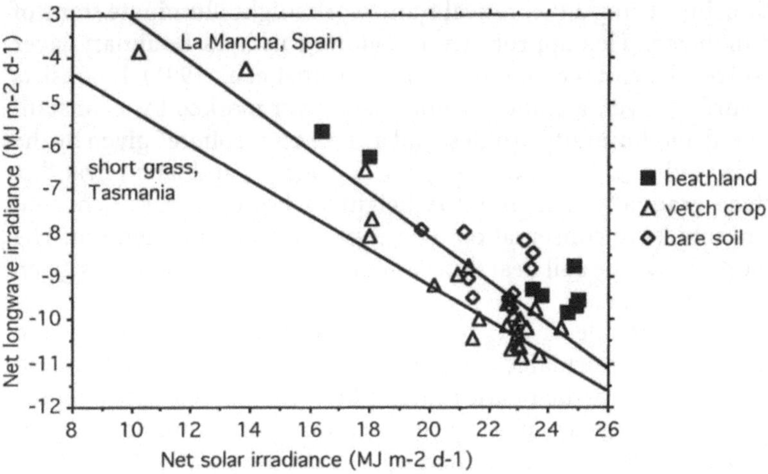

Fig. 10. Comparison of daily measurements (MJ m^{-2} day^{-1}) of net longwave irradiance (L_*) and net shortwave irradiance (K_*) for sites in Tasmania, Australia (short grass) and La Mancha, Spain (bare soil; heathland; vetch crop). The *regression lines* given by Eqs. (47) and (48) are shown

here. Figure 10 presents the results of daily measurements of net solar and net longwave irradiance in Tasmania, Australia and La Mancha, Spain.

The Tasmanian data represent measurements taken over short, cut grass in Cressy during the period February 1988 to October 1988. A total of 66 days were analysed and the results are summarised in Eq. (47) and shown as the bottom line in Fig. 10.

$$L_* = -0.4029K_* - 1.226\left(\text{MJ}\,\text{m}^{-2}\,\text{day}^{-1}\right), \tag{47}$$

with $r^2 = 0.914$ and SEE = 0.95.

The La Mancha data encompass a variety of surfaces from natural heathland to vetch crop and bare soil; all measurements were taken during June 1991. Applying a linear regression to the La Mancha data gives the relationship

$$L_* = -0.471K_* + 1.203\text{MJ}\,\text{m}^{-2}\,\text{day}^{-1}, \tag{48}$$

with $r^2 = 0.748$ and SEE = 0.818.

Results show consistent relationships between individual surfaces, but a much larger error if a single relationship is applied to the entire data set. In particular the Tasmanian relationship obtained for short cut grass has a substantially lower intercept than the La Mancha data for most of the range of net solar irradiance measurements. The heathland surface shows the highest intercept, followed by bare soil and vetch crop.

It is evident that these measurements are linked to the entire surface-energy balance by their dependence on surface temperature. An examination of L_*/K_* relationships using a theoretical framework might illuminate some of the processes involved. This approach was followed using a boundary-layer model which solves the surface-energy balance (Brunet et al. 1991). It consists of a turbulent surface layer, a planetary boundary layer marked by a constant temperature, wind and humidity profiles, and a "free atmosphere" given by the 850 hPa airsonde readings. As the day progresses, sensible and latent heat flux from the surface gradually increases the height of both the turbulent and planetary layers subject to conservation of energy, mass and momentum. The model also assumes that the soil heat flux is a conservative term and it is given as 10% of the net radiation.

An important advantage of this type of relationship is that the air and surface temperatures are calculated implicitly in response to external boundary conditions. In this study the relationship of Swinbank (1963) was adopted to represent incoming longwave irradiance. In addition, transmission of solar radiation through clouds was represented by a transmissivity factor of 0.34, which is typical of low clouds (List 1968).

The procedure followed was to select a set of parameters describing surface conditions (aerodynamic roughness, albedo, percent saturation at the surface) and upper boundary conditions (850 hPa geostrophic wind, environmental lapse rate, humidity and wind gradients). Solar radiation and cloud cover are externally forced into the system which then predicts the characteristics of the boundary layer every half hour.

The results for a set of conditions typical of the December month in Tasmania are presented in Fig. 11. All surfaces have an albedo of 10% and

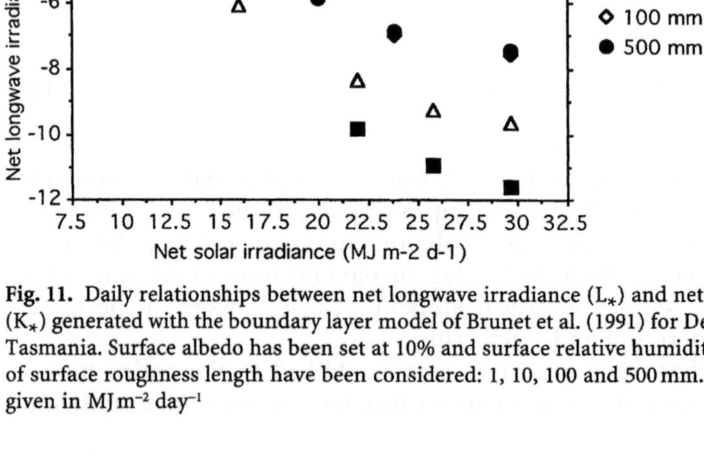

Fig. 11. Daily relationships between net longwave irradiance (L_*) and net shortwave irradiance (K_*) generated with the boundary layer model of Brunet et al. (1991) for December conditions in Tasmania. Surface albedo has been set at 10% and surface relative humidity at 10%. Four values of surface roughness length have been considered: 1, 10, 100 and 500 mm. Irradiance values are given in MJ m^{-2} day^{-1}

surface relative humidity has been taken as 10% of saturation. The roughness length has been allowed to vary from 1 to 500 mm for varying levels of cloud cover. The results presented show the importance of roughness in explaining the daily K_*/L_* relationships. Although surface humidity and wind conditions are also important, their effects were nearly one order of magnitude less effective than the role of roughness in generating differences. Clearly, this term is important in that it affects the ability of the atmosphere to dissipate the net radiation energy absorbed by the surface. It is likely that the lower intercept for short grass in Fig. 10 largely reflects the lower roughness of the surface, although the pattern for bare soil does not agree with this hypothesis.

A further illustration of this process is provided by data collected above a dry eucalypt forest canopy near Hobart, Tasmania. Thirty-three days of daily K_* and L_* values were obtained for the months of December 1992 and January 1993. The results exhibit considerable scatter as shown in the relationship

$$L_* = -0.293K_* + 0.167\left(\mathrm{MJ\,m^{-2}\,day^{-1}}\right), \tag{49}$$

with $r^2 = 0.45$ and SEE = 2.05.

The three regression relationships have been plotted in the same graph (Fig. 12). It is evident that on the basis of surface roughness the data may be grouped into three distinct sets, with the lowest roughness (grass) exhibiting the largest deficit in net longwave irradiance. This is followed by the two vegetated surfaces in La Mancha, (bare surface was not considered) occupying the middle niche, and with the lowest deficit being obtained above the forest surface.

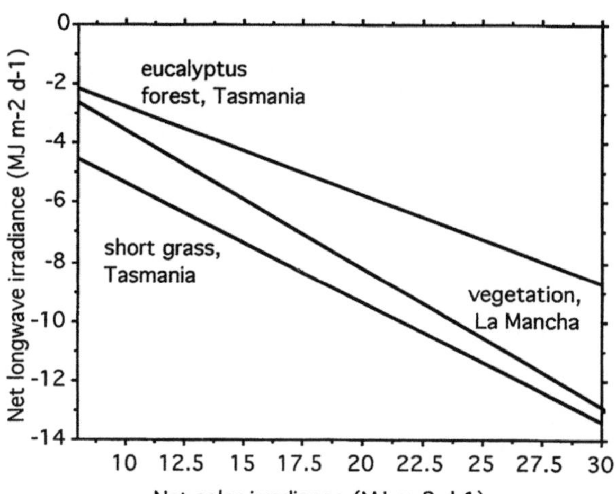

Fig. 12. Daily relationships between net longwave irradiance (L_*) and net shortwave irradiance (K_*) obtained for short grass in Tasmania [Eq. (47)], heathland and vetch crop in La Mancha, Spain [Eq. (48)] and an eucalypt forest near Hobart, Tasmania [Eq. (49)]

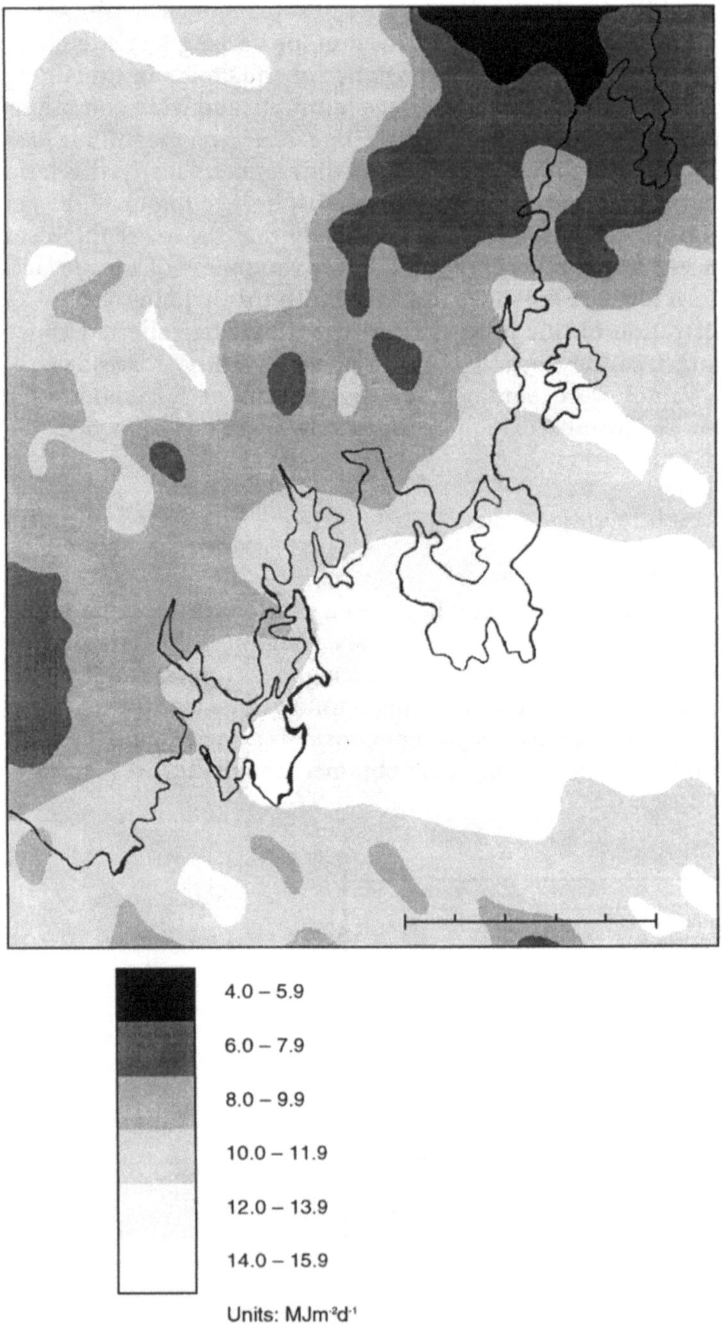

Fig. 13. Mean daily net radiation (MJ m^{-2} day^{-1}) for the Hobart, Tasmania region for 18 January 1988. The image represents a 2° × 2° region centred on Hobart Airport. A distance scale with 10-km intervals is shown at the *bottom right*

The above relationships indicate that surface albedo and surface roughness are two important parameters determining the L_*/K_* ratio. Soil-moisture effects, which might vary seasonally, may also be important. Satellite data may be used to map Q_* for a given L_*/K_* ratio if one assumes little variability in surface texture, roughness and soil-moisture status in the region.

Once L_* is known, the net radiation may be obtained simply from the relationship

$$Q_* = K_* + L_* . \tag{50}$$

The relationship for short grass expressed by Eqs. (47) and (50) may be used to map the net radiation at the surface in southeastern Tasmania for 18 January 1988. GMS satellite data have been used and the estimation of K_d and α_g has already been discussed in previous sections. It also is important to keep in mind that the results are truly applicable to short grass, a surface type which is widely represented in this region.

The results are shown in Fig. 13. The geometric details of the image are similar to those of Fig. 5, which contains an outline of the Tasmanian coastline and Hobart Airport in the geometric centre. Cloud cover affects K_d which in turn causes high variability in Q_*, ranging from 5 to 15 MJ m^{-2} day^{-1}. Lowest values are observed in the Eastern Tiers which run parallel to the Tasmanian east coast. A secondary low region is also obtained along the mountainous to the southwest. Values increase inland towards the Central Plateau and Midlands regions. Although only a single day is presented here, the pattern is characteristic of the cloudiness regime in Tasmania, with higher levels along the coast and southwest.

6 Discussion and Conclusions

Net radiant energy Q_* is a critical component of the energy balance because it drives the physical and biological processes at and near the land surface.

On a large scale, Q_* defines the rate at which land and oceans exchange heat and moisture with the atmosphere, the rate of warming (or cooling) of the air, soil and water layers close to the surface and the distribution of land and sea surface temperatures at the interface. This is essential information in general circulation models and in the development and interpretations of theories of climatic control and climatic change.

At a regional scale, the primary forcing of the atmospheric boundary layer and its surface layer, within which most life processes occur, is through the absorption of solar radiation at the land surface. This results in diurnal and seasonal variation in the exchange of turbulent latent and sensible heat fluxes between the soil surface, vegetation and the atmosphere; in the soil heat flux; and in the surface temperature. Thus the radiation balance, through the energy budget, controls the heating of the air, and hence the microclimate of plants and animals.

There is a wide range of applications at regional and local scale which require estimates of the temporal and spatial variation in surface-radiation balance and energy budget. These include the estimation of evaporation and transpiration in water balance studies, catchment hydrology, agricultural meteorology and bioclimatology; crop productivity studies in agronomy; as well as various engineering activities including building and urban design problems and indoor climate control.

Throughout the review we emphasise the use of satellite-based methods for regional rather than global scale mapping of the radiation-budget components. Particular emphasis is placed on the use of operational meteorological satellites because of the frequency and accessibility of the data as well as their relatively low cost. We discuss a range of approaches developed for the estimation of solar irradiance (K_d) from satellite-based remote sensing. These fall into three broad groups: methods based on physical principles which account for scattering, absorption and reflection processes; statistical and empirical techniques which relate brightness to insolation; and methods which combine physical and statistical approaches.

Records of the components of the Earth's radiation budget derived from satellite data are not yet sufficient for long-term trend detection. However, significant progress is made towards developing global-scale climatologies for planetary albedo, surface albedo, solar iradiance and net radiation. Satellite-derived radiation budget data are now linked to regional and large-scale circulations, and they are used for identifying seasonal cycles and general patterns of variability (see Carleton 1991). Major progress has also been made in the use of surface solar-irradiance estimates for crop-yield prediction (e.g. AgRISTARS). Much of this progress is due to improvements in satellite coverage, advances in satellite systems and greater diversity in the spectral sensitivity of satellite sensors. However, the relatively coarse resolution of the geostationary and polar-orbiting satellites used in deriving surface radiation budget components places a limit on many regional and local-scale applications.

Carleton (1991) has identified several critical issues which affect obtaining radiation budget information from satellite data. These include appropriate spatial and temporal resolution; differences between satellite systems; narrow-band to broad-band conversion; drift in sensor calibrations (sensor degradation) and differences in sensor field of view. There are also differences between satellite and ground truth due to inaccuracies in satellite-Earth co-location; differences in spatial coverage (assumptions of spatial homogeneity); non-synchronized satellite and ground data; differences in spectral responses of the radiometers and systematic errors due to temporal sampling.

Satellite information used in bioclimatology and global climate change includes radiation-budget components (albedo, irradiance, NDVI) and surface temperatures. This information is used for the estimation of evaporation from the land surface and in estimating soil moisture status and biomass production (as reviewed by Kalma and Calder 1994). A key role of satellite data in climate modelling is in the parameterization of land-surface conditions. Remote-sensing applications in hydrology include the estimation of evaporation, the compilation of inventories of water resources, the assessment of land cover and the monitoring of soil moisture.

Below, we review the use of satellite-derived surface radiation budget data in one particular application: the estimation of potential and actual regional-scale evaporation from the land surface. Such estimates are needed in bioclimatology and hydrology.

Equilibrium evaporation ($\lambda_v E_{eq}$) (see Brutsaert 1982) has been defined as the evaporation from extensive wet surfaces in advection-free conditions which is controlled by the difference between net radiation Q_* and the soil heat flux Q_g. It is given by

$$\lambda_v E_{eq} = (\Delta/\Delta+\gamma)[Q_* - Q_g], \tag{51}$$

where λ_v is the latent heat of vaporization, Δ is the slope of the saturation vapour pressure /temperature curve, and λ is the psychrometric constant.

Priestley and Taylor (1972) proposed that evaporation from a wet surface under conditions of minimal evaporation may be estimated from

$$\lambda_v E_{PT} = \alpha(\Delta/\Delta+\gamma)Q_*,$$

where $\alpha = 1.26$ according to measurements in a wide range of environments (Brutsaert 1982). It is included to account for the aerodynamic component of the Penman combination equation. Shuttleworth (1991) argues that on physical grounds the vapour pressure gradient between the land surface and the atmosphere, the main component of the aerodynamic term, should be proportional to net radiation and inversely proportional to wind speed. An added consideration in favour of the Priestley-Taylor relation is that the radiation component in the Penman equation is typically five to six times larger than the aerodynamic component (Shuttleworth 1991). On the other hand, advection effects may introduce substantial differences to the vapour pressure gradient and require α values well in excess of 1.26.

A map of Priestley-Taylor potential evaporation $\lambda_v E_{PT}$ for Tasmania for 18 January 1988 is shown in Fig. 14. It has been obtained by applying Eq. (51) to the net radiation map shown in Fig. 13. It is evident that the variability in net radiation introduces differences in $\lambda_v E_{PT}$ which are larger than can be attributed to variations in the aerodynamic component due to advection differences. If it is assumed, as in Fig. 13, that the values apply to short paddock grass, it is shown that minimum values of 2 mm day^{-1} are obtained in the northeast and southwest quadrants of the image. By contrast, values exceeding 4.5 mm day^{-1} are obtained in the Central Plateau area.

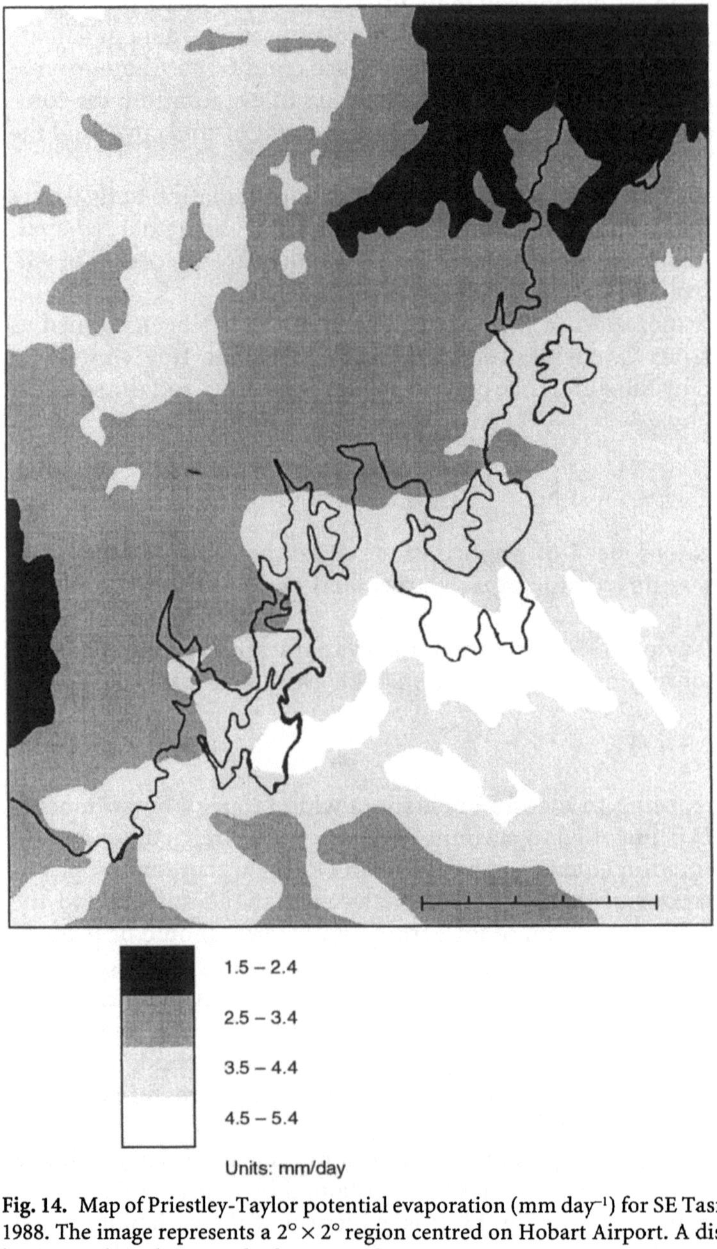

Fig. 14. Map of Priestley-Taylor potential evaporation (mm day^{-1}) for SE Tasmania on 18 January 1988. The image represents a 2° × 2° region centred on Hobart Airport. A distance scale with 10-km intervals is shown at the *bottom right*

Thus $\lambda_v E_{PT}$ may be obtained from estimates of Q_* based on remote sensing techniques, with an appropriate value for α in Eq. (52) and a correction for the soil heat flux Q_g. However, since we are dealing with a wet surface, Q_g is expected to be small and can perhaps be neglected. Significant progress has also been made in estimating daily values of regional-scale actual evaporation from the surface by combining net radiation measurements with remotely sensed thermal imagery.

Many workers (e.g., Seguin and Itier 1983; Jackson et al. 1977) have used an empirical relation between daily totals of actual evaporation $\lambda_v E_a$ and net radiation Q_* which incorporates the difference between remotely sensed ground surface temperature T_g and ambient air temperature T_a near midday or at the time of its maximum:

$$\lambda_v E_a = Q_* - b\left(T_g - T_a\right)_{1pm.} \tag{53}$$

Here b is an empirical constant, and Q_* and T_g may be obtained from remote sensing. Certain assumptions need to be made about the spatial variation in ambient air temperature T_a. The second RHS term of Eq. (53) is an estimate of the sensible heat flux between the land surface and the adjacent air and Eq. (53) thus expresses the surface energy balance over a day. Expressions similar to Eq. (53) have been used by Lagouarde and Brunet (1989), Lagouarde (1991) and Kalma et al. (1991) with some success, especially over short vegetation.

Nieuwenhuis (1986) estimated the daily total of actual evapotranspiration E_a as a fraction of the daily total of potential evapotranspiration E_p from

$$E_a / E_p = 1 - b'\left(T_g - T_g'\right)_i, \tag{54}$$

where $(T_g - T_g')_i$ is the ground surface temperature difference at time i between actual and potential evaporation conditions. Potential evapotranspiration PET may be obtained from the Priestley-Taylor expression as outlined earlier. The calibration constant b′ is expressed as a function of wind speed (u) with

$$b' = c + du, \tag{55}$$

where c and d are empirical constants.

Menenti (1984) related spatial differences in evaporation (ΔE_a) to spatial differences in surface albedo ($\Delta\alpha$) and in surface temperature (ΔT_g) with

$$\Delta E_a = p + q(\Delta\alpha) + r\left(\Delta T_g\right), \tag{56}$$

where p, q and r are empirically determined constants. Both $\Delta\alpha$ and ΔT_g may be obtained with remote sensing.

Some published work has concentrated on the one-dimensional boundary layer, dealing with land–atmosphere interactions for a given set of boundary conditions (Brunet et al. 1991). Other studies have used remotely sensed surface-temperatures maps in combination with simple surface heat balance models to derive the regional evapotranspiration field (Price 1980; Carlson et al. 1984; Taconet et al. 1986; Tarpley 1994). Another approach has been to use the visible and near infrared bands of NOAA–AVHRR to derive indices of dryness which may be related to soil dryness status and evapotranspiration (Jupp et al. 1992; McVicar et al. 1992).

In conclusion, it should be stated that although the field of satellite climatology is over thirty years old, applications of satellite data to environmental problems have not kept up with the vast advances in satellite spectral and spatial resolution, computing power and image processing capabilities. The problem probably arises because of a lack of knowledge by potential users who could greatly benefit from the data. There is also the added consideration that the data were, until recently, difficult to obtain and costly. This is no longer the case. At present much of the data, especially from the operational (meteorological) satellites, is frequent and freely available. It may now be accessed from existing computer networks. We hope that this review has illustrated some uses and applications of satellite data.

Much work needs to be done in the various areas that have been discussed. Substantial advancement is only possible by developing and coordinating ground measurement programs which could be related to the satellite data. Some progress has been made in several recent large-scale integrated field experiments such as EFEDA and Hapex-Sahel.

In the field of solar radiation, the problem rests with aerosol variability: ultraviolet radiation studies need reliable high-resolution surface measurements; surface albedo maps rely on broad-band and bidirectional conversions which still need examination; and the net radiation field is linked to the surface temperature and as a result, to the rest of the surface energy balance. It is only when these issues are addressed that a truly operational system will be obtained which will be capable of providing the relevant climatologies.

Acknowledgements. We gratefully acknowledge the assistance provided by John Parry, Peter Cinquini and John Patterson who helped in processing the satellite data. Robert Slack helped to produce the diagrams. Critical comments by four anonymous reviewers of an earlier draft of this paper have been most helpful in revising the manuscript.

References

Barker HW, Davies JA (1989) Surface albedo estimates from Nimbus-7 ERB data and a two-stream approximation of the radiative transfer equation. J Climate 2:409–418

Barker KS, Smith RC, Green AES (1980) Middle ultraviolet radiation reaching the ocean surface. Photochem Photobiol 32:367–374

Barrett EC, Curtis LF (1992) Introduction to environmental remote sensing. Chapman and Hall, London, 426 pp

Barton IJ, Robertson DF (1975) Measurements of erythemally effective ultraviolet radiation and total ozone content. Nature 258:68–69

Bittar A, Mackenzie RL (1990) Spectral ultraviolet intensity measurements at 45° S: 1980 and 1988. J Geophys Res 95:5597–5603

Brakke TW, Kanemasu E (1981) Insolation estimation from satellite measurements of global radiation. Remote Sensing Environ 11:157–167

Breon F, Frouin R, Gautier C (1994) Global shortwave energy budget at the earth's surface from ERBE observations. J Climate 7:309–324

Brest CL, Goward SN (1987) Deriving surface albedo from narrow-band satellite data. Int J Remote Sensing 8:351–367

Briegleb BP, Minnis P, Ramanathan V, Harrison E (1986) Comparison of regional clear sky albedos inferred from satellite observations and model calculations. J Clim Appl Meteorol 25:214–226

Brunet Y, Nunez M, Lagouarde JP (1991) A simple method for estimating regional evapotranspiration from infrared surface temperature data. ISPRS J Photogramm Remote Sensing 46:311–327

Brutsaert WH (1982) Evaporation into the atmosphere: theory, history and applications. Reidel, Dordrecht, 299 pp

Bullrich K (1964) Scattered radiation in the atmosphere and the natural aerosol. Adv Geophys 10:99–260

Bureau of Meteorology (1990) The Bureau of Meteorology's ozone program. Bur Meteorol Rpt, Melbourne, 23 pp

Calvet JJ, Viswanadham Y (1993) Amazon forest radiation budget from satellite data. J Appl Meteorol 32:844–847

Cano D, Monget JM, Albuisson M, Guillard H, Regas N, Wald L (1986) A method for the determination of the global solar radiation from meteorological satellite data. Solar Energy 37:31–39

Carleton AM (1991) Satellite remote sensing in climatology. Belhaven, London, 291 pp

Carlson TN, Dodd JK, Benjamin SG, Cooper JN (1981) Satellite estimation of the surface energy balance, moisture availability and thermal inertia. J Clim Appl Meteorol 20:67–87

Carlson TN, Rose FG, Perry EM (1984) Regional-scale estimates of surface moisture availability from GOES infrared satellite measurements. Agronomy J 76:972–979

Cess RD, Dutton EG, Deluisi JJ, Jiang F (1991) Determining surface solar absorption from broadband satellite measurements for clear skies: comparison with surface measurements. J Climate 4:236–247

Cess RD, Nemesure S, Dutton EG, Deluisi JJ, Potter GL, Morcrette JJ (1993) The impact of clouds on the shortwave radiation budget of the surface atmosphere system: interfacing measurements and models. J Climate 6:1764–1772

Chen TS, Ohring G (1984) On the relationship between clear sky planetary and surface albedos. J Atmos Sci 41:156–158

Chen E, Allen LH, Bartholic JF, Bill RG, Sutherland RA (1979) Satellite-sensed winter nocturnal temperature patterns of the Everglades agricultural area. J Appl Meteorol 18:992–1002

Chertock B, Frouin R, Gautier C (1992) A technique for global monitoring of net solar irradiance at the ocean surface. Part II. Validation. J Appl Meteorol 31:1067–1083

Coulson KL (1959) Characteristics of the radiation emerging from the top of a Rayleigh atmosphere, 1 and 2. Planet Space Sci 1:256–284

Darnell WL, Gupta SK, Staylor WF (1983) Downward longwave radiation at the surface from satellite measurements. J Clim Appl Meteorol 22:1956–1960

Darnell WL, Staylor WF, Gupta SK, Denn FM (1988) Estimation of surface insolation using sun-synchronous satellite data. J Climate 1:820–835

Dave JV, Halpern P (1976) Effects of changes in ozone amount on the ultraviolet radiation received at sea level of a model atmosphere. Atmos Environ 10:547–555

Davies JA, McKay DC (1982) Estimating solar irradiance and components. Solar Energy 29:55–64

Davies JA, Schertzer WS, Nunez M (1975) Estimating global solar radiation. Bound Layer Meteorol 9:33–52

Dedieu G, Deschamps PY, Kerr YH (1987) Satellite estimation of solar irradiance at the surface of the earth and of surface albedo using a physical model applied to Meteosat data. J Clim Appl Meteorol 26:79–87

Deschamps PY, Herman M, Tanre D (1981) Influence de l'atmosphere en teledetection des ressources terrestres. Modelisation et possibilites de corrections. Spectral Signatures of Objects in Remote Sensing INRA, Versailles, pp 543–558
Diak GR, Gautier C (1983) Improvements to a simple physical model for estimating insolation from GOES data. J Clim Appl Meteorol 22:505–508
Diekmann FJ, Hopp S, Rieland M, Benesch W, Czeplak G, Kasten F (1988) An operational estimate of global solar irradiance at ground level from Meteosat data: results from 1985 to 1987. Meteorol Rundsch 41:65–79
Elwood JM, Lee JAH, Walter SD, Mo T, Green AES (1974) Relationship of melanoma and other skin cancer mortality to latitude and ultraviolet radiation in the United States and Canada. Int J Epidemiol 3:325–332
Elwood MJ, Swerdlow AJ, Cox B (1989) Trends in incidence and mortality from cutaneous melanoma in England and Wales. In: Refshauge W (ed) The ozone layer and health. Trans Menzies Found 15:131–136
Farmer CB , Toon GC, Schaper PW, Blavier JF, Lowes LL (1987) Stratospheric trace gases in the spring 1986 Antarctic atmosphere. Nature 329:126–130
Fisher MS, Kripke ML (1982) Suppressor T lymphocytes control the development of primary skin cancers in ultraviolet irradiated mice. Science 216:1133–1134
Frederick JE, Lubin D (1988) The budget of biologically active ultraviolet radiation in the earth-atmosphere system. J Geophys Res 93 (D4):3825–3832
Frouin R, Chertock B (1992) A technique for global monitoring of net solar irradiance at the ocean surface. Part I: Model. J Appl Meteorol 31:1056–1066
Frouin R, Gautier C, Katsaros KB, Lind RJ (1988) A comparison of satellite and empirical formula techniques for estimating insolation over the oceans. J Appl Meteorol 27:1016–1023
Gates DM, Keegan HJ, Schleter JC, Weidner VR (1965) Spectral properties of plants. Appl Optics 4:11–20
Gautier C, Diak G, Masse S (1980) A simple physical model to estimate incident solar radiation at the surface from GOES satellite data. J Appl Meteorol 19:1005–1012
Gautier C, Frouin R (1985) Satellite-derived ocean surface radiation fluxes. In: Deepak A, Fleming HE, Chahine MT (eds) Advances in remote sensing retrieval methods. Deepak, Hampton, Virginia, pp 311–329
Green, AES, Sawada T, Shettle EP (1974) The middle ultraviolet reaching the ground. Photochem Photobiol 19:251–259
Green AES, Cross KR, Smith LA (1980) Improved analytic characterization of ultraviolet skylight. Photochem Photobiol 31:59–65
Gupta SK (1989) A parameterization for longwave surface radiation from sun-synchronous satellite data. J Climate 2:305–320
Gupta SK, Darnell WL, Wilber AC (1992) A parameterization for longwave surface radiation from satellite data: recent improvements. J Appl Meteorol 31:1361–1367
Gutman G, Ohring G, Tarpley D, Ambroziak R (1989) Albedo of the US Great Plains as determined from NOAA-9 AVHRR data. J Climate 2:609–617
Hay JE (1981) The mesoscale distribution of solar radiation at the earth's surface and the ability of satellites to resolve it. In: Bahm RJ (ed) Satellites and forecasting of solar radiation. Proc 1st Worksh on Terrestrial solar resource forecasting and on use of satellites for terrestrial solar resource assessment, 2–5 Feb 1981, Int Sol Energy Soc, Newark, Delaware, 144 pp
Hay JE (1984) An assessment of the mesoscale variability of solar radiation at the earth's surface. Solar Energy 32:425–434
Hay JE, Suckling PW (1979) An assessment of the networks for measuring and modelling solar radiation in British Columbia and adjacent areas of western Canada. Can Geogr 33:222–238
Hinds MW, Kolonel LN (1980) Malignant melanoma of the skin: Hawaii, 1960–1977. Cancer 45:811–817
Hiser HW, Senn HV (1980) Mesoscale mapping of available solar energy at the earth's surface by use of satellites. Solar Energy 24:129–141

Holman CDJ, James IR, Gattey PH, Armstrong BK (1980) An analysis of trends in mortality from malignant melanoma of the skin in Australia. Int J Cancer 26:303–309

Hounam CE (1963) Estimates of solar radiation over Australia. Aust Meteorol Mag 43:1–14

Jackson RD, Reginato RJ, Idso SB (1977) Wheat canopy temperature: a practical tool for evaluating water requirements. Water Resour Res 13:651–656

Jupp DLB, McVicar TR, Walker J, Held A, Kalma JD, McDonald E (1992) Remote sensing of change in components of the water balance of the Murray Darling Basin using satellite imaged and spatially registered environmental data. CSIRO Div Water Resour, Consultancy Rep No 92/31

Kalma JD, Calder IR (1994) Land surface processes in large-scale hydrology. Operational Hydrology Report No 40, World Meteorol Org, Geneva, WMO-No 803, 60 pp

Kalma JD, Sivapalan M, Alksnis H (1991) Evaporation from pasture. Proc Int Hydrology and water resources Symp, Perth, October 1991, pp 681–686

Kalma JD, Laughlin GP, Green AA, O'Brien MT (1986) Minimum temperature surveys based on near-surface air temperature measurements and airborne scanner data. J Climatol 6:413–430

Kalma JD, Byrne GF, Johnson ME, Laughlin GP (1983) Frost mapping in southern Victoria: an assessment of HCMM thermal imagery. J Climatol 3:1–19

Kerschgens M, Pilz U, Raschke E (1978) A modified two-stream approximation for computations of the solar radiation budget in a cloudy atmosphere. Tellus 30:429–435

Kirkpatrick J, Dickinson KJM (1984) Vegetation map of Tasmania. 1:500 000. Forestry Commission Tasmania, Hobart

Kraemer KH, Lee MM, Scotts J (1984) DNA repair protects against cutaneous and internal neoplasia: evidence from xeroderma pigmentosum. Carcinogenesis 5:511–514

Lacis AA, Hansen JE (1974) A parameterisation for absorption of solar radiation in the earth's atmosphere. J Atmos Sci 31:118–133

Lagouarde JP (1991) Use of NOAA AVHRR data combined with an agrometeorological model for evaporation mapping. Int J Remote Sensing 12:1853–1864

Lagouarde JP, Brunet Y (1989) Spatial integration of surface latent heat flux and evaporation mapping. Adv Space Res 9:259–264

Lagouarde JP, Brunet Y, Kerr Y, Imbernon J (1991) Estimating the daily upward longwave radiation from NOAA-AVHRR data for mapping net radiation. Adv Space Res 11:151–161

Lazlo I, Jacobowitz H, Gruber A (1988) The relative merits of narrow-band channels for estimating broad-band albedos. J Atmos Ocean Technol 5:757–773

Li Z, Leighton HG, Masuda K, Takashima T (1993a) Estimation of shortwave flux absorbed at the surface from TOA reflected flux. J Climate 6:317–330

Li Z, Leighton HG, Cess RD (1993b) Surface net solar radiation estimated from satellite measurements: comparison with tower observations. J Climate 6:1764–1772

Linacre ET (1992) Climate data and resources. Routledge, London, 366 pp

List R (1968) Smithsonian meteorological tables. Smithsonian Institution, Washington

Madrid CR (ed) (1978) The NIMBUS-7 user's guide. Prepared by the Landsat/Nimbus Project, Goddard Space Flight Center, NASA, Beltsville Maryland, 263 pp

McVicar TR, Jupp DLB, Yang X, Tian G (1992) Linking regional water balance models with remote sensing. Proc 13th Asian Conf on Remote sensing, Ulaanbaatar, Mongolia, 7–11 Oct 1992, B.6:1–6

Mekler Y, Joseph JH (1983) Direct determination of surface albedos from satellite imagery. J Clim Appl Meteorol 22:530–536

Menenti M (1984) Physical aspects and determination of evaporation in deserts applying remote sensing techniques. Rep No 10, ICW, Wageningen

Michael KJ, Nunez M (1991) Derivation of ocean-atmosphere heat fluxes in a tropical environment using satellite and surface data. J Climatol 11:559–575

Minnis P, Harrison EF (1984) Diurnal variability of regional cloud and clear-sky radiative parameters derived from GOES data. Parts I, II and III. J Clim Appl Meteorol 23:993–1051

Monteith JL, Unsworth MH (1990) Principles of environmental physics. Edward Arnold, London, 291 pp

Monteith JL, Szeicz G (1961) The radiation balance of bare soil and vegetation. Q J Roy Meteorol Soc 87:159–170

Moran MS, Jackson RD, Hart GF, Slater PN, Bartell RJ, Biggar SF, Gellman DI, Santer RP (1990) Obtaining surface reflectance factors from atmospheric and view angle corrected SPOT-1 HRV data. Remote Sensing Environ 32:203–214

Morcrette J, Deschamps P (1986) Downward longwave radiation at the surface in clear-sky atmospheres: comparisons of measured, satellite-derived and calculated fluxes. ISCLP: Proc Int Conf Rome, Italy, 1985; ESA-SP 248, ESA, Paris, pp 257–262

Moser W, Raschke E (1983) Determination of global radiation and of cloudiness from Meteosat imaging data. Ann Meteorol (New Ser) 18:161–163

Moser W, Raschke E (1984) Incident solar radiation over Europe estimated from Meteosat data. J Clim Appl Meteorol 23:166–170

Nagaraja Rao CR, Takashima T, Bradley WA, Lee TY (1984) Near ultraviolet radiation at the earth's surface: measurements and model comparisons. Tellus 36B:286–293

Nieuwenhuis GJA (1986) Integration of remote sensing with a soil water simulation model (SWATRE). In: Hydrological applications of space technology. Proc Cocoa Beach Worksh, Florida, Aug 1985. IAHS Publ 160:119–128

Nunez M (1978) The radiation index of dryness in Tasmania. Aust Geogr Stud 16:126–135

Nunez M (1987) A satellite-based solar energy monitoring system for Tasmania, Australia. Solar Energy 39:439–444

Nunez M (1990a) Solar energy statistics for Australian capital cities. Solar Energy 44:343–354

Nunez M (1990b) Satellite estimation of regional solar energy statistics for Australian capital cities. Meteorological Study No 39, B Meteorol, AGPS, Canberra, 314 pp

Nunez M (1990c) Mapping ultraviolet radiation in the Australian region. 59th ANZAAS Congr, 14–16 Feb 1990, Hobart, Univ Tasmania Printers, p 148

Nunez M (1993) The development of a satellite-based insolation model for the tropical western Pacific Ocean. Int J Climatol 13:607–627

Nunez M, Hart TL, Kalma JD (1984) Estimating solar radiation in a tropical environment using satellite data. J Climatol 4:573–585

Nunez M, Skirving WJ, Viney NR (1987) A technique for estimating regional surface albedos using geostationary satellite data. J Climatol 7:1–11

Nunez M, Forgan B, Roy C (1994) Estimating ultraviolet radiation at the earth's surface. Int J Biometeorol 38:5–17

Oke TR (1988) Boundary layer climates, Methuen, London, 435 pp

Otterman J (1984) Albedo of a forest modeled as a plane with dense protrusions. J Clim Appl Meteorol 23:297–307

Otterman J, Tucker CJ (1985) Satellite measurements of surface albedo and temperatures in semi-deserts. J Clim Appl Meteorol 24:228–235

Paltridge GW (1988) Spectral and total albedo to solar radiation of ice and water clouds-experimental results from ASPIRE. Atmosfera 1:5–16

Paltridge GW, Platt CMR (1976) Radiative processes in meteorology and climatology. Elsevier, Amsterdam, 318 pp

Paltridge GW, Proctor D (1976) Monthly mean solar radiation statistics for Australia. Solar Energy 18:235–243

Pierrehumbert CL (1972) Precipitable water statistics, Australia. Monthly statistics of precipitable water between surface and 400 mb at 2300 GMT, 1958–1969. Bur Meteorol Tech Rep, 66 pp

Pinker RT, Ewing JA (1985) Modeling surface solar radiation: model formulation and validation. J Clim Appl Meteorol 24:389–401

Pinker RT, Ewing JA, Tarpley JD (1985) The relationship between the planetary and surface net radiation. J Clim Appl Meteorol 24:1262–1268

Pinker RT, Lazlo I (1992) Modelling surface solar irradiance for satellite applications on a global scale. J Appl Meteorol 31:194–211

Pinker RT, Tarpley JD (1988) The relationship between the planetary and surface net radiation: an update. J Appl Meteorol 27:957–964
Pinty B, Ramond D (1987) A method for the estimation of broad-band surface albedo from a geostationary satellite. J Clim Appl Meteorol 26:1709–1722
Pinty B, Szejwach G (1985) A new technique for inferring surface albedo from satellite observations. J Clim Appl Meteorol 24:741–750
Pinty B, Szejwach G, Stum J (1985) Surface albedo over the Sahel from Meteosat radiances. J Clim Appl Meteorol 24:108–113
Price JC (1980) The potential of remotely sensed thermal infrared data to infer surface soil moisture and evaporation. Water Resour Res 16:787–795
Priestley CHB, Taylor RJ (1972) On the assessment of surface heat flux and evaporation using large scale parameters. Monthly Weather Rev 100:81–92
Rao PK, Holmes SJ, Anderson RK, Winston JS, Lehr PE (1990) Weather satellites: systems, data and environmental applications. Am Meteorol Soc Publ, 503 pp
Raschke E, Preuss HJ (1979) The determination of the solar radiation budget at the Earth's surface from satellite measurements. Meteorol Rundsch 32:18–28
Rawlins F (1989) Aircraft measurements of the solar absorption by broken cloud fields: a field study. Q J R Meteorol Soc 115:365–382
Rimoczi-Paal A (1983) A determination of global radiation from satellite pictures and meteorological data. Solar Energy 31:79–84
Roberts A (1991) Human exposure to ultraiolet radiation under all sky conditions in Hobart, Tasmania. Honours Thesis, Dept Geography and Environ Stud, University of Tasmania, 67 pp
Roy CR, Gies HP, Elliott G (1989) The ARL solar ultraviolet radiation measurement programme. In: Refshauge W (ed) The ozone layer and health. Trans Menzies Found 15:71–76
Saunders RW (1990) The determination of broad-band surface albedo from AVHRR visible and near-infrared radiances. Int J Remote Sensing 11:49–67
Schmetz P, Schmetz J, Raschke E (1986) Estimation of daytime downward longwave radiation at the surface from satellite and grid point data. Theor Appl Climatol 37:136–149
Scotto J, Cotton G, Urbach F, Berger D, Fears T (1988) Biologically effective ultraviolet radiation: surface measurements in the United States, 1974 to 1985. Science 239:762–764
Seguin B, Itier B (1983) Using mid-day surface temperatures to estimate daily evaporation from satellite thermal IR data. Int J Remote Sensing 4:371–383
Sellers PJ, Rasool SI, Bolle HJ (1990) A review of satellite data algorithms to studies of the land surface. Bull Am Meteorol Soc 71:1429–1447
Sellers WD (1965) Physical climatology. University of Chicago Press, Chicago, 272 pp
Shettle EP, Green AES (1974) Multiple scattering calculation of the middle ultraviolet reaching the ground. Appl Optics 13:1567–1581
Shuttleworth WJ (1991) Evaporation models in hydrology. In: Schmugge TJ, Andre JC (eds) Land surface evaporation: measurement and parameterization. Springer, Berlin Heidelberg New York, pp 93–120
Skirving WJ (1984) Estimates of surface albedo using geostationary meteorological satellite data. Honours Thesis, Dept Geography, University of Tasmania, 83 pp
Smith WL, Woolf HM (1984) Geostationary satellite sounder (VAS) observations of longwave radiation flux. WMO/WCRP WCP-70, WMO, Geneva, 36 pp
Stanhill G, Hofstede GJ, Kalma JD (1966) Radiation balance of natural and agricultural vegetation. Q J R Meteorol Soc 92:128–140
Stanhill G, Israeli M, Rosenzweig D (1973) The solar radiation balance of scrub forest and pasture on the Carmel Mountain, Israel: a comparative study. Ecology 54:819–828
Staylor F, Suttles JT (1986) Reflection and emission models for deserts derived from Nimbus-7 ERB scanner measurements. J Clim Appl Meteorol 25:196–202
Stolarski RS, Krueger AJ, Schoeberl MR, McPeters RD, Newman PA, Alpert JC (1986) Nimbus 7 satellite measurements of the springtime Antarctic ozone decrease. Nature 322 (28 Aug):808–811

Stuhlmann R, Rieland M, Raschke E (1990) An improvement in the IGMK model to derive total and diffuse solar radiation at the surface from satellite data. J Appl Meteorol 29:586–603

Stull R (1991) An introduction to boundary layer meteorology. Kluwer, Dordrecht, 666 pp

Stum J, Pinty B, Ramond D (1985) A parameterization of the broad-band conversion factors for Meteosat visible radiances. J Clim Appl Meteorol 24:1377–1382

Suckling PW, Hay JE (1976) Modelling direct, diffuse and total solar radiation for cloudless days. Atmosphere 14:298–308

Suckling PW, Hay JE (1977) A cloud layer-sunshine duration model for estimating direct, diffuse and total solar radiation. Atmosphere 15:194–207

Swinbank WC (1963) Long-wave radiation from clear skies. Q J R Meteorol Soc 89:339–348

Taconet O, Bernard R, Vidal-Madjar D (1986) Evapotranspiration over an agricultural region using a surface flux/temperature model based on NOAA-AVHRR data. J Clim Appl Meterol 25:284–307

Tanré D, Deroo C, Duhaut P, Herman M, Morcrette JJ, Perbos J, Deschamps PY (1985) Effects atmosphériques en télédétection-logiciel de simulation du signal satellitaire dans le spectre solaire. Proc 3rd Int Colloq on Spectral signatures of objects in remote sensing ESPA SP-247, pp 315–319

Tanré D, Herman M, Deschamps PY, De Leffe A (1979) Atmospheric modeling for space measurements of ground reflectances, including bidirectional properties. Appl Optics 18:3587–3594

Tarpley JD (1979) Estimating incident solar radiation at the surface from geostationary satellite data. J Appl Meteorol 18:1172–1181

Tarpley JD (1994) Monthly evapotranspiration from satellite and conventional meteorological observations. J Climate 7:704–713

Taylor VR, Stowe LL (1984) Atlas of reflectance patterns from uniform earth and cloud surfaces. NOAA Tech Rep NESDIS 10, Washington, DC, 66 pp

Tuzet A, Moser W, Raschke E (1984) Estimating global solar radiation at the surface from Meteosat data in the Sahel region. J Rech Atmos 18:31–39

Uchijima Z (1976) Maize and rice. In: Monteith JL (ed) Vegetation and the atmosphere, vol 2. Case studies. Academic Press, London, pp 33–64

Valiente J, Nunez M, Lopez E, Moreno J (1995) Narrowband to broadband conversion for Meteosat visible channel and broad-band albedo using both AVHRR-1 and -2 channels. Int J Remote Sensing 16:1147–1166

Vulis IL Cess RD (1989) Interpretation of surface and planetary directional albedos for vegetated regions. J Climate 2:986–996

Welch RM, Cox SK, Davis JM (1980) Solar radiation and clouds. Meteorol Monogr 39, Am Meteorol Soc, Boston, 96 pp

Wilson RG, Petzold DE (1972) Daily solar radiation differences between stations in southern Canada. McGill Univ, Climate Bull 11:15–22

Worrest RC (1982) Review of literature concerning the impact of UV-B radiation on marine organisms. In: Calkins J (ed) The role of solar ultraviolet radiatiation in marine organisms. Plenum Press, New York, p 429

Wydick JE, Davis P, Gruber A (1987) Estimation of broad-band planetary albedo from operational narrow-band satellite measurements. NOAA Tech Rep NESDIS 27, 32 pp

Mathematical Models in Crop Bioclimatology in the Former USSR (History, Achievements and Prospects)

O.D. Sirotenko[1]

1 Development of Methods of Mathematical Simulation of the Effects of Hydrometeorological Conditions on Crop Production

Agriculture and related scientific disciplines are undergoing a rapid evolution. Mathematics, rather than being just a useful tool for describing research results and correlating them with theoretical concepts, has found growing support as the only possible approach to some problems. In terms of science methodology the current situation may be defined as the commencement of the maturity stage with the "normal" ways of conducting scientific research being dominant. Three development stages may be discerned in the ways in which we describe and explain relationships between the physical environment and crop production. In the first stage descriptive approaches prevailed, with relationships between weather (climatic) conditions and crop production being explained on the basis of agronomy and plant physiology concepts and qualitative laws of physics. This descriptive approach was replaced with an empirical statistical one which emphasized mostly the search for direct relations between input and output of a black box – the Crop–weather system. A specific date can be pinpointed when the former USSR initiated development of the third, current, stage in resolving crop–environment problems. In 1964 three papers were published by two groups of authors: Budyko and Gandin (Budyko 1956; Budyko and Gandin 1964) on the one hand, and Budagovsky, Nichiporovich and Ross (Budagovsky et al. 1964) on the other. These papers laid the groundwork for a quantitative theory of crop production, resulting in the change from a static to a dynamic presentation of crop–environment relationships.

1.1 Static Models, Empirical-Statistical Approach

A static model is a mathematical structure which explicitly does not include time. In addition, the system's behavioral characteristics which are essentially time-dependent have to be discarded. Since everything in this world is changing, either rapidly or slowly, static models are always an approximation but are

[1] (Russian Research Institute of Agricultural Meteorology, Obninsk, Russia

sometimes very useful due to their simplicity. Static models for agriculture are simultaneously empirical-statistical models which use a minimum of a priori information about the physical mechanisms of processes. Let us begin analyzing them with the investigations of methodological importance.

1.1.1 Methodological Studies

The book by Obukhov (Obukhov 1949) published in 1949 summarized research results extending back to the beginning of this century. It may be considered a basic work in applying mathematical statistics methods to investigations of crop-production–weather relations, comparable in significance with the well-known work by R. Fisher (Fisher 1925).

Imagine the essence of the problem which Obukhov approached. To investigate the effects of weather on winter-rye yields in dry areas near the Volga, he first had to separate the contributions made by autumn–winter and spring–summer periods. To this end, a regression equation was written relating yields to visual estimates of the winter-rye crop state obtained at snow melt. The equation enables calculation of the time series of crop yields purified from autumn and winter effects. The spring and summer crop growing season is split into nine subperiods and calculation is made of coefficients of correlation between the purified crop yield series and meterological factors (Table 1).

How may the total effects of temperature and rainfall be taken into account for all nine stages of spring and summer growth? It is clear that in such a case the simple summation of meteorological parameters would not be applicable and a reliable regression equation cannot be written using 19 independent variables for 23 observations. To reduce dimensionality, Obukhov suggested an effective, although certainly not optimal, approach. Weights were attributed to each subperiod's temperature and rainfall, proportional to the square of partial coefficient of correlation between the yield estimate and

Table 1. Coefficients of correlation between purified crop yield series and meteorlogical factors

Growth stage	Rain (total mm)	Mean temperature (°C)
1. Commencement of growth	0.02	0.39
2. Vigorous growth	0.18	0.20
3. 〃	0.54	–0.28
4. Stem production	0.81	–0.53
5. 〃	0.55	–0.58
6. Heading	0.26	–0.10
7. Flowering	0.36	–0.41
8. Grain filling	0.19	–0.34
9. Maturity	0.04	–0.34

temperature at constant rainfall, and between the yield estimate and rainfall at constant temperature. As a result, 18 indicators were replaced by 3.

A special discussion of the multidimensionality problem is presented in Sirotenko (1969). The reasoning behind it is as follows. Crop response to hydrometeorological factors which are random time functions undergoes a regular variation during plant ontogenesis whose mean time may be assumed to equal 100 days. How many hydrometeorological factors can be taken into account? At least three: air temperature, a moisture indicator, and solar radiation. Thus, to solve the problem, if only in a linear approximation (without considering the interaction and cumulative impact of the environmental factors), regression equations involving 200 to 300 coefficients have to be identified. This is ultimately impracticable, since the corresponding field test would run for millennia. It will be shown below that this problem can be solved in a satisfactory enough way within the framework of the dynamic approach. (Then the number of coefficients to be estimated would not exceed five to ten.) Although the empirical-statistical approach is not as effective, it may also be used to relieve the burden of "dimensionality damnation". Let us consider the course of any meteorological parameter x throughout the growing season as a discrete function of time x_j assigned in $j = 1, 2, \ldots, n$ at equidistant points (where n is the number of days, 10-day periods, or other time intervals making up the growing season). At any degree of accuracy, this function may be presented by expansion into a series. To this end, orthogonal Chebyshev polynomials, or, which would be more effective, the eigenvectors of covariance matrix may be conveniently used. As a result, a transition may be effected from the set $x_1, x_2, \ldots, x_n$ to a new set of predictors $z_1, z_2, \ldots, z_p$, the latter being linear combinations of basic variables. In this case, as a rule, $p \ll n$. The variables $z_1, z_2, \ldots, z_p$ are orthogonal, as distinct from the basic predictors, thus enabling an easier elaboration of a regression equation. The greater the correlation of the basic set of predictors, the better results this approach will yield. Its effectiveness is quite adequate, provided that the basic variables represent the same physical quantity, with predicted data being uniformly "extended" over the entire aggregate of the basic variables, rather than being centered on a small subgroup (Sirotenko 1969, 1971).

The use of a priori information for elaborating regression equations essentially boils down to listing the set of independent variables. Therefore the natural potential for improving the accuracy of such relationships lies in considering various additional data.

Zabelin discussed an interesting illustration of this idea (Ulanova and Zabelin 1990). To forecast cereal crop yields y in Kazakhstan, predictors such as rainfall totals and mean temperatures for the third 10-day period in May and three 10-day periods in June (R_1, R_2, R_3, R_4 and t_1, t_2, t_3, t_4 respectively) were selected. By the least squares method a regression equation was obtained:

$$y = -0.105R_1 - 0.91R_2 + 1.294R_3 + 0.500R_4 + 0.559t_1 - 0.904t_2 + 1.38t_3 - 0.508t_4. \tag{1}$$

In Eq. (1) all predictors are presented in a standardized scale. Under conditions of deficient moisture, the rainfall factors should be nonnegative, and the temperature factors nonpositive. In Eq. (1) these conditions are often violated due to close linear relationships among predictors. To obtain acceptable results, the ridge regression method was used. At the minimum value of a ridge parameter k = 0.3, the following relationship was obtained:

$$y = 0.04R_1 + 0.070R_2 + 0.038R_3 + 0.288R_4 - 0.036t_1 - 0.044t_2 - 0.249t_3 - 0.015t_4. \quad (2)$$

The difference between Eqs. (1) and (2) needs no comment. When both relationships were checked against independent data for 4 y, prediction errors with Eq. (2) were only half those with Eq. (1).

The choice of a nonlinear model structure would be the next step on the way to a fuller use of a priori data. Satyr (1973) described the relationship between air temperature, soil moisture, and biomass production as an asymmetric "hill" whose top crowns the point representing an optimum combination of the above factors. The "hill" was described with the help of a bivariate log-normal distribution formula whose 5 parameters were made dependent on plant development rates.

Dmitrenko (1976) proposed the best-known statistical model. It was designed for calculating and forecasting the yields of winter wheat and rye, spring barley, maize, potato, sugar beet and sunflower. A formula was used to express the relationship between winter cereal crop yields and influencing factors as

$$y = Y_j(1-P)f(k)\eta_8(W)\alpha_8 + \sum_{i=1}^{n} \eta_i(T_i)\eta_i^*(R_i)\alpha_i, \quad (3)$$

where Y is the annual statistical maximum yield, P is plant density, f(k) is the tillering index, $\eta_i(T)$, $\eta_i^*(R)$ and $\eta_8(W)$ are crop production functions for air temperature T, rainfall R, and water storage W at the beginning of plant growth, and i is the number of the growing-season subperiod. These functions are bell-shaped and can vary from 0 to 1.

1.1.2 Applied Results

The methods designed by Ulanova to forecast winter wheat yields (Ulanova 1963, 1975, 1977) seem to be the most popular. They are based on two initial factors: soil water stored in a 1-m-thick topsoil layer at the beginning of wheat plant growth, and the number of tillers per 1 m^2 in spring, characterizing the crop state upon overwintering.

Similar relationships using additional data on air temperature and rainfall for selected time periods prior to the date of forecast announcement were found for the Ukraine, Central Black Soil Region and northern Caucasus

(Derkach 1978), Georgia and Azerbaijan (Tsertsvadze and Meladze 1979), Armenia (Mkrtchyan and Arustamyan 1975), Tajikistan (Zainutdinov 1978), Kazakhstan (Maslovskaya 1974), and southern Russia (Svisyuk 1971).

For spring cereal crops, many yield-forecasting methods were also designed using, as a rule, the same predictors as for winter cereals. However, the validity of these methods is usually less than for winter cereal crops because the inertia factors (soil-water and crop-state characteristics) at the same forecast time exert less impact on the ultimate yields of spring crops. A monograph has been written devoted to the study of soil-water storage as a crop yield predictor (Mescherskaya et al. 1982).

Methods for forecasting spring cereal crop yields were designed for the Volga and Central Black Soil Regions (Kirilicheva 1980), Lower Volga and northern Caucasus (Svisyuk 1971, 1991), western Siberia (steppe and wooded steppe) (Ponomarev 1974), Trans-Urals (Kozhushnyan and Uzyanova 1966), western Kazakhstan (Fedyushina 1980), Turgai province of Kazakhstan (Fedyushina and Zavyalova 1980), Kulunda Zone of western Siberia (Shchenina 1978), Kemerov region (Tyukalo 1980), Krasnoyarsk area (Andropova and Dobretsov 1977), and Russia's Far East (Kresnyanskaya 1977).

Note that all the above-discussed crop-yield forecasting methods were designed specifically for steppe and wooded steppe zones where crop production often suffers from moisture deficit. In the forest, nonchernozem, soil zone, the soils usually are wetted each year to field capacity, thus soil moisture storage in spring can no longer be assumed to be an effective predictor there. This accounts for the fact that the few methods of crop yield forecasting elaborated for the forest zone had, as a rule, lower validity (Polevoy and Myzina 1976).

1.2 Dynamic Models, Theory

As was observed above, multidimensionality is the main problem in designing statistical models of the crop–weather system. It can be basically handled by a change-over to the dynamic (plant growth-oriented) presentation of the relationships. Let M(t) be the crop biomass considered as the function of time t. To estimate it, a difference equation may be written:

$$M^{j+1} = M^{j} + F\left(M^{j}, X^{j}, A^{j}\right)\Delta t, \tag{4}$$

where $j = 0, 1, \ldots, N$; X is the vector characterizing the current state of the crop environment, A^j is the vector of model parameters which may vary with time; Δt is the time step of e.g., 1 day; F is the predetermined functional. The number of model parameters k (vector A's dimensionality) does not depend on either N or Δt. This makes up the basic difference between the dynamic model Eq. (4) and empirical statistical relationships. The dimensionality A is determined solely by the degree of detail in the mathematical description of the crop production processes. Practice has shown that an agroecosystem model may be comprehensive enough if the dimensionality value A is not greater than 10. Thus a changeover to the dynamic description would result in reducing by at least 100 times the number of parameters to be determined. In practice the gain would be even better, since Eq. (4) is nonlinear.

Finally, Eq. (4) enables, through M^j, the preceding plant ontogenesis to be taken into account when describing crop response to the current hydrometeorological conditions.

It was emphasized above that in the USSR the works by Nichiporovich, Budagovsky, Ross, Budyko and Gandin published in 1964 laid the foundations for the theory of agroecosystem photosynthetic productivity. Summing up, their research program was essentially aimed at sequentially fulfilling three tasks:

1. Establishing the photosynthesis–respiration relationship between the CO_2 exchange of an individual leaf and the governing factors of environment and plant parameters.
2. Establishing the relationship between the whole canopy gas exchange and the governing factors, i.e., reduce to integration over the space coordinate of equations describing gas exchange within selected canopy layers.
3. Establishing the relationship between the canopy gas exchange rate and current biomass production formulated as a set of differential equations; integrate this set of equations over time, to obtain the ultimate state of crop biomass.

To implement this program, Ross and Bichele (Kan et al. 1989) designed a static model of canopy photosynthesis distinguished by a fuller description of the crop radiation regime, and Budyko and Gandin (Budyko 1964; Budyko and Gandin 1964) designed a model featuring a detailed description of the turbulent transfer of CO_2 and other substances within the canopy. Around the same time, Budagovsky published a monograph (Budagovsky 1964) which was a significant step in the development of the theory of evapotranspiration as a process which is closely associated with agroecosystem productivity. In the USSR, Budagovsky's contribution has been valued as highly as that of Penman. Somewhat earlier a monograph was published by Budyko (1956) summing up years of theoretical work and experiments on developing methods for calculating the heat balance of underlying surfaces, including crop fields.

Planetary-boundary-layer theory developed by Obukhov and Monin (cf. Zilitinkevich 1970) should also be referred to as fundamental. The same applies to work made by Chudnovsky (1948, 1976) who developed an approach involving a composite mathematical description of the heat regime at the soil–air interface.

These efforts led, on the one hand, to the establishment of renowned scientific schools of ground-air and soil physics comprising plant habitat. On the other hand, these investigations resulted in the development of a crop production theory and contributed to the elaboration of agroecosystem dynamic models. In this connection, mention should be made of an exceptionally stimulating role of the efforts undertaken by de Wit et al. (1971) which led to the ELCROS model.

Now let us briefly discuss the logical structure of the crop production theory as it is currently understood (Sirotenko 1981, 1984). A crop is regarded as an open developing system – green machine – which takes up energy and substrates from the environment and produces organic materials. According to some estimates, using 2000 kg of carbon dioxide from the atmosphere and 250–400 kg of minerals from soil, a plant can produce 1000 kg of organic matter, taking up 500–1000 tonnes of water in the process.

A mathematical structure representing the above description is made up of two related sets of equations:

1. An energy and mass exchange set describing the transfer of solar radiation, heat, moisture, CO_2, and other materials at the atmosphere–soil interface, where plant matter parameters act as distributed sources (sinks) of the materials transported, and water flows serve to additionally bind the atmospheric and soil media "from the inside".
2. A biodynamic set describing plant growth and development, and the dynamics of plant biomass, leaf area, and crop geometry, as well as plant parameter variations during ontogenesis.

The first of these sets determines the rates of the green-machine operation depending on the current ambient conditions, the second prescribes the changes the machine will undergo in time.

Therefore, the task of calculating ultimate crop yields under known ambient conditions may be looked upon as a comprehensible, although complicated, problem of mathematical physics. It is with regard to this problem that we will analyze relevant studies of far-ranging topics.

2 Simulating Energy and Mass Exchange for the Soil–Plant–Atmosphere System

2.1 Defining the Main Problem

Let us formulate a set of energy and mass exchange equations for the soil–plant–atmosphere (SPA) system. Assuming, as usual, the flows of all substances in the ground air to be quasi-steady state, the following equations may be written for a homogeneous canopy, to describe transfer processes for heat, water vapour, and carbon dioxide (Sirotenko 1981).

In these equations, a, p, s are the subscripts defining values for atmosphere, plants, and soil respectively; $Z_1 = Z + \Pi_1$; T_a, T_p, T_s are temperature values; q_a, q_s are specific humidity values for atmosphere and soil air; q_p is the saturating humidity at temperature T_p and the plant tissue's water potential ψ_p;

k is the turbulent exchange coefficient, S is the leaf-area specific density ($cm^2 cm^{-3}$); C is the CO_2 content of air ($g cm^{-3}$); D_T is the heat and water exchange coefficient ($cm s^{-1}$); Ω is the CO_2 exchange rate per unit leaf area ($g s^{-1} cm^{-2}$); I, $I^{\uparrow}$, $I_{\downarrow}$ are the total, upward, and incident fluxes of shortwave radiation; F, $F^{\uparrow}$, $F_{\downarrow}$ are the respective fluxes of longwave radiation; ρ_a, C_p are the air density and heat capacity; χ is the latent heat of vaporization; σ is the Stefan–Boltzmann constant; ω, γ_F are the coefficients; θ_S is the bulk moisture of soil ($cm^3 m^{-3}$); K_s is the hydraulic conductivity of soil ($cm s^{-1}$); ω_S is the soil-water potential (cm); ε_1 is the evaporation (condensation) rate per unit soil volume (cm^3 water cm^{-3} soil $\cdot s^{-1}$); ε_2 is the root water-uptake rate per unit soil volume (cm^3 water cm^{-3} soil $\cdot s^{-1}$); V_θ is soil porosity; ζ_s is the convolution index; D_v is the coefficient of water vapour diffusion in air; ρ_w is water density; C_s, λ_s are the volumetric water capacity and heat conductivity of soil, respectively; θ_p is the volumetric moisture content of plant tissue (cm^3 water cm^{-3} plant biomass); α_p is the plant biomass specific density (cm^3 plant biomass cm^{-3} volume); β_p is the ratio cross section of root vessels or stems to total area of their cross sections at a given z ($cm^2 cm^{-2}$); K_p is the hydraulic conductivity index of xylem ($cm s^{-1}$); ψ_p is the plant tissue water potential (cm).

$$\frac{d}{dz}\left[K\frac{dT_a}{dz}\right]+SD_T\left[T_p-T_a\right]=0 \tag{5}$$

$$\frac{d}{dz}\left[K\frac{dq_a}{dz}\right]+SD_q\left[q_p-q_a\right]=0 \tag{6}$$

$$\frac{d}{dz}\left[K\frac{dc}{dz}\right]+S\Omega=0. \tag{7}$$

An equation of the heat balance of plant biomass may be written in a steady-state approximation, for crops with thin leaves and not very bulky stems, as

$$\frac{d}{dz}\left[I+F\right]-\rho_a C_p D_T S\left[T_p-T_a\right]-\chi\rho_a D_q S\left[q_p-q_a\right]=0. \tag{8}$$

Equation (8) contains two unknown functions I and F – the fluxes of shortwave and longwave radiation. To calculate them, let us use the simplest equations of radiation exchange in a two-flow approximation (Dubov et al. 1987):

$$\frac{dI^{\uparrow}}{dz}=\frac{\omega}{2}SI_{\downarrow}+\left[1-\frac{\omega}{2}\right]SI_{\uparrow},\quad \frac{dI^{\downarrow}}{dz}=\left[1-\frac{\omega}{2}\right]SI_{\downarrow}+\frac{\omega}{2}SI^{\uparrow}, \tag{9}$$

$$\frac{dF^{\uparrow}}{dz}=-S\gamma_F F^{\uparrow}+\sigma T_p{}^4 S,\quad \frac{dF^{\downarrow}}{dz}=S\gamma_F F_{\downarrow}-\sigma T_p{}^4 S. \tag{10}$$

As the next step, let us write equations for the second medium of plant environment – soil. The most widely used form of mathematical description of a one-dimensional transport of water in the zone of aeration affected by capillary and gravitational forces is represented by the following equation:

$$\frac{\delta\theta_S}{\delta t} = \frac{\delta}{\delta z} K_S \frac{\delta(\psi_S - z_1)}{\delta z} - \varepsilon_1 - \varepsilon_2. \tag{11}$$

Obviously, Eq. (11) can describe only the isothermal processes of soil-water transport. An equation of isothermal water-vapour transfer through the porous soil medium, with a source term describing changes of phase for water, may be written as

$$\frac{\delta}{\delta t}\left[\rho_a (V_\theta - \theta_s) q_s\right] = \frac{\delta}{\delta z} D_v \xi_s \rho_a \left[V_\theta - \theta_s\right] \frac{\delta q_s}{\delta_z} + \varepsilon_1 \rho_w. \tag{12}$$

Let us write an equation of heat transfer in soil in such a way as to calculate the energy aspect of water phase changes as

$$\frac{\delta}{\delta t}\left[C_s T_s\right] = \frac{\delta}{\delta z} \lambda_s \frac{\delta T_s}{\delta z} - \chi \rho_w \varepsilon_1, \tag{13}$$

where C_s and λ_s are assumed to be the functions not only of the vertical coordinate, but of soil moisture as well.

To close the above set of equations, another relationship is needed which could describe water transport through the plant vessels of the root–leaf pothway, such as (Sirotenko 1981):

$$\frac{\delta}{\delta t}\left[\alpha_p \theta_p\right] = \frac{\delta}{\delta z}\left[\beta_p K_p \frac{\delta(\psi_p - Z_1)}{\delta z}\right] + \varepsilon. \tag{14}$$

To integrate the set of Eqs. (5) to (14), boundary conditions must be specified. Assume these to be relationships between the flows of heat, water, carbon dioxide, and radiation (I and F), and their values measured (predetermined) for undisturbed atmosphere above the crop, the values of temperature and moisture at the lower boundary of the simulated soil layer, continuity conditions of air temperature and humidity, heat and water-vapour balances at the interface (soil surface), as well as some other quite obvious relations.

This set of equations, then, represents a "normal" definition of an integrated energy and mass-exchange problem for the SPA system. Similar integrated problems, including different modifications for the two joining media with plants inserted, were studied by numerical methods by Sirotenko, Menzhulin, Poluektov with co workers and students (cf. monographs of Sirotenko 1981; Polucktov 1991 and papers of Menzhulin 1977; Savvateyev

1979; Boiko and Sirotenko 1985). The SPA integrated heat exchange is dealt with in monographs (Kurtener and Chudnovsky 1969; Chudnovsky 1976).

Menzhulin (1977) and Savvateev (1979) elaborated the most detailed, in terms of biology, model of SPA water exchange. However, the simulations made by running the model are of limited applicability since it lacks even rough, as they may be, experimental data on a large number of constants used in the model formulas. This model, in our opinion, indicates where a limit is to be set on a detailed description of the physiological aspects of water exchange. It may not be advisable to go beyond this limit, given today's level of experimental data provision and, chiefly, today's insight into the processes simulated.

2.2 Energy and Mass Exchange in the Canopy

Budyko, Gandin, Budagovsky, Poluektov, and others (Budyko and Gandin 1968; Menzhulin 1970, 1973, 1974; Bondarenko et al. 1982) analyzed problems of energy and mass exchange for the aboveground part of the crop [i.e., the main equations are Eqs. (5) to (8)]. Their efforts were aimed mainly at simplifying the problem and obtaining analytical solutions. The results will be discussed in the relevant section.

The canopy radiation regime is described by Eqs. (9) and (10). In the former USSR, Ross, Nielson, and Tooming conducted the most significant investigations on this issue. The results were summed up by Ross (1975) which presented a comprehensive description of the problems encountered in studying radiation absorbtion, scattering, transmission, and emission by crops. Tooming, in his monograph (Tooming 1977), also discussed applied aspects of the theory. These studies could provide a satisfactory response to the crop-production simulation requirements, forming a basis for future simulation developments, primarily for the calculation of shortwave radiation. Calculation of the longwave radiation regimes of crops, however, involves several unresolved problems (for a description, see Poluektov 1991).

No definition of the crop aerodynamic regime was given in the problem statement [Eqs. (5) to (14)] although Eqs. (5) to (7) include the function k(z) – the turbulent exchange coefficient – and D_T and D_q, when expressed through resistance, depend on wind velocity. Thus,

$$D_T = \frac{1}{r_a}, \quad D_q = \frac{1}{r_a + r_s}. \tag{15}$$

where r_s is the stomatal resistance and r_a is the resistance of the leaf boundary layer which is inversely proportional to the local wind velocity.

Therefore, the functions k(z) and u(z) should be specified in order to solve Eqs. (5) to (14). That might be done in several ways. The task of calculating the crop water and heat regimes was dealt with using a linear approximation of these functions (Asaturov 1969; Budyko and Gandin 1968), and k(z) was dealt with by an exponential approximation of k(z) (Budagovsky 1964; Sirotenko and Boiko 1976). Calculation of k(z) and u(z) with the help of a physically based model would seem to be more attractive. The one used most often is a model of the crop aerodynamic regime for the case of neutral stratification designed by Menzhulin (1970, 1973), despite all the criticism (Budagovsky 1980; Dubov et al. 1987) of this semiempirical model. Nevertheless, since most crop canopies (as is true at least for cereals) are usually adequately ventilated, the accuracy with which k(z) and u(z) are assigned exerts comparatively little effect on calculations of water and heat regimes. Therefore, in most cases, for practical purposes, any reasonable representation of these values suffices. Budagovsky (1980) and Dubov et al. (1987) give the most complete and qualified overviews of the studies of turbulent exchange in canopies.

As is shown above, the formula for calculating D_q includes stomatal resistance r_s. As a rule, in the available models, r_s is given as a purely empirical function of environmental factors, such as radiation I, temperature T, water potential ψ_e, etc. A satisfactory solution can hardly be obtained by this approach. Moldau mode a most significant contribution to resolving this important and complicated problem. On the basis of this own experimental results he proposed a stomatal regulation model (Moldau 1973) and substantiated a concept of stomata as universal controls of mass exchange enabling an agreement between the plant requirements and current ambient conditions (Moldau 1977).

2.3 Heat and Water Transport in Soil

As Chudnovsky (1976) pointed out, a complete definition of a model for the soil-temperature field should consist of setting up and resolving a set of four initial equations for heat transport by way of conduction, convection, radiation, and moisture. Nevertheless there is an effective means of bypassing these predicaments and fulfilling numerous practical tasks, viz. an equivalent heat conductivity model which is confined to just one equation [Eq. (13)] with a complicated (due to the above-mentioned factors) dependence of the heat-transport coefficients on soil-water and other soil characteristics. Chudnovsky and his disciples (Kurtener and Chudnovsky 1969; Chudnovsky 1976) determined the relationships between C_s, λ_s, and the governing factors required for solving Eq. (13) for many soils in the USSR.

The water regime is of basic importance for plant life. No crop production model can be constructed which does not feature water, on the contrary – there are examples of models with the only dynamic unit being represented by soil-water transport equations. When describing the water regime, two basically different processes should be distinguished, viz. soil wetting and drying. During the growing season, when viewed as a whole, wet periods (following watering and heavy rains) account for a moderate fraction of the total time. The drying regime is of much greater importance for plant life. In principle, Eqs.

(11) and (12) can describe both regimes. However, taking into account the difference in the importance of these regimes, as well as the fact that their characteristic times essentially differ for crop-production simulation purposes, the main emphasis should be placed on the description of the slower process of drying. That would enable an essentially longer time step for the numerical solution of the problem. The wetting regime, in its turn, can be parameterized using the following straightforward scheme: the rainfall total (or irrigation norm) is added to the water storage value for the topsoil layer; if the resulting total is greater than field capacity, then the excess water enters the second layer, and the procedure is repeated for the latter as well as all other layers.

To solve the water-transfer equation, two additional functions should be available – the relationships between the hydraulic conductivity k_s and the soil-water potential ψ_s, and the extent of soil saturation. Denisov (1968, 1978) made a theoretical substantiation of these relationships. A large number of studies have been conducted to approximate these relationships from experimental data, and to establish correlations between the formula coefficients and easily measured soil parameters such as density, texture, humus content, etc. Globus (1987) analyzed and summarized 24 models for the volumetric-moisture–hydraulic-potential relationship and 28 models for soil-water conductivity. Zoslavsky et al. (1988) presented a description of a software package which enables (using a dialog mode) on the basis of the available data, selection of the most convenient model and estimation of its parameters. Pavlova et al. (1986) describe interesting proposals for the generalization of the dependence of soil-water potential and soil-water conductivity on moisture content. The presence of roots in soil is treated as the vertical coordinate-distributed function of sink or source – ε_2 in Eq. (11). The available functional expressions for ε_2 may be grouped into two categories: microscopic and macroscopic. Models falling into the first category treat the root system as a set of cylinders of a permanent radius and usually use an analytical solution of the equation for radial water flow. Tupchiev et al. (1973) discuss the applicability of asymptotic methods to solving equations of this type. An advantage of the microscopic models lies in the convenient formalization of processes of soil-water uptake. However, assignment of boundary conditions at the root surface, as well as some other tasks, still remain problematic.

The macroscopic representation of function ε_1 assumes that the functioning of the entire root system should be considered as it absorbs water from a certain soil volume. In this representation ε_2 is usually assigned as

$$\varepsilon_2 = k(\psi_s)\frac{\psi_p - \psi_s}{\beta(z)}, \tag{16}$$

where $\beta(z)$ is a proportionality factor treated in different ways by different authors. In some investigations $\beta(z)$ was obtained as the ratio of the root weight for a given layer to the total root weight, or was assigned as an exponential function depending on the current total weight of the root system. It is not quite clear if the roots absorbing water from saturated soil layers can transport it via xylem, and then yield it to drier layers. Some scientists, e.g., Poluektov (1991), assumes it to be possible, others do not, thus believing that $\varepsilon_2 \geq 0$ (Sirotenko 1981).

The next task is to estimate ε_1, i.e., assign the dependence of internal soil evaporation (condensation) on the determining factors. Dandaron (1974a,b) and others made experimental and theoretical studies of the kinetics of water-vapour sorption and desorption by soil under changes in the temperature of a desiccated layer. One should agree with Budagovsky (1980) that to date no satisfactory model has been constructed which would permit description of evaporation from drying soil although several semiempirical approaches have been proposed enabling successful solutions of selected problems. These include the desiccated layer concept developed by Budagovsky (1964, 1980). Novak (1980) presented an overview of the state of the art in studying water and water-vapour transport in soil under conditions of evaporation.

2.4 Parameterization of Energy and Mass Exchange Models

An energy and mass-exchange calculation scheme for applied problems should be simple and precise, i.e., a compromise is to be found between "the quicksand of overcomplication and the pitfall of oversimplification". Some hope for designing such schemes was pinned on analytical solutions of stationary equations for transfers inside the canopy (Budagovsky 1964; Budyko and Gaudin 1968; Asaturov 1969; Sirotenko and Boiko 1976). Unfortunately, this hope did not come true: the solutions obtained were rather cumbersome and had to be paid for by making uncertain assumptions. Thus for most such schemes, plant biomass T_p was assumed to be constant, and T_a and q_a were found as functions of z. Obviously, the differences $T_p - T_a$ and $q_p - q_a$ which determine the rate of plant biomass–air exchange in this instance could not be estimated accurately enough.

Most specialists seem now to support an idea that the rational way to simplify the problem of energy and mass exchange would be to abandon the canopy's vertical coordinate, keeping it for soil, i.e., the sought model is to be a steady-state, single-layer model ("big leaf") for the air medium, and a nonsteady-state, multilayer model for the soil medium. The best-known schemes of this type were designed under the supervision of Budagovsky (Budagovsky and Lozinskaya 1976a,b) and Sirotenko (Sirotenko and Mokievsky 1990). Let us discuss the latter by way of example.

The following set of balance equations makes up the basis for the calculation scheme:

$$
\begin{aligned}
XE_p &= R_p - H_p, \\
XE_s &= R_s - H_s - G, \\
Q_R &= E_p,
\end{aligned}
\tag{17}
$$

where X is the latent heat of vaporization, E_p is the transpiration rate, R_p and R_s are the canopy and soil-surface radiation balances respectively; H_p and H_s are the heat turbulent fluxes from plants and soil to the atmosphere; E_s is the soil-surface evaporation, G is the heat flux into the soil, and Q_R is the root system's water absorption rate.

The set of Eq. (17) describes the energy- and mass-exchange processes in a steady-state approximation. The first of them is the canopy heat-balance equation, the second is the soil-surface heat-balance equation, and the third is for the crop water balance.

Expressions for heat and water flows may be written as

$$H_p = (1-K)\rho C_p \frac{T_p - T_a}{r_a}, \tag{18}$$

$$H_s = K\rho C_p \frac{T_s - T_a}{r_a}, \tag{19}$$

$$E_p = (1-K)\rho \frac{q_p - q_a}{r_a - r_{st}}, \tag{20}$$

$$E_s = K\rho \frac{q_s - q_a}{r_a + r_{ss}}, \tag{21}$$

Resistance values of r_s, r_{ss} and r_a are calculated on the basis of semi-empirical formulas depending on leaf-water potential, soil moisture, and air velocity above the crop respectively. $K = 0.7^L$, where L is the leaf area index.

Although the set Eq. (17) is nonlinear, the use of linearization for the Stefan–Boltzmann law and the Magnus formula enabled its analytical solution to be found and the explicit formulas obtained for calculating the values of E_p, E_s, T_e and T_s to be averaged over the daytime period.

In order to follow the simulation scheme one needs to know maximum and minimum air temperatures, mean daily deficit of air humidity, air velocity at 15.00 h, and daily sunshine hours. To start the simulation, soil moisture content by layers, crop biomass in dry mass units and plant height are assigned for the initial date. Knowledge of soil hydrophysical characteristics is also required.

The model was calibrated for cereal crops. It was checked against the most precise experimental data currently available – measurements obtained at the

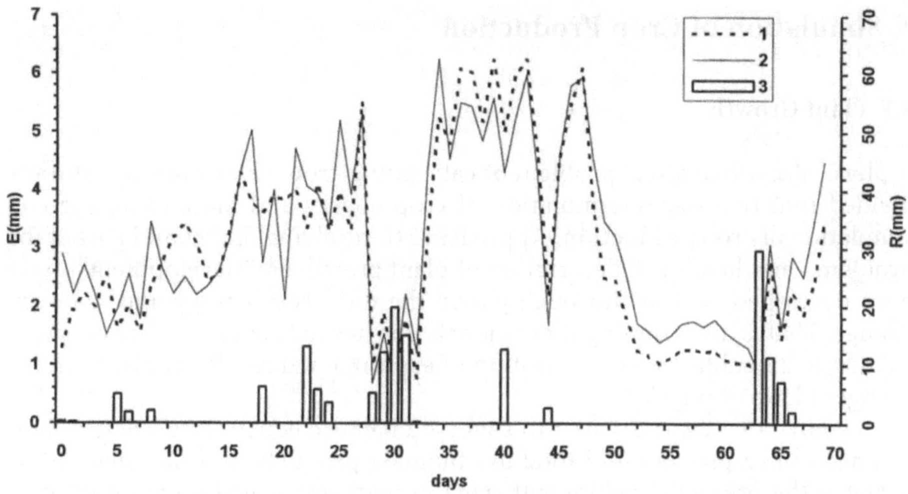

Fig. 1. Dynamics of actual (*1*) and simulated (*2*) daily values of evapotranspiration *E* and rainfall *R* (*3*) for 18 May–26 July 1973 (measurements were obtained at a large soil pan at Lake Valday, Novgorod Region, Russia)

large-scale hydraulic soil pan at Lake Valday (Novgorod Region, Russia) for a 70-day growing period of spring barley. The dynamics of actual and simulated daily evapotranspiration values for the period from 18 May (emergence) to 26 July (maturity) for spring barley are shown in Fig. 1. The scheme was also tested for cereal crops and peas under Moldavian conditions (Direktorenko 1989). The coefficient of correlation between measured and simulated values was 0.91. According to Kharchenko et al. (1970), the USSR's best-substantiated heat balance-based method of calculating evapotranspiration designed by Kharchenko allows an estimation validity of ca. 28% for monthly and 10-day values. The error for the method described above is half as much, being respectively 11 and 16%. The results obtained indicate that this method has great potential, which nevertheless needs to be tested with additional experimental data.

As easy-to-use simulation scheme could be constructed, in particular, assuming the heat flux into the soil to be a permanent fraction of the radiation balance. A more complex model was designed, free of that assumption, taking into account the daily course of energy- and mass-exchange processes (Boiko et al. 1990). A comparison of the simulation results obtained with the simpler model (Sirotenko and Mokievsky 1990) and the more complex one (Boiko et al. 1990) based on the barley evapotranspiration data for the Lake Valday soil pan showed that the use of hourly energy- and mass-exchange data did not improve the simulation accuracy. This conclusion seems to indicate that there is a certain complexity limit whose violation could result in reducing the effectiveness of the model as a tool for resolving applied problems.

3 Simulation of Crop Production

3.1 Plant Growth

A physical-mathematical problem of calculating crop yields may be naturally divided into two stages; simulation of crop energy and mass exchange and simulation of crop production. Approaches to resolving the second part of the problem – mathematical description of plant growth and development – have been developed far less thoroughly than the models of energy and mass exchange. That is, in designing these models one has to lean upon mere concepts leading to a qualitative generalization of experimental results, rather than the stringent laws of physics.

Ivanov (1946) was the first to suggest a balance equation to relate photosynthesis to respiration and total dry biomass production. Ross (1966, 1967) initiated the present development stage in mathematical simulation of plant growth when he proposed a set of equations which served as a basis for designing pilot dynamic models of crop production:

$$\frac{dm_j}{dt} = \varepsilon_\phi \sum_{i=1}^{4} A_{ij}\overline{\Phi}_{ci} - \varepsilon_R \overline{R}_{ci} - V_j + M\sum_{i-1}^{4} B_{ij}, \tag{22}$$

where i and j are the plant organs (leaves, stems, roots, and reproductive organs); m_j is the dry mass of the j-th organ; ε_ϕ, ε_R are the photosynthesis and respiration effectiveness indexes (g dry mass g^{-1} CO_2); Φ_{ci} and R_{ci} are the total daily photosynthesis and total daily respiration; V_j is the dry plant biomass lost per day due to the leaf fall ($g\,d^{-1}$); M is the plant's total dry mass; A_{ij} is the fraction of the i-th organ's daily production of "fresh" assimilates exported into the j-th organ in 24 hours; B_{ij} is the exchange of "old" assimilates (in $g\,d^{-1}$) between the i-th and j-th organs, related to 1 g of whole plant dry biomass. A_{ij} and B_{ij} are presented as "growth matrices". During the vegetative period when leaves act as the main centers of growth, use is made of growth functions A_j, while the import–export functions B_j are employed during the reproductive period. Sepp and Tooming (1982), modified the equation written by Ross as

$$\frac{\Delta m_j}{\Delta t} = \varepsilon A_i(t)\left[\int_t^{t+1}\int_0^{L(t)} (\Phi_1 - R_1)dLdT - \sum_{i=2}^{n} R_{ci}\right] + MB_j, \tag{23}$$

where $\sum_{i=2}^{n} R_{ci}$ is the total daily respiration of all the organs except the leaves.

There has been an attempt to improve Ross' growth functions (Galyamin and Siptits 1977): it was suggested that they be calculated in relation to the aboveground plant biomass production, rather than to the total biomass production. "Idealized" curves were plotted for the functions of vegetative growth of wheat leaves, stems, and ears related to the aboveground biomass production which, in the author's opinion, do not depend on the ambient conditions (Baranovsky et al. 1978). Galyamin and Siptits (1977) attempted to calculate the dependence of the root growth function on the water regime and nitrogen content of soil with the help of nonlinear regression equations.

To simplify the determination and application of growth functions, Abashina (1979) suggested that they be represented exponentially. Polevoy (1983) related growth functions to effective temperatures and the dates of plant growth stage commencement and used, as the basis for their determination, a three-parameter logistic curve of plant biomass production dynamics.

The growth function method is widely used due to its simplicity and because it presents an opportunity to take advantage of the standard biometric measurements. Its significant limitation is that it is closed upon itself: experimental data are used to construct the model which is designed essentially to establish relationships for these same experimental data. The use of more or less stringently predetermined growth functions limits the possibility of simulating the plant response to environmental conditions. Abandoning the enforced partitioning of assimilates would enable a more detailed consideration of environmental effects on plant photosynthesis and respiration on the one hand, and growth and development on the other.

The type of model in which the partitioning and redistribution of plastic materials in plant ontogenesis are shaped on the basis of the current demand and supply ratio may be referred to as autoregulatory to distinguish such models from the ones which address the task with the help of growth functions. In the former USSR, Bichele, Moldau, and Ross were the first to pioneer a design of this kind (Bichele et al. 1980b). Growth is simulated in such a model using the concept of a reserve pool of available assimilates, which stipulates that photosynthates initially are pooled in the reserve and only then can be used for growth which is treated as a process of turning the reserves into the structural biomass. The model consists of two parts (shoots and roots) and the flow rate of reserve carbohydrates between them is assumed to be proportional to the carbohydrate concentration difference.

Sirotenko and Abashina (1985) appear to present the most complete model of the autoregulatory category. Dry plant biomass (m) is divided into four components: soluble sugars (mc), reserve carbohydrates (ms), mobile nitrogen reserve (mN), and structural matter (mg):

$$m = mc + ms + mN + mg. \tag{24}$$

Figure 2 illustrates a formalized representation of a plant's main physiological relationships. The diagram includes two sources of the reserve of soluble sugars – photosynthesis and breakdown of structural materials – and two sinks – carbohydrate losses for synthesizing new structures and meeting all energy requirements of the organism. In addition, the diagram distinguishes the routinely occurring processes of soluble sugars being channeled to the carbohydrate reserve pools and back.

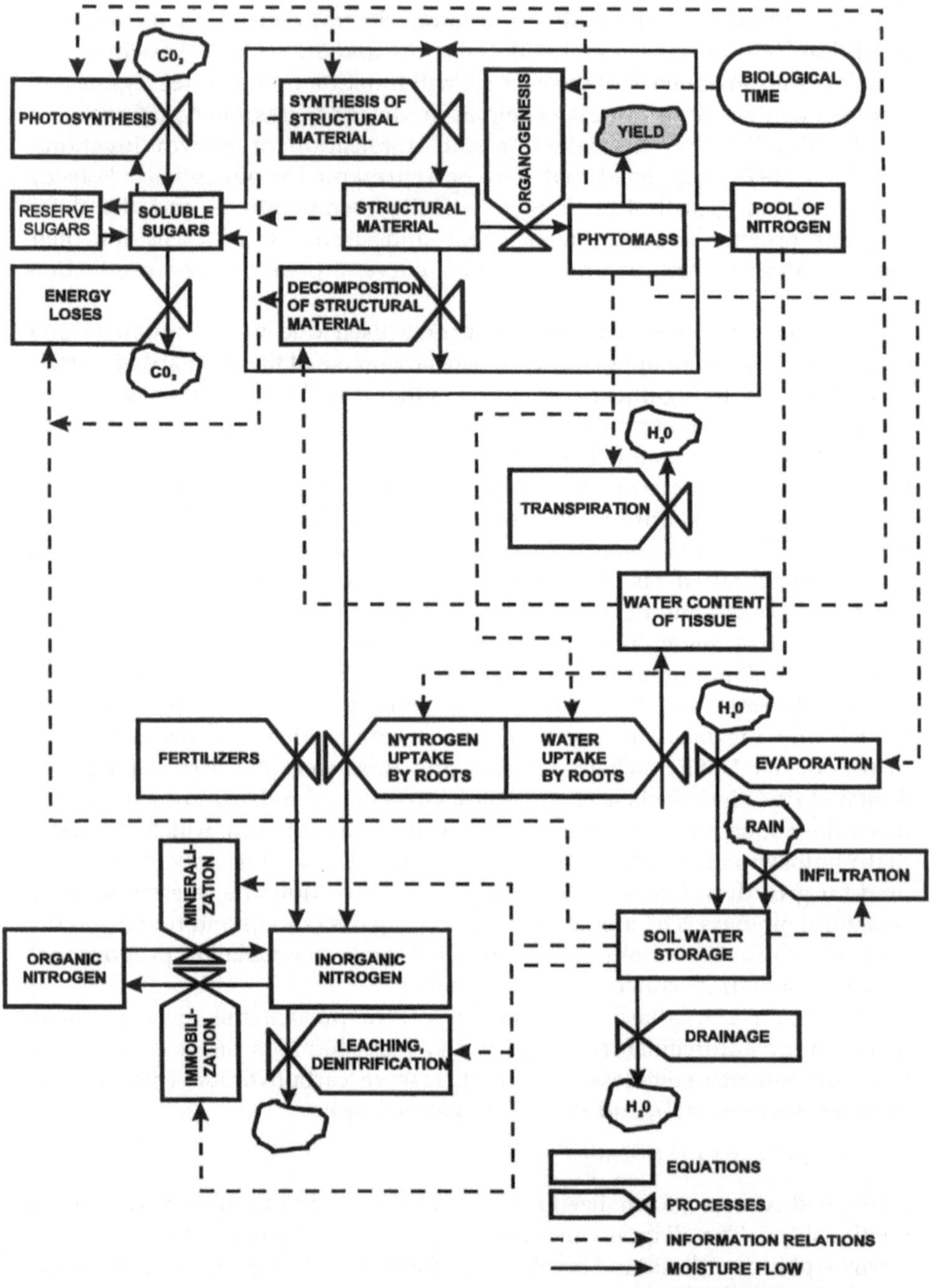

Fig. 2. Relational diagram, of energy and mass exchange processes simulated using a set of differential equations (23)–(26)

The plant nitrogen cycle is arranged in a simpler way since nitrogen-containing substances play the part of structural material and do not have to act as an energy supplier at the same time, as carbohydrates do.

These considerations allow the description of the plant biomass components to be made with the following set of differential equations:

$$\frac{d(mc)}{dt} = \nu_1\Phi - \nu_2(G-Q) + K_{sc}ms - K_{sc}mc - D, \tag{25}$$

$$\frac{d(ms)}{dt} = K_{cs}mc - K_{sc}ms, \tag{26}$$

$$\frac{d(mN)}{dt} = P - \nu_3(G-Q), \tag{27}$$

$$\frac{d(mg_{ij})}{dt} = G_{ij} - Q_{ij}, \tag{28}$$

where Φ is the total gross photosynthesis (g CO_2 s^{-1}); G is the structural biomass production rate in a growing plant (g structural biomass s^{-1}); Q is the structural matter breakdown rate (g structural biomass s^{-1}); D is the total respiratory loss of carbohydrates for plant energy requirements (g CH_2O s^{-1}); P is the rate of nitrogen uptake by roots (g N s^{-1}); K_{sc}, K_{cs} are constants (C^{-1}); i is the number of the group where a given plant organ belongs; and j is the plant organ number in the order of appearance.

Equations (25) to (27) represent the dynamics of soluble sugars, reserve carbohydrates, and the mobile nitrogen pool, respectively, and the subsystem Eq. (28) represents the dynamics of structural biomass of individual plant organs whose number changes in the ontogenesis.

For spring wheat, to which the model may be oriented, by way of example, the following plant-organ groups may be distinguished: (1) kernel, (2) embryo, (3) main root, (4) pairs of seminal roots, (5) stems, (6) ears, (7) leaves of the first (main) stem, (8) leaves of the second stem, etc.

In most cases, the number of equations [Eq. (28)], which must equal the total number of organs in an "average" plant, would not be greater than a few dozen.

Plant development is simulated by assigning an algorithm specifying the time of switching on and off each equation in the subsystem Eq. (28). To this end the function $\lambda(t)$ is assigned which characterizes how well the growth processes are provided with the substrates required. The production of new roots and tillers is linked to the condition of $\lambda(t) \geq \lambda_{max}$. A similar rule is applied to determine the time when every next i-th leaf is to start growming – the next leaf in turn will start growing only when its predecessor has changed from importing to exporting assimulates with the export rate being greater than the minimum level. However, the number of plant organs in this model does not increase monotonously. As the new organs appear, those that were produced earlier die. To determine the time of an organ's death, use is made of the degree of degradation of its structural biomass expressed as a ratio of the current structural biomass mg_{ij} to the maximum one $\max(mg_{ij})$. An organ dies (a corresponding equation is deleted) as soon as the above ratio falls below a predetermined critical level. A similar rule is used for determining the moment when an organ "yellows", i.e., the time when its photosynthesis stops.

Gulyaev (Gulyaev et al. 1983) proposed a relational diagram for a whole plant regulatory system of the "source-sink" type similar to the one in Fig. 2. To simulate the reproductive development period, he distinguished two stages. At the first stage the structural elements of reproductive organs take shape, and autoregulation is performed of the potential capacity of a reproductive organ (the number of kernel sites formed in a wheat ear is determined) according to the available photosynthetic potential of a plant. At the second stage, the kernel biomass grows and the ultimate weight will be determined by the supply of "fresh" assimilates to support the potential grain biomass accumulation rate, and by the amount of reserve materials in vegetative organs.

3.2 Photosynthesis and Respiration

The literature on simulating photosynthesis and photorespiration is quite voluminous – the subject is discussed in practically all monographs on mathematical simulation (Tooming 1977; Bichele et al. 1980a; Sirotenko 1981; Poluektov 1991). Therefore here it is worth noting only that in the best-developed dynamic models the carbon dioxide gas exchange of a leaf is simulated with the help of the well-known semiempirical formula by Chartier. The derivation of the most complete version of that formula with an explicit account for respiration is given in Poluektov (1991).

To make use of Chartier's formula in the autoregulatory dynamic models, an additional relationship between the photosynthetic rate and the assimilate content of plant foliage needs to be estimated. (As an approximation of experimental data, it is to be found in Bichele et al. 1980a.)

Laisk (1977) developed an original mathematical model of the Calvin cycle. By mathematically analysing the kinetics of the photosynthesis pentosphosphate cycle it was found which cycle stages would include limiting responses under different environmental conditions. Photorespiration kinetics were accounted for by an assumption that carbon dioxide and oxygen compete for the acceptor ribulosediphosphate. Laisk's model is of an explanatory type and seems to be too complicated to be used in dynamic modeling.

Let us consider the studies of dark respiration as a part in crop production. Simultaneously, McCree (1971), and Tooming (1970) were the first to propose (for higher plants) the equations of two-component respiration which took into account the losses for respiration associated with structural growth and maintenance. Respiration was regarded as a function of whole plant photosynthesis and dry mass (McCree 1971), or foliage only (Tooming 1970). Kuperman and Khitrovo (1977), Khitrovo and Kuperman (1980), and Kuperman et al. (1983) discuss the parameter evaluation in simulating respiration for different levels of nitrogen supply. The authors conclude that under an improved nitrogen supply the value of the maintenance respiration index adjusted to unit dry biomass should increase, whereas the unit protein-adjusted index should decrease. Studies made by Semikhatova et al. (1979) and Moldau (1984) on the role of maintenance respiration in adaptative restructuring of plants to lower temperatures (Semikhatova et al. 1979) and water deficit (Moldau 1984) established the rule that the losses for maintenance respiration are greater for adverse weather conditions.

The wild species of wheat, millet, sorghum, and cotton were definitely proved to be ahead of their modern cultivars in terms of the maximum rate of light-saturated CO_2 exchange per unit leaf area. Comprehension of this discovery, as we view it, should provide a formidable impact on the development of mathematical simulations of crop production. The plant growth rate and productivity, subdued as they may be under photosynthetically unfavourable conditions, are not able indeed to exceed a certain limit when the conditions get better. That limit appears to be determined by the growth processes. Therefore it seems to be quite likely that under conditions which may be unfavourable for plant growth although adequate for intensive photosynthesis, the latter is nevertheless inhibited. Plants do not store excessive amounts of unused assimilates. This aspect undermines the concept of the independence and main role of photosynthesis in crop production, and reduces the level of its consideration to that of assimilate transport and growth processes. It is from the same viewpoint that we need to regard the role of photorespiration: under conditions when the carbon supply is limiting, a decrease of the effective photosynthetic rate by as much as one third due to photorespiration appears to retard the growth rate as well; if the growth processes are limiting, then photorespiration can hardly exert an adverse impact on plant productivity (Laisk 1977).

It appears that the model designers should be encouraged to shift their attention from the mere description of observed relationships between the carbon dioxide gas exchange and environmental factors toward gaining an insight into the mechanisms providing for the integration of the whole plant physiological processes. The most interesting proposals on the issue were made by Semikhatova (1980) on integrating the processes of photosynthesis and respiration; Kuperman (1972) on developing concepts of autoregulation of coordination between the plant's levels of nitrogen and light supply; and Mokronosov (1982, 1983) on substantiating a qualitative theory of the donor–acceptor relations in the plant ontogenesis.

4 Applied Dynamic Models and Their Use

4.1 Applied Dynamic Models

The USSR's first dynamic models of crop production were developed by Galyamin and Siptits (1977) (Moscow) for soil and water management applications, and Sirotenko and Boiko (1976) (Obninsk) for use in agricultural meteorology. Later similar research was initiated under the leadership of Poluektov (Bondarenko et al. 1982) in St. Peterburg, Kan (1981) in Novocherkassk, and Tooming (1984) in Estonia. Currently Polevoy (1983) (Odessa, Ukraine) is also involved in this area of research.

By way of example, let us discuss one of these developments – the model designed by the author with E.V. Abashina as a coworker – which is a set of conventional differential equations, such as

$$\frac{dm_p}{dt} = G_p - D_p - q_p - P_p, \tag{29}$$

$$\frac{dW_i}{dt} = q_{i-1} - q_i - TR_i - \sigma_i E_s, \tag{30}$$

$$\frac{dN_k}{dt} = H_k + U - h_k + V_{k-1} - A_k, \tag{31}$$

where m_p is the p-th organ's weight (mg cm^{-2}), $p \in$ l, s, ls, r, k, z; t is the time, in days; G_p, D_p, q_p, P_p are the rates of growth, respiration, breakdown, and fall of plant tissues [mg/(cm·d)]; W_i is the water stored in the i-th soil layer [mm (i = 1.15)]; q_{i-1} and q_i are the water flows across the top and bottom boundaries of the i-th layer (mm/d); for the first layer the flow across the top boundary (q_o) is equal to effective rainfall ($R_{3\Phi}$); TR_i is the water lost to transpiration from the i-th soil layer (mm/d); σ_i is a logical variable equal to 1 at i = 1 and to zero in all other cases; E_a is the soil-surface evaporation, (mm/d); N_k is the nitrate content of the k-th soil layer, k = 1.3 (mg/cm^2); H_k is the mineralization rate of mobilizable organic nitrogen [mg/(cm^2·d)]; U is the fertilizer amount, with an account for immobilization [mg/(cm^2·d)]; V_{k-1} and V_k are the flows of mineral N with water across the boundaries of the k-th layer, $V_o = 0$ [mg/(cm^2·d)]; h_k is the denitrification rate of mineral N [mg/(cm^2·d)]; and A_k is the rate of N absorption by plants from the k-th soil layer [mg/(cm^2·d)].

The model assumes that in the growth process all carbohydrates produced for a day will be transformed into the structural biomass:

$$G_p = \alpha_p \left(F + \sum_{p=1}^{6} q_p \right), \tag{32}$$

where α_p denotes the growth functions which determine what fraction of the carbohydrate reserve pool should be spent for the p-th organ's growth, $\alpha \geq 0$, $\sum_{p=1}^{6} \alpha_p = 1$, F is the daily total gas exchange of green plant organs calculated by the Chartier formula.

This mathematical structure served as a basis for developing a number of model versions. It features a nitrogen part and a feedback loop to calculate the repeated use of carbohydrates exported from the ageing plant tissues which disintegrate. Routinely obtained weather data are used as model input; temperature, air humidity deficit, hours of sunshine, and daily rainfall totals. The model is run at a 1-day time step.

The model was calibrated for spring barley, spring and winter wheat (Sirotenko et al. 1985; Dolgy-Trach 1987), potato (Ustinova et al. 1990), tomato (Kharchevskaya 1986), green pea (Direktorenko 1989), alfalfa (Daus 1991), maize (Bulgaria; Sirotenko et al. 1982), and sunflower (Bulgaria; Dukov 1987).

The model for grain crops was tested for a wide variety of soil and climatic conditions of the Russian Plain. The model adequacy may be evaluated using the data shown in Figs. 3 and 4. Figure 3 indicates that the model is able to adequately enough monitor the dynamics of arable soil-water content for soils of different genesis. The model enables a satisfactory simulation of significant year-to-year crop-yield variability of spring cereals typical of arid conditions (cf. Fig. 4). Insufficient field observations are available to establish the validity of the nitrogen part, although selected model checks showed satisfactory results (Abashina 1979).

Let us briefly characterize other dynamic models designed in the USSR. Galyamin's model (Galyamin 1980) was developed for irrigation scheduling applications and was parameterized for spring wheat and maize. The model is run at a daily time step. The biological part is based on growth functions. Account is taken of the heat regime and soil-water effects on the plant development rate. The model lacks a part for computing soil-water dynamics and a part for available N dynamics in soil. In addition to weather data and model needs actual 10-day data on the 1-m-thick soil layer's water content, which makes the model applicability questionable.

Polevoy's model (Polevoy 1983) structurally resembles Galyamin's. The original aspects of the way in which growth functions are defined in the model are described above. Another distinction is a 10-day time step dramatically reducing the costs of preparing input meteorological data. However, in such a case the physically based interpretation of the model parameters becomes problematic, and the model acquires an excessively localized nature.

Sepp and Tooming (1982) designed the dynamic model of potato crop production, POMOD. This model, as distinct from the model CROP-WEATHER-3, has a better developed part for computing the crop radiation regime, although it is significantly inferior in terms of the detailed description of the crop water regime. Photosynthesis is simulated at a 1-h time step, and biomass at a 1-day time step. Model validity was estimated by checking against the long-term data collected at several agrometeorological stations

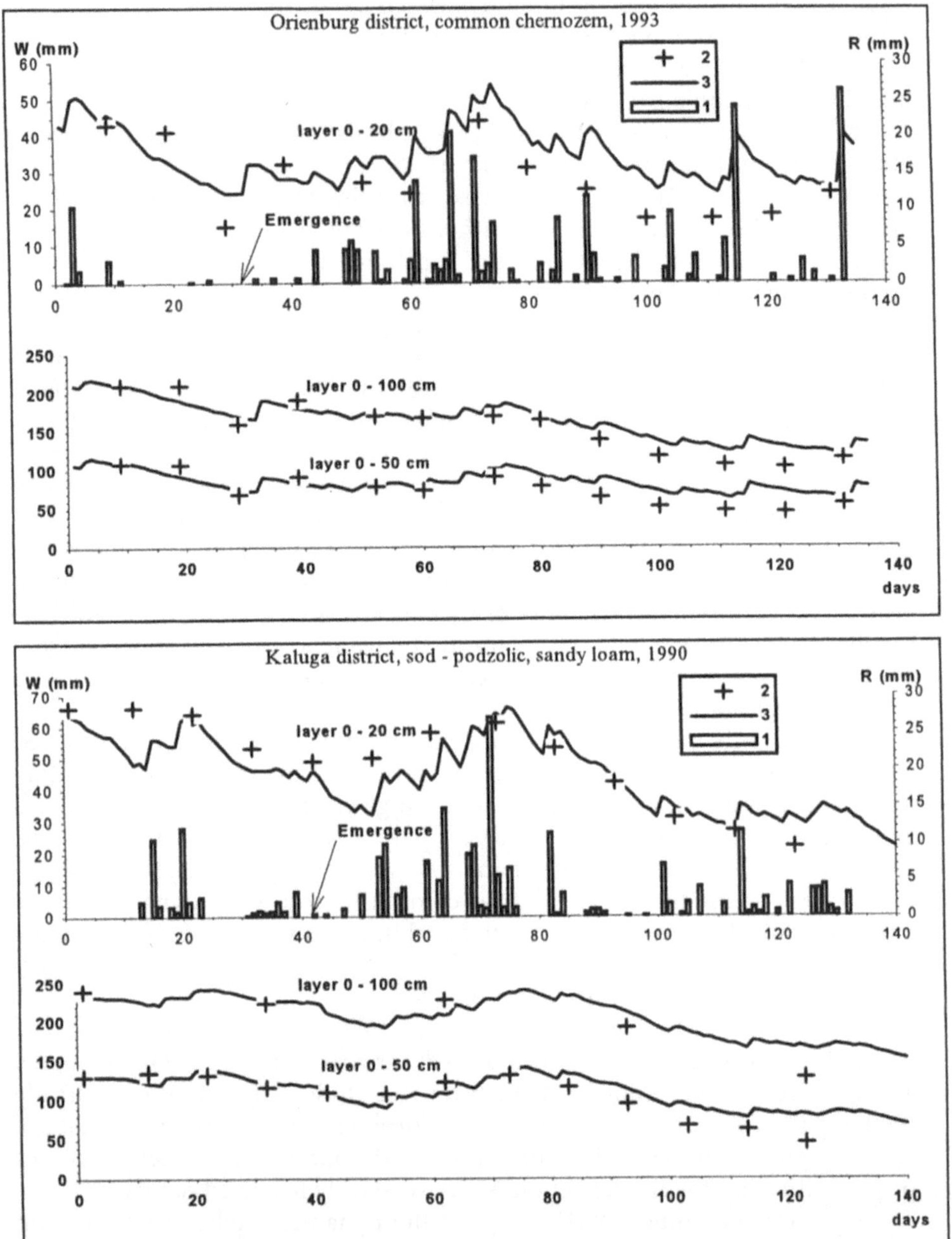

Fig. 3. Measured (*2*) and calulated (*3*) soil moisture storage *W* in the layers 0–20 cm, 0–50 cm, 0–100 cm and precipitation (*R*) (*1*)

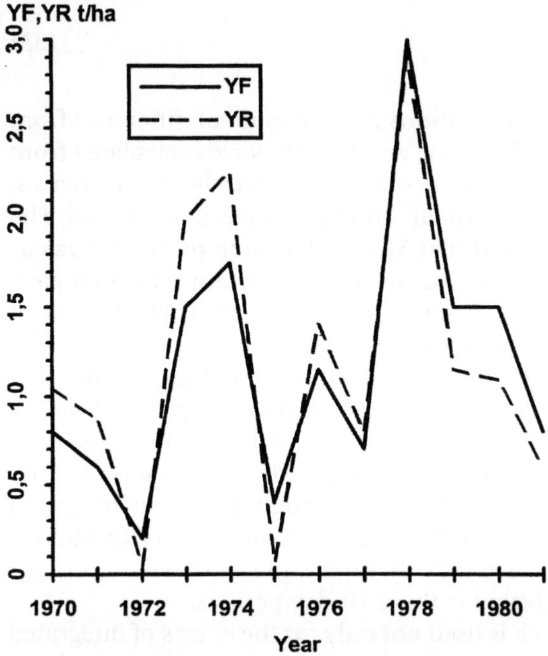

Fig. 4. Dynamics of actual (*YF*) and calculated (*YR*) yield of spring wheat for 1970–1981 for agrometeorological station Yershov (Saratov District)

in Estonia (coefficient of correlation between simulated and actual data was ca. 0.7–0.8).

Poluektov (1991) developed a large-scale basic model of an agroecosystem including a part for calculating the dynamics of the soil-nitrogen regime (Chuvashina 1983) which was identified for several crops, in particular spring wheat (Poluektov et al. 1984).

Kan et al. (1984) have developed dynamic models primarily for applied purposes. In recent years they have increased the number of the model parts by adding submodels of the dynamics of soil phosphorus, symbiotic nitrogen fixation, and the quality of crop yield (Kan 1981). The same authors designed crop-production models for soybean, maize, and winter wheat for rain-fed and irrigated areas.

4.2 The Use of Models

4.2.1 Diagnosing Agrometeorological Conditions

An agriculturalist always faces this problem as he tries to estimate what effects antecedent weather conditions (previous weeks, 10-day periods, a season) might have on future crop yields. Let us represent the measure of differentiating the estimated period's conditions from the reference ones as a ratio, such as

$$Y_k = \frac{y}{y_k} 100\%, \quad (33)$$

where Y_k is the estimated weather conditions (the measure of difference from the k-th reference conditions), and y and y_k are the crop yields calculated from the estimated period's actual and reference data respectively. As references, the normal, last year's conditions, "optimal" conditions, etc. may be used. The results (Sirotenko et al. 1985) showed that Y_k as a deviation of the two calculated values is practically independent of the way the future meteorological conditions have been assigned. This makes it possible to dissociate the task of estimation from the accuracy of crop yield forecasting.

The top rows in Table 2 (Sirotenko et al. 1985) represent the dynamics of forecast yields for two years; estimated weather conditions are given below the yield. Table 2 indicates that in 1977 the weather (including the preceding winter and autumn) in terms of favorability to spring wheat crops was just 70% of the normal compared to 202% in 1978. The growing season is estimated as 59% for 1977 and 144% for 1978 respectively. In 1977 the weather conditions for May 11–20 and June 1–10 were especially severe; the expected crop yields declined by 21 and 45% respectively for these 10-day periods.

A similar estimating approach is used not only for the effects of integrated meteorological conditions, but also for delineating the role of selected environmental factors. A dynamic model of this type may be used as a kind of measuring device, integrating the effects of plant habitat conditions on crop productivity.

4.2.2 Forecasting Crop Yields

The involved nature of the task of forecasting crop yields is associated with the lack of valid long-term weather forecasts. When dynamic models are used to obtain a crop-yield forecast, mean long-term weather data are used instead of the unknown values of future weather conditions. The use of dynamic models for crop-yield forecasting based on this choice of future weather has led to a certain ammount of disappointment. The accuracy of crop-yield estimates thus obtained didn't appear to have any essential improvement as compared with conventional regression schemes. Limited space does not allow us to discuss this interesting issue at great length. Note only that the way in which the future weather is assigned may play a crucial role. Since dynamic models are nonlinear, the use of smoothed "normals" as a forecast leads to biased crop-yield estimates. When a smoothed air-temperature curve is "disturbed" by a random value with a mean of 0 °C and a 4 °C standard deviation, crop yields are reduced by 9% (Sirotenko and Pavlova 1988). To eliminate such an effect, a model was proposed using the Monte-Carlo approach to generate an ensemble of annual realizations of the daily set of weather parameters corresponding to the predetermined climatic normals (Sirotenko and Pavlova

Table . Dynamics of predicted crop yields and estimates for spring wheat growth and development (% of normal) in 1977–1978, calculated from date for Ershov agrometeorological station, Saratov District

Year	Index	Emergence	Final dates of periods estiamted May 10	May 20	May 31	June 10	June 20	June 30	July 10	July 20	July 31	Harvest
1977	Yield ($q\,ha^{-1}$)	13.5	13.5	10.7	10.5	5.8	6.3	9.4	7.9	8.0	–	8.0
	Estimates: 10-day periods	–	100	79	98	55	109	149	84	101	–	–
	Current growing season	100	100	79	78	43	47	70	58	59	–	59
	Current farming year	117	117	93	91	50	55	82	69	70	–	70
1978	Yield ($q\,ha^{-1}$)	16.1	–	16.5	16.8	16.9	15.7	17.0	20.3	21.5	23.3	23.2
	Estimates: 10-day periods	–	–	102	102	101	93	108	119	106	108	–
	Current growing season	100	–	102	104	105	97	106	126	133	145	144
	Current farming year	140	–	143	146	147	137	148	177	187	203	202

1988). The model features long-range statistical relations, calculating auto- and cross-correlation relations of daily values of solar radiation, temperature, and air humidity. Discriminant analysis is used to synchronize the time of rainfall with variations of other meteorological values.

Sirotenko and Abashina (1985) compared different modes of assigning future weather data to obtain dynamic model-based crop-yield forecasts. Using data for 31 years, wheat yields were calculated assigning, at the commencement of heading, mean long-term values, the data for each year from the available sample (i.e., 31 × 31 simulations were made), and finally, data for an analogue year based on similarity to the current-year conditions. The coefficients of correlation between the crop-yield series estimated using this approach and the yield series estimated from the actual data were 0.76, 0.80, and 0.83 respectively. Thus, the use of the analogue approach to assigning the future weather seems preferable. Such was the result of developing a computer-based system for forecasting alfalfa yields (Daus 1991).

4.2.3 Managing Crop Production

Every year the task of crop production management becomes increasingly more urgent due to the requirements of saving resources and environmental considerations whose significance grows as farming becomes more intensive. Dynamic models (systems of conventional differential equations) make it possible, for the first time, to obtain a mathematically correct definition for problems of crop management. The issue may be discussed using an example of computing an optimum irrigation schedule formulated by Galyamin (1980) as follows.

Define the determinate discrete function $\varphi(\tau)$, $\tau \in [\underline{\tau}, \overline{\tau}]$ satisfying the constraints

$$\sum \varphi(\tau) \le \Phi, \quad 0 \le \varphi(\tau) \le \varphi^o,$$

so that

$$m(\overline{\tau}) \to \max,$$

where Φ is the irrigation norm value, i.e., the total amount of water to be delivered in time $\overline{\tau} - \underline{\tau}$ to the irrigated crop field; φ^o is the daily maximum amount of water; $\underline{\tau}, \overline{\tau}$ are the dates of planting and harvest respectively.

Sirotenko et al. (1987) resolved the optimum irrigation problem using another definition. Define the function $\varphi(\tau)$, $\tau \in [\underline{\tau}, \overline{\tau}]$, satisfying the constraints

$$0 \le \varphi(\tau) \le \varphi^o, \quad m(\tau) = m_0$$

so that

$$\sum \varphi(\tau) \rightarrow \min.$$

The first definition of the problem assumes that an irrigation regime can be found which would maximize crop yields under the predetermined irrigation norm, whereas the second one assumes that the minimum irrigation regime be sought for obtaining the predetermined crop yield level. The latter option seems to be preferable in view of environmental considerations. Figure 5 shows the minimum irrigation regimes found for three levels of yield of wheat – 2, 3, and 3.5 t ha^{-1}. At the prescribed yield levels of 2 and 3.5 t ha^{-1}, water is applied only during critical periods – generative plant organ production – and only after this restriction is met is the need of irrigation on other dates considered (cf. Fig. 5).

This method is heuristic and cannot guarantee an exact solution. Shlyachkova (1990) describes the exact solution of a problem of optimum crop management, maximizing the function representing the accumulated total crop biomass. To this end the Pontryagin principle is used which enables a complete solution of the problem for a simplified agroecosystem model.

4.2.4 Estimating Soil and Climatic Resources

The development of mathematical crop-production models has led to the revision of methods of estimating natural soil and climatic land resources. Tooming was among the first who readily used these new opportunities and suggested the reference-yield method (Tooming 1976, 1982), enabling a comparision of the potential yield (PY), actually possible yield (APY), and harvested yield. PY is defined as the yield under ideal weather conditions determined by the incoming PAR and biological characteristics of a cultivar. APY is the possible yield maximum under actual weather conditions. PY and APY any be obtained under ideal soil conditions. This idea (conceptually resembling the four plant production levels suggested by de Wit) has been used by numerous authors with varying modifications. However, as noted in (Tooming 1982), PY is an abstract concept since it is still not clear which meteorological conditions should be accepted as ideal for crop yield production.

To make the above approach more specific, a system of reference yields was proposed (Table 3; Sirotenko et al. 1990).

The simulation is made in a simple way: NY is estimated using the actual daily data for soil-water storage and mobilizable N content of soil solution. To estimate SDY, simulated water-storage data are substituted with "optimum" values. To estimate CDY, current values of nitrate concentration are substituted with "optimum" values. Finally, MPY is estimated using actual weather data and optimum soil-water and nitrogen values. Table 4 shows the results of calculating the four levels of spring wheat yield for the Krasnodar region,

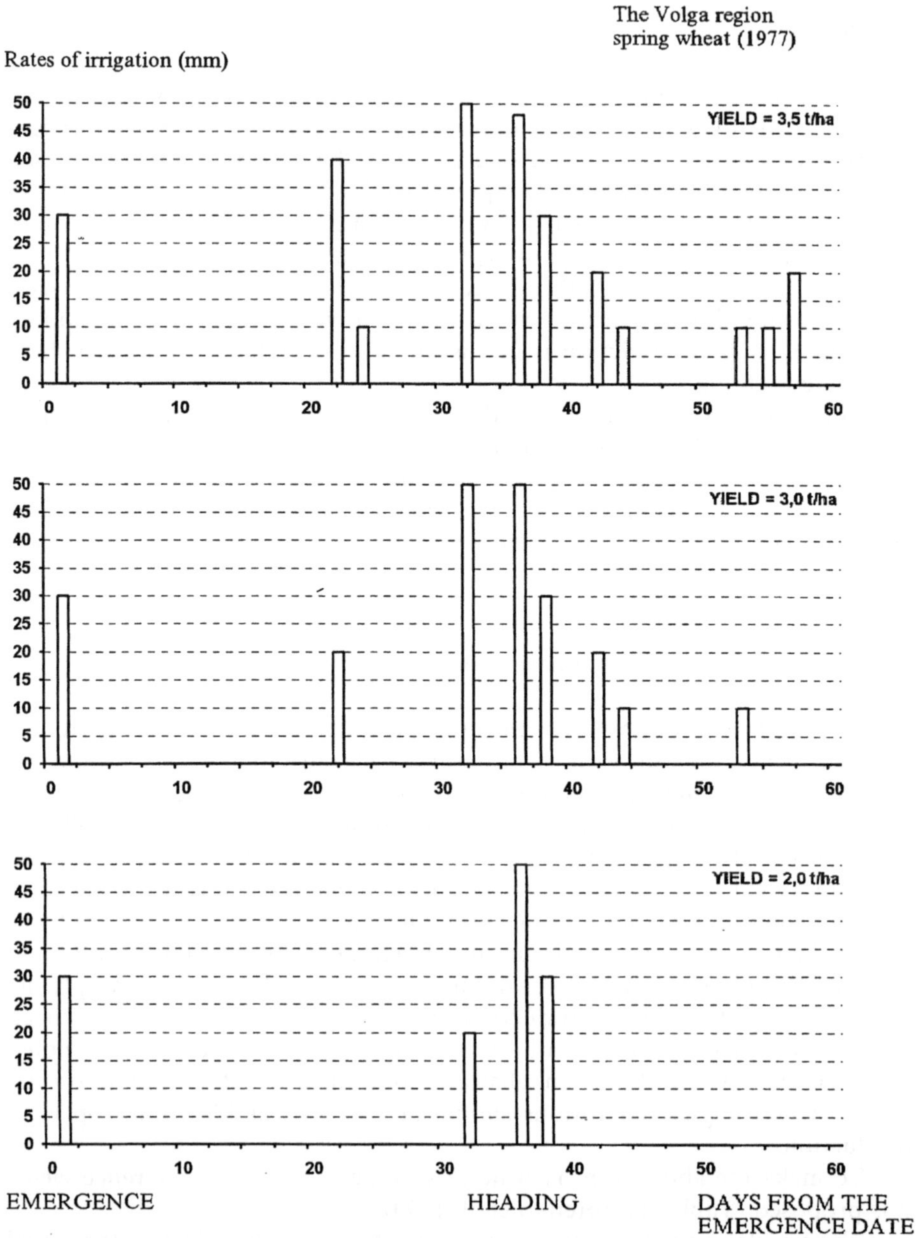

Fig. 5. Plots of the minimal irrigations (dates and rates) needed to obtain the desired yield

Russia. Thus, optimization of the water regime would raise the production by 1 $t\,ha^{-1}$ (SDY – NY), that of N supply by 0.2 $t\,ha^{-1}$ (CDY – NY), and both by 1.3 $t\,ha^{-1}$ (MPY – NY). Note that the resultant increase due to the integrated optimization of the water and nutrition regime would be 0.1 $t\,ha^{-1}$ greater than

Table 3. A proposed system of reference yields

Regimes assigned	Water	Optimum	Actual	Optimum	Actual
	Nutrition	Optimum	Optimum	Actual	Actual
Plant production levels		Maximum possible (MPY)	Climatically determined (CDY)	Soil-determined (SDY)	Nominal (NY)

Table 4. Results of calculating four levels of spring wheat yield for the Krasnodar Region, Russia

Reference yield	NY	SDY	CDY	MPY
Yield ($t ha^{-1}$)	2.6	3.6	2.8	3.9

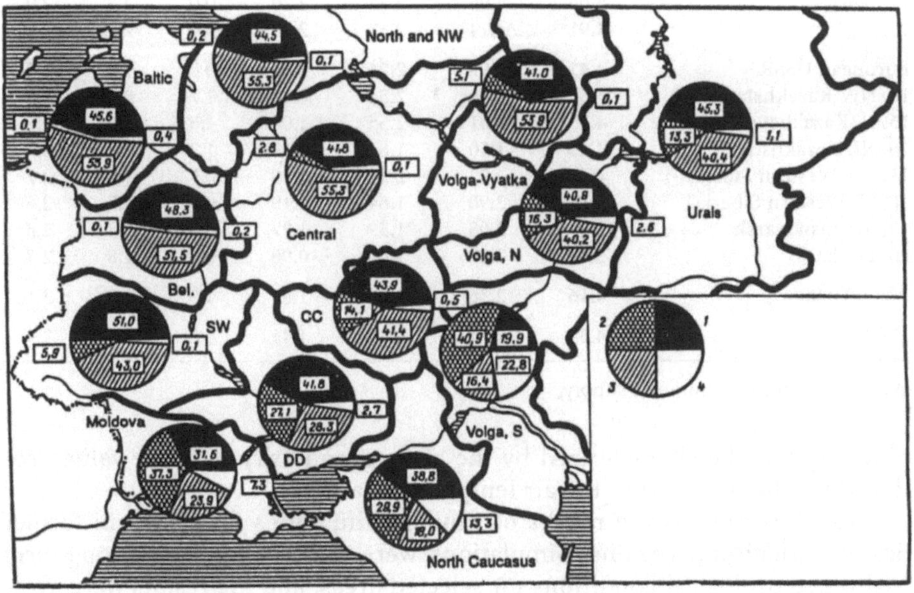

Fig. 6. Degree of utilization (*1*) and possibilities of transformation of soil-climatic resources of the European former USSR (in % of radiation-thermal potential, MPY) because of (*2*) optimization of the water regime; (*3*) optimization of mineral nutrition (fertilization); (*4*) additional effect from a combination of these factors

the sum total of the effects of separate optimizations, which may be attributed to the interaction of the simulated factors.

This conclusion was comfirmed by the data, simulated in a similar manner, on the extent of using the soil and climatic resources of the area: its bioclimatic potential could be significantly raised through optimization of the water regime and/or mineral sypply (cf. Fig. 6). In this case the bioclimatic

Table 5. The simulated current values of potential yields of spring wheat and land improvement response to chemicals and/or water management practices for the former USSR area

	Potential yields, $t\,ha^{-1}$				Yield increase with optimization, $t\,ha^{-1}$		
Economic regions	MPY	CDY	SDY	NY	W	N	W, N
1. North, northwest	3.91	3.82	1.20	1.17	0.0	2.6	2.7
2. Central	4.16	3.66	1.53	1.41	0.1	2.3	2.8
3. Volga-Vyatka	3.92	3.22	1.41	1.30	0.1	1.9	2.6
4. Central Black Soil	4.27	2.73	1.94	1.29	0.6	1.4	3.0
5. North Volga	4.23	2.44	1.76	1.14	0.6	1.3	3.1
6. South Volga	3.75	1.27	1.99	0.81	1.2	0.5	3.0
7. N. Caucasus	3.84	2.12	2.52	1.66	0.8	0.5	2.2
8. Ural	4.46	2.64	2.14	1.31	0.8	1.3	3.1
9. SW Ukraine	4.96	4.22	2.62	2.32	0.3	1.9	2.6
10. Donets-Dnieper (Ukr.)	4.87	2.79	3.27	1.98	1.3	0.8	2.9
11. S. Ukraine, Moldova	5.49	2.86	3.51	1.95	1.6	0.9	3.6
12. Baltic	5.06	4.96	2.22	2.20	0.0	2.8	2.9
13. Byelorussia	4.91	4.74	2.33	2.33	0.0	2.4	2.6
European USSR	4.41	2.94	2.24	1.55	0.7	1.4	2.9
14. NW Kazakhstan	3.52	0.89	2.34	0.71	1.7	0.2	2.8
15. N. Kazakhstan	4.73	2.01	2.55	1.04	1.5	1.0	3.7
16. NE Kazakhstan	4.40	1.50	1.84	0.69	1.2	0.8	3.7
17. SW Western Siberia	4.10	2.92	2.11	1.39	0.7	1.5	2.7
18. SE Western Siberia	3.81	2.90	1.64	1.19	0.5	1.7	2.6
19. S. Krasnoyarsk	3.87	3.08	1.59	1.27	0.3	1.8	2.6
20. Far East	3.63	2.77	1.40	0.98	0.4	1.8	2.7
Asian USSR	4.16	2.36	2.02	1.07	1.0	1.3	3.1
USSR	4.34	2.77	2.18	1.41	0.8	1.4	2.9

W, Water regime; N, mineral supply.

potential may be characterized by the aboveground dry biomass value produced for the period with the air temperature above 5 °C.

Table 5 contains the results of similar simulations for the main former Soviet agricultural regions. Simulations were made using mean long-term hydrometeorological conditions for selected areas, and the results were averaged over 20 economic regions. The simulation system Crop-Weather (Sirotenko et al. 1990), designed under the author's supervision, was used. The tabulated data deserve individual analysis. For illustration, Fig. 6 shows how farm productivity can be increased by changing agricultural practices in different geographical regions. Note the large reserve increases in yield and that huge resources are in store which could effect higher levels of farm productivity: optimization of mineral nutrition is able to double crop yields (from 1.4 to 2.8 $t\,ha^{-1}$), optimization of water regime can increase crop yields by 1.5 times (from 1.4 to 2.2 $t\,ha^{-1}$), and, finally, integrated optimization could triple crop yields.

4.2.5 Estimating Climate Change Effects

The task of estimating the future climate change effects on agriculture appears to be complex with many aspects. Despite a landslide increase in the number of publications and several reviews (Warrick et al. 1989; Parry 1990), this problem has not been adequately solved yet due to the weakness and excessive

Table 6. Forecast of spring cereal crop yields as a percentage of present-day values under GCM scenarios and the paleoscenario (Eemian Interglacial)

Economic regions	Modern yields t ha^{-1}	GFDL $CO_2 \times 1$ (%)	GFDL $CO_2 \times 2$ (%)	UKMO $CO_2 \times 1$ (%)	UKMO $CO_2 \times 2$ (%)	EI $CO_2 \times 1$ (%)	EI $CO_2 \times 2$ (%)
1. North, northwest	1.17	−25	20	−21	30	0	65
2. Central	1.41	−34	0	−30	9	0	53
3. Volga-Vyatka	1.30	−38	−5	−26	13	0	50
4. Central Black Soil	1.29	−46	−16	−36	−3	3	51
5. North Volga	1.14	−48	−20	−33	0	5	50
6. South Volga	0.81	−38	0	−31	10	17	79
7. N. Caucasus	1.66	−36	−6	−28	1	43	90
8. Ural	1.31	−48	−20	−37	−4	8	54
9. SW Ukraine	2.32	−41	−17	−36	−9	−2	39
10. Donets-Dnieper (Ukr.)	1.98	−41	−16	−32	−5	30	69
11. S. Ukraine, Moldova	1.95	−28	3	−21	11	53	106
12. Baltic	2.20	−23	10	−21	17	−8	38
13. Byelorussia	2.33	−36	−7	−33	−1	−4	37
European USSR	1.55	−39	−9	−31	2	13	61
14. NW Kazakhstan	0.71	−48	−11	−46	−6	28	93
15. N. Kazakhstan	1.04	−43	−10	−42	−8	18	72
16. NE Kazakhstan	0.69	−43	−5	−54	−22	8	69
17. SW Western Siberia	1.39	−35	−2	−40	−7	9	52
18. SE Western Siberia	1.19	−40	−4	−47	−15	3	55
19. S. Krasnoyarsk	1.27	−36	−2	−37	−5	−5	38
20. Far East	0.98	−34	5	−36	2	−2	55
Asian USSR	1.07	−40	−5	−43	−9	9	62
USSR	1.41	−40	−8	−35	−1	12	62

GFDL scenario, summertime temperatures increase by an average of 5.2 °C, rainfall increases by 15%.
UKMO scenario, summertime temperatures increase by an average of 4.2 °C, rainfall increase by 25%.
EI scenario, Eemian Interglacial, climate wetter and milder than at present.
$CO_2 \times 1$, CO_2 levels at present-day values.
$CO_2 \times 2$, CO_2 levels double present-day values.

empiricism of the analytical tools being used. In our opinion, sustained progress in this field might result from the development of national and, later, global simulation systems based on crop-production models and large data bases. Such a system for the former USSR's area was created in the Institute of Agricultural Meteorology, Obninsk, Russia (Sirotenko et al. 1990). The system's simulation fragment is presented in Table 6. To interpret the simulation results, note that the scenarios GFDL and UKMO assume that summertime temperatures increase in the given area by 5.2 and 4.2 °C on average, respectively, whereas the paleoclimatic scenario which corresponds to the Eemian Interglacial (EI) optimum assumes the summertime temperatures to drop by 0.5 °C (with a winter-time increase by 7.2 °C) and the annual rainfall to be by 120 mm larger. Under the scenarios GFDL and UKMO the rainfall will be, respectively, 15 and 25% greater, although that cannot compensate for the greater losses of water through evaporation. In response to the changes in climatic conditions crop yields are going to change too – under the scenarios GFDL and UKMO they will be lower by 39 and 31% respectively, and 13% higher for EI since the latter assumes a milder and wetter climate. Direct effects of the CO_2 increase may be assumed to reduce the losses attributed to the worsening climatic conditions.

5 The Future of Crop Production Simulation

5.1 Can Simulation Progress?

The question of developing criteria for model evaluation and comparison arises as a result of the exponentially growing number of mathematical simulation studies of crop production. In the Russian language literature this issue was first addressed by Bichele et al. (1980a) who presented a table evaluating 18 models by 23 indicators using a five-point appraisal approach. The inventory of dynamic models was set out by Sirotenko (1981) with an extensive model classification. Giving their due to the above attempts at generalization of what had been achieved, one should note the inherent limitations of this approach since conventional scientific publications do not seem to be able to provide for an in-depth objective outlook in terms of model quality and development.

The wide applicability of mathematical techniques and computers has turned them into a kind of fetish. A number of recent papers have voiced concern about an oversimplified reduction approach to the simulation of complex systems. Nalimov (1983) states, for example, that "construction of models in ecology with the help of differential equations looks like a parody of physics". This may be an overstated argument, although the call to critically analyze the fundamentals of ecological simulation seems to be highly topical. Nalimov's opinion is that simulation does actually pool together the efforts of different specialists; however, this has not yet resulted in any novel insight into the essence of ecological phenomena. Longer computations associated with a formal complication of models, despite the hope some authors pin on it, cannot by itself bring about a better understanding of the studied nonlinear problems.

A tendency has taken shape, viz. that of littering science with "prefabricated" models which can substitute each other but which have not been checked against independent actual data. This seems to be a common limitation for most models. Often, model validity is tested using too little information which cannot reflect year-to-year variability of weather conditions. Thus Yusupov (1990) compares simulated and actual cotton yields for 5 years. These data might appear adequate for a preliminary conclusion on the model quality. However, it was found that the interannual variability of actual crop yields under irrigated conditions does not exceed observation errors. It is clear that given such materials, it is not possible to make a valid check of a model, i.e., a specially planned experiment is needed – an issue which is discussed below.

The criticism of crop-production mathematical simulation may be summed up as follows: most models being created lack either the originality necessary to make them of interest for theory, or the specificity necessary for them to be useful in practice. Einstein's quotation is well known; that experiment will remain a decisive criterion of the physical suitability of a mathematical statement, even though the creative beginning belongs to mathematics. Thus human reason initially has to construct, independently, the required forms, before we will be able to find them in nature. In simulating crop production it is not often that mathematics plays the role of "creative beginning".

Frequently, the models proposed are just an impoverished formalization of the ideas put forward by biologists a long time ago and well described in textbooks. Summing up, these problems prevent us from giving a unique positive answer to the question asked in the section heading.

Nevertheless, examples can be given for which the beginning of creation may be attributed to the ideas originating from mathematics. The most impressive example is the principle of maximum productivity of photosynthesizing systems, according to which the plant adaptation in the plant community is aimed at achieving the potential CO_2 gas exchange under the given ambient conditions. Proceeding from this principle, Tooming suggested the notion of IDA (Irradiation Density of Adaptation) which determines competitive interactions in plant communities. The introduction of IDA stimulated physiologists to conduct corresponding experiments. It was shown that a low IDA level is a prerequisite for both photosynthesis and plant growth in the community, whereas at high IDA, plant communities would not be likely to appear on Earth. The spectrum of ideas, whose starting point was the variation problem of plant-community potential productivity stimulated the development of a new line of investigation.

Another argument to be voiced in discussing the issue of mathematical simulation significance is the explanation of the Issyk-Kul phenomenon. In the former USSR area, at the Prszewalski cultivar testing grounds near Lake Issyk-

Kul, 1718 m above the sea, record breaking cereal yields (ca. 11–12 t ha^{-1}) have been common. As reasons for this phenomenon, higher natural radiation, local soil composition, etc. were named. In 1986, spring wheat yields were 8.5 t ha^{-1}, whereas the simulated yield data computed using the Crop-Weather-3 model (which had not been adjusted) was 7.8 t ha^{-1}. This is very good agreement, particularly when one considers that prior to this case simulated yields, even for optimum soil-moisture and mineral-supply conditions, had not exceeded 5.5 t ha^{-1}. The high yields at Lake Issyk-Kul thus may be accounted for by purely meteorological reasons: moderate temperatures, high solar radiation, and a long growing period. With the unlimited soil water of 1986, calculated yields amounted to 12 t ha^{-1} which constitutes a record-breaking result.

5.2 Experimental Support to Simulation

It is a commonly opinion held that experimental support is the main cause retarding the construction of dynamic models. Nevertheless there has been little progress in dealing with the problem. In this respect only two investigations appear to be interesting enough: Poluektov (1991) observes that the currently predominant concept of a field test, the formation of which was infuenced by R. Fisher's studies, is statically,rather than dynamically, oriented. The main predicament in studing agroecosystems is associated with their adaptive characteristics, i.e., the set of the processes which enable plant survival under stress. Therefore, states Poluektov, the primary task should be the selection of a test version which would correspond to conditions resulting in the highest productivity. Then, by the consecutive imposition of stress impacts, one should study their effects at different plant development stages. The proposed scheme includes nine field test versions with different combinations of water stress in the growing and reproductive development periods, coupled to three grades of N stress (nil, low, high). This approach however doesn't take into account the fact that many ambient factors (air temperature, solar radiation, etc.) cannot be field-controlled, i.e., for most cases what we have at hand is a passive, rather than a planned experiment.

Assume that $x_1, x_2, \ldots x_n$ are the meteorological factors influencing crop yields y. Sirotenko and Varcheva (1992) showed that climate, by annually generating a new line in the matrix

$$X = \begin{bmatrix} x_{11}, & x_{12}, & \ldots, & x_{1n} \\ x_{21}, & x_{22}, & \ldots, & x_{2n} \\ \ldots, & \ldots, & \ldots, & \ldots \\ x_{N1}, & x_{N2}, & \ldots, & x_{Nn} \end{bmatrix}$$

thus assigns a potential test plan for any site on the Earth's surface (where N is the number of a current year of observation). Following the modern theory of

Table 7. Mean estimated regression coefficients for air temperature and growing season rainfall for the five best and five worst plans selected from the sliding 10-year observation series for 1951–1987

	det (X*X)	Air temperature $\bar{\theta}_1$	Growing season rainfall $\bar{\theta}_2$
Best plans	3714	–0.70	0.07
Worst plans	907	1.80	0.03

experimenthal planning, the volume of θ the regression coefficient scattering ellipsoid θ may be accepted as an effectiveness criterion of the plan X

$$\hat{\theta} = (X * X)^{-1} X * Y, \tag{34}$$

which is inversely proportional to the covariant matrix determinant C = (X*X), The superscript * denotes the matrix transposition operation. Therefore, the test plan will be optimal (such plans are referred to as D-optimal) if the determinant C is maximum.

Consider an example. For the Yershov testing ground (dry area at the Volga) for 37 years (1951–1987), the following estimated regression coefficients were obtained, relating spring-wheat yields with air temperature and growing season rainfall:

$$\hat{\theta}_1 = -0.91 \text{ quintal ha}^{-1}{}^{\circ}\text{C}^{-1}, \quad \hat{\theta}_2 = 0.08 \text{ quintal ha}^{-1}\text{ mm}^{-1}.$$

For comparison, Table 7 gives the mean estimates of the same coefficients for the five "best" and five "worst" years selected from a host of plans for the sliding 10-year observation series for 1951–1987.

The estimated regression coefficients for temperature under the "worst" plans have a wrong sign, and are strongly underrated for rainfall.

Figure 7 demonstrates the dynamics of the estimated detC for 7-year plans for nearly 100 years. The figure shows that the year quality may change dramatically if a year is substituted, varying widely – by more than 2 orders of magnitude. Based on the simulation results it may be concluded that success in identifying the parameters of both static and dynamic models depends strongly on the composition of hydrometeorological conditions in those years which made up the sample. Rather often, one can find periods (cf. Fig. 7) for which meteorological characteristics detC $\approx$ 0 thus making the parameter identification absolutely invalid. May this not be an explanation why some models are insensitive and inadequate, especially for crop-yield forecasting purposes? The above approach enables an objective evaluation of information content in the observation materials available to a model designer, by comparing the implemented random plan with the D-optimal one. In addition, the designer has an opportunity to compute the combinations $x_1, x_2, \ldots, x_n$ maximizing detC thus assigning the test conditions N + 1, N + 2, etc. to fill gaps in

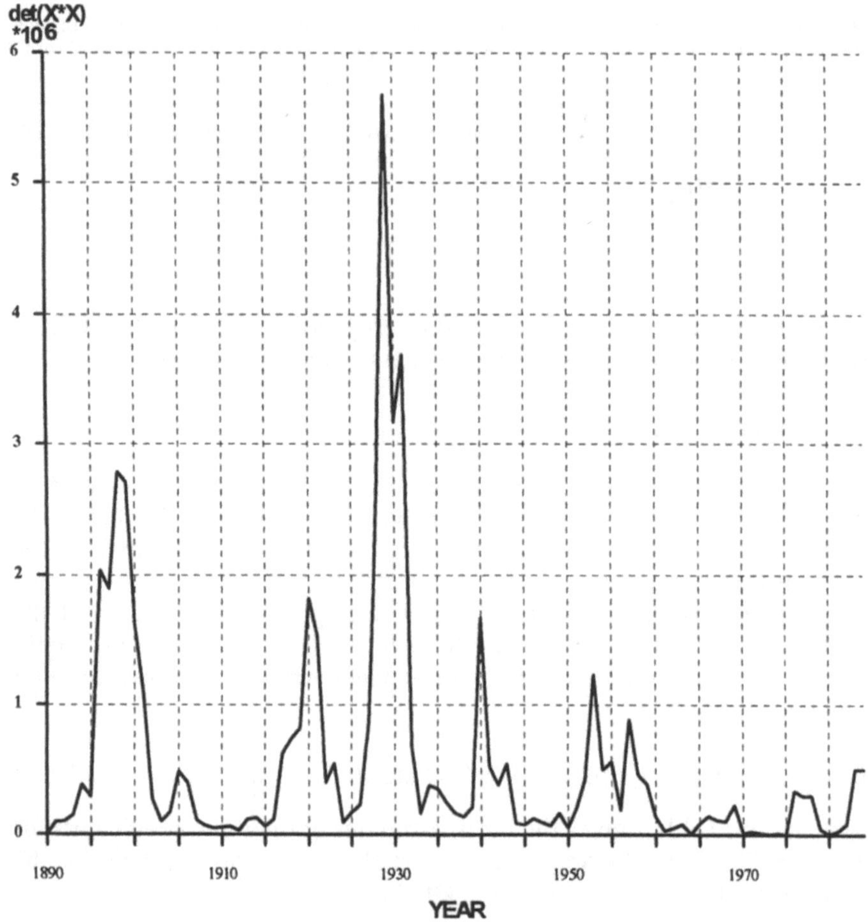

Fig. 7. Dynamics of the determinant of a four-dimensional covariance matrix *det* (X^*X) calculated from 7-year-long sliding periods ($x = x_1, x_2, x_3, x_4$) are monthly values of air temperature and rainfall for June–July

the available data. In other words, the possibility arises of combining a passive experiment (i.e., conventional field test) with an active one implemented by controlling plant-water supply and nutrition (cf. preceding sections).

The approach discussed may provide a basis for optimizing the planning of geographic experiments and monitoring systems, wherein the spatial variability of climate should substitute for its time variability.

5.3 Onward to Next-Generation Models

This subsection, summing up, deals with the analysis of development trends in the mathematical simulation of crop production. In this connection, let us first of all address the unsolved problems found in the physical aspects of the issue. These include: (1) the well-known problem of closing the hydrodynamics equation set describing the canopy's turbulent regime; (2) the problem of correctly representing the soil's internal evaporation and root-water uptake; (3) the problem of assigning internal boundary conditions in the integrated problem of energy and mass exchange of the plant habitat media, etc. It is safe to say, however, that biological aspects, rather than physical ones, limit the development of crop-simulation mathematical models. Practically all of the biological problems relate to performance of the plant community as a kind of primitive biochemical reactor. The primacy of metabolism kinetics is abandoned only when attempting to simulate plant developmental morphology which in no way proceeds from crop biomass dynamics. The laws of nature do not fit into a rigid scheme – the faster the metabolic rate, the faster the developmental morphology, and the faster the plant development rates. The hypothesis that as soon as a simulated concentration reaches some critical value, new plant organs are initiated may be more flexible. However, both this assumption and the concept of reserve pool by and large cannot embrace all the diversity of the plant autoregulation system. Therefore of fundamental importance for the development of mathematical simulations seem to be the efforts which justify the existence of a second, metabolically independent, program in plants – the developmental morphology and/or ontogenesis program (Mokronosov 1983). In this connection the studies made by Abashina and Nemchenko (1990) are of considerable interest – they suggested a model of hormonal regulation of plant growth. It may be possible that calculation of plant hormone regulatory functions will provide an additional basis for the ontogenesis program in the next generation models. Interesting suggestions on plant growth and development were made by Dobrachev (1982), Kan (1989), Gulyaev (1983), Moldau (1984) and others. In these studies, the use of the principle of optimum appears to be very promising (its origin may be traced back to Darwin's theory of natural selection). However, in practice this principle has been only applied to writing the statement of a local optimization procedure. Thus, according to Tarko and Sudullaev (1984), newly formed biomass is partitioned among the cotton plant leaves, stems, and roots so that the maximum biomass accretion rate should be guaranteed at the next time step, provided there are no changes in the environment. The optimum behavior should result, it seems, not at the expense of "fitting" such procedures into the model, but from formalizing the hypothesis which would embody the behavioral mechanism.

Currently it seems impossible to foretell the shape of mathematical structures in which the plant growth and development programs will be implemented in the next generation models. It will suffice to state only, that they will look as different from present models as warm-blooded mammals from cold-blooded reptiles. Thanks to the calculation of the adaptive aspect of plant growth and development, model behavior must become much more complex and "live" without essentially complicating the models themselves.

In addition to these new approaches, there are other promising advances in model development. For example, Kan (1992) showed that nonhomogeneity of a plant population and microecology were significant for simulating crop production. When calculated, these aspects can make the models look stochastic and will open new vistas for agroecosystem studies.

Drawing an analogy with the development of modern physics, the most impressive event in crop simulation must be the "large unification" of models of soil fertility and production of vitally (most) important crops, aimed at establishing computerized systems which would continuonsly simulate the behavior of agronomic systems for as much as 1–2 (or more) decades ahead for different cropping patterns, land amelioration, and chemical application strategies, taking into account the future changes in climate and chemical composition of the atmosphere.

In conclusion I consider it essential to add that after submitting this study for printing, the monograph of A.P. Boyko "Research of the hydrometeorological regime of plant envi ronments by means of numerical methods" (St. Petersburg, 1993) was published.

The research has been conducted on the basis of same results of numerical experiments made with the help of the models of energy and mass exchange in the system "soil–plant–atmosphere". The models with distributed and lumped parameters are discussed. Considerable attention in the monograph is paid to the development and analysis of the economic numerical techniques and algorithms.

References

Abashina EV (1979) A method to account for nitrogen nutrition of vegetation in dynamic models designed for assessment of agrometeorological conditions for yield of cereal crops. Collect Works IEM (Institute of Experimental Meteorology) 13 (91):101–119

Abashina EV, Nemcnenko OA (1990) Simulation submodel for metabolism of phytohormones. Collect Works ARRIAM (All-Russian Research Institute of Agricultural Meteorology) 26:53–58

Andropova TM, Dobretsov AN (1977) Prediction of spring wheat yield under the conditions of the Krasnoyarsky kray. Chem Agric 11:64–66

Asaturov ML (1969) An addition to the theory of the water–heat regime of vegetation cover. Collect Works Main Geophys Observ 249:65–71

Baranovsky PM, Galyamin EP, Filimonov MS et al. (1978) Issues of management of crop yield under irrigation. All-Russian Inst Irrigated Farming, Yolgograd, 128 pp

Bichele S. Moldau H, Ross Y (1980a) Mathematical modelling of plant transpiration and photosynthesis under soil moisture stress. Hydrometeoizdat, Leningrad, 223 p

Bichele I, Moldau H, Ross Y (1980b) A submodel of assimilation and growth of plants under water deficit. Tartu, Acad Sci Estonia A-5:21

Boiko AP, Sirotenko OD (1985) Modelling of energy and mass exchange of soil-plant-atmosphere system under deficit of soil moisture. Collect Works ARRIAM, Hydrometeoizdat, Leningrad 10:3–26

Boiko AP, Safarov SG, Sirotenko OD (1990) Modelling of daily variations in total evaporation from agricultural crops. Collect Works ARRIAM 26:22–33

Bondarenko NF, Zhukovsky EE, Mushkin IG (1982) Modelling productivity of agroecosystems. Hydrometeoizdat, Leningrad, 262 pp
Budagovsky AI (1964) Evaporation of soil moisture. Nauka, Moscow, 243 pp
Budagovsky AI (1980) Evaporation of soil waters. In: Nerpin SV (ed) Physics of soil waters. Nauka, Moscow, pp 13–95
Budagovsky AI, Lozinskaya FA (1976a) A system of equations for heat and moisture exchange in vegetation cover. Water Resour 1:78–94
Budagovsky AI, Lozinskaya EA (1976b) Theory of total evaporation and its applications. Water Resour 2:34–55
Budagovsky AI, Nichiporovich AA, Ross YK (1964) Quantitative theory of photosynthesis and its use for solving scientific and applied problems of physical geography. Izv USSR Acad Sci Ser Geogr 6:13–27
Budyko MI (1956) Heat balance at the Earth's surface. Hydrometeoizdat, Leningrad
Budyko MI (1964) An addition to the theory of the influence of climate factors on photosynthesis. Proc USSR Acad Sci 158 (N2):331–334
Budyko MI, Gandin LS (1964) About taking into account characteristics of physics of the atmosphere in agrometeorological research. Meteorol Hydrol 11:3–11
Budyko MI, Gandin LS (1968) An addition to the theory of heat regime of vegetation cover. Collect Works Main Geophys Observ 229:69–73
Chudnovsky AF (1948) Physics of heat-exchange in soil. Gostechizdat, Moskow
Chudnovsky AF (1976) Thermal physics of soils. Nauka, Moscow, 352 pp
Chuvashina NB (1983) Development of the method to calculate dynamics of content of mineral forms of nitrogen in soil under agricultural crops on the basis of mathematical modelling. Author's abstract of dissertation for Candidate in agriculture, Agrophysicol Res Inst, Leningrad, 19 pp
Dandaron ZD (1974a) Approximate calculation of evaporation from soil with a dry layer. Meteorol Hydrol 2:166–172
Dandaron ZD (1974b) Effect of negative change in water evaporation from soil with a dry layer. Meteorol Hydrol 7:66–74
Daus YV (1991) Development of a mathematical model for water-heat regime and productivity of lucerne for the purposes of agrometeorological provision of agriculture. Author's abstract of dissertation for Candidate of Geographic Science, OHMI-Hydrometeorol Inst, Odessa, 16 pp
Denisov YM (1968) Transfer of moisture and heat in soil. Collect works SANIGMI 39 (54):3–19
Denisov YM (1978) Mathematical model for moisture, heat and salt transfer in soils. Meteorol Hydrol 3:71–79
Derkach LN (1978) Impact of meteorological conditions on the yield of new variaties of winter wheat. Collect Works USSR Hydrometeocentre 193:99–108
De Witt CT, Brouwer R, Penning de Vries FWT (1971) A dynamic model of plant and crop growth. In: Wareing PP, Cooper JP (eds) Potential Crop Production. A case study. Heinemann Education Books, London, pp 117–142
Direktorenko VM (1989) Calculation methods for assessing agrometeorological conditions for peacrop and programming its yield in Moldavia. Author's abstract of dissertation for Candidate of Geographic Science, OHMI-Hydrometerol Inst, Odessa, 14 pp
Dmitrenko VP (1976) Assessment of impact of air temperature Omitrenko and precipitation on yield formation of major cereal crops. Guidebook. Hydrometeoizdat, Leningrad, 48 pp
Dobrachev YP (1982) Structure of autoregulatory model for plant growth AURORA. Dynamic modelling in agrometeorology. Hydrometeoizdat, Leningrad, pp 17–27
Dolgy-Trach VA (1987) Application of model WEATHER-YIELD-3 to calculate winter wheat yield. Collect Works ARRIAM 21:54–66
Dubov AS, Bykova LP, Marunich SV (1987) Turbulence in vegetation cover. Hydrometeoizdat, Leningrad, 182 pp
Dukov R (1987) Analysis of multi-year variations in sunflower production and the qualitative estimation of agrometeorological conditions of its yield with the help of a dynamical model. Author's Abstr, BAH, Sofia, 31 pp

Fedyushina DP (1980) A method for prediction of spring wheat yield for the western Kazakhstan. Collect Works KazNIGMI 70:12–22
Fedyushina DP, Zavyalova LA (1980) A method for prediction of spring wheat yield for the Turgaiskaya region. Collect Works KazNIGMI 70:23–27
Fisher RA (1925) The influence of rainfall on the yield of wheat at Rothamsted. Philos Trans R Soc Ser B 213:89–142
Galyamin EP (1980) Optimization of operational distribution of water resources in irrigation. Hydrometeoizdat, Leningrad, 217 pp
Galyamin EP, Siptits OS (1977) A dynamic model for the productive process of maize and its applications for optimization of water regime. Collect Works IEM 8:114–123
Globus AM (1987) Soil - hydrophysical provision for agroecological mathematical models - Hydrometeoizdat, Leningrad, 427 pp
Gulyaev VI (ed) (1983) Photosynthesis and production process. Naukova dumka, Kiev, 14 pp
Gulyaev BI, Ilyashek EM, Mitrofanov BA (1983) Photosynthesis and production process. Naukova dumka, Kiev, 144 pp
Ivanov LA (1996) Light and moisture in the life of our wood species. V Timiryazev reading. Izv USSR Acad Sci Moscow, 60 pp
Kan NA (1981) System of simulation modelling of agrocenoses. In: Stolyarov AI (ed) High-intensity use of irrigated lands. Novocherkassk, South Institute of Hydrotechniques and heclamation, pp 149–154
Kan NA (1989) Problems in the methodology of simulation modelling of agrocenoses. Inst Soil Sci Photosynthi, Pushchino, 31 pp
Kan NA (1992) Orginogenesis and morphological structure of cereal crops in mathematical models of agroecosystems. Hydrometeoizdat, St. Petersburg, 133 pp
Kan NA, Chuprinin VG, Sorochan KI (1984) Mathematical support for the system of simulation modelling of agrocenoses. In: Stolyarov AI (ed) Regimes of irrigation, water supply and agrotechnology of agricultural crops on the Northern Caucasus. Novocherkassk, South Institute of Hydrotechniques and heclamation, pp 67–80
Kharchenko et al. (1970) Recommendations for calculating evaporation from natural lands and agricultural fields by the heat-water balance method. Hydrometeoizdat, Leningrad, 65 pp
Kharchevskaya NF (1986) Calculation of tomato yields using a dynamic model "weather-yield". Collect Works ARRIAM 21:48–54
Khitrovo EV, Kuperman IA (1980) Influence of mineral nutrition on the relation between characteristics of respiration gas exchange of plants calculated on the basis of protein nitrogen and dry biomass. Physiol Biochem Cult Plants 12 (3):269–279
Kirilicheva KV (1980) Assessment of the effect of temperature and precipitation on the yield of spring wheat. Collect Works USSR Hydrometeocentre 124:39–46
Kozhushnyan EA, Uzyanova VP (1966) On the issue of prediction of spring wheat yield. Collect Works Sverdlovskaya Hydrometeorol Serv 5:27–32
Kresnyankaya VP (1977) Dependence of spring wheat yield on agrometeorological conditions of the growing period in the Primorsky kray. Collect Works DVNIGMI 59:9–11
Kuperman IA (1972) About nitrogen and light uptake ratio control for higher terrestrial plants. Physiological mechanisms of adaptation and resistence of plants. Nauka, Norosibirsk, pp 5–34
Kuperman IA, Khitrovo EV (1977) Respiration gas exchange as an element of the production process of plants. Nauka, Novosibirsk, 184 pp
Kuperman IA, Khitrovo EV, Maslova IY (1983) About investigation of causes of plant productivity reduction at nitrogen excess. Physiol Biochem Cult Plants 15 (N5):419–426
Kurtener DA, Chudnovsky AF (1969) Calculation and control of thermal regime in open and protected ground. Hydrometeoizdat, Leningrad, 299 pp
Laisk AH (1977) Kinetics of photosynthesis and photorespiration C_3 plants. Nauka, Moscow, 195 pp

Maslovaskaya AD (1974) A method of long-term agrometeorological prediction of republican averaged yield of winter wheat in Kazakhstan. Collect Works KazNIGMI 47:117-130

McCree KJ (1971) An equation for the rate of respiration of white clover plants grown under controlled conditions. In: Setlik I (ed) Production and measurement of photosynthetic productivity. PUDOC, Wageningen, pp 221–229

Menzhulin GV (1970) An addition to the methodology of calculating meteorological regime of vegetation cover. Meteorol Hydrol 2:92–99

Menzhulin GV (1973) An addition to the theory of stationary meteorological regime in vegetation cover. Collect Works Main Geophys Observ 227:20–28

Menzhulin GV (1974) Modelling of meteorological regime of vegetation cover. Collect Works Main Geophys Observ 318:5–34

Menzhulin GV (1977) Modelling moisture exchange and transpiration in the soil-plant-atmosphere system. Collect Works State Hydrol Inst 247:36–44

Mescherskaya AV, Boldyreva NA, Shavaeva ND (1982) Mean regional reserves of productive moisture in soil and hight of snow cover. Statistical analysis and examples of use. Hydrometeoizdat, Leningrad, 243 pp

Mkrtchyan RS, Arustamyan SA (1975) Methodological guidance on long-term agrometeorological prediction of winter wheat yield in the Armyanskaya SSR. ZakNIGMI, Erevan, 27 pp

Mokronosov AT (1982) Donor-acceptor relations in plant onthogenesis. In: Nichiporovich AA (ed) Physiology of photosynthesis. Nauka, Moscow, pp 235–250

Mokronosov AT (1983) Photosynthesis function and integrity of a plant organism. Timiryazev readings XLII USSR Acad Sci, Moscow, Nauka, 63 pp

Moldau AA (1973) Influence of water deficit on stomatal resistance (Mathematical model). Izv Acad Sci Estonian SSR Ser Biol 22 (N4):348–357

Moldau AA (1977) Stomata - universal regulators of photosynthesis. Physiol Plant 24 (5):969–975

Moldau AA (1984) Autoregulation of production process of plants at water deficit. Author's abstract of dissertation for the degree of Doctor of biological sciences. Inst Plant Physiol, Moscow, 44 pp

Nalimov VV (1983) Analysis of principles of environmental forecast. Quest Philos 1:108–117

Novak V (1980) Movement of water in the soil during evaporation. In: Nerpin SV (ed) Physics of soil waters. Nauka, Moscow

Obukhov VM (1949) Yield and meteorlogical factors. Gosplanizdat, Moscow, 316 pp

Pavlova VN, Sirotenko OD (1985) About use of dynamic models for assessing the effect of possible changes and climatic variations on yield of agricultural crops. Collect Works AURIAM 10:81–90

Pavlova KK, Morozova NS, Lavrov SA (1986) An addition to the definition of moisture transfer parameters of frozen soils. Proc State Hydrometeorol Inst 308

Parry M (1990) Climate change and world agriculture. Earthscan, London, 157 pp

Polevoy AN (1983) Theory and calculation of crop productivity. Hydrometeoizdat, Leningrad, 175 pp

Polevoy AN, Myzina TI (1976) Methodological guidance on compiling an agrometeorological forecast of region-averaged yield of spring barley in the Nechernozemnyz zone of the European part of USSR. Hydrometeoizdat, Moscow, 39 pp

Poluektov RA (1991) Agroecosystem dynamics simulation. Hydrometeoizdat, Leningrad, 312 pp

Poluektov RA et al. (1984) A basic simulation model of wheat productivity. Wiss Z Humbolt-Univ Berl Math Naturwiss Reihe XXXIII:339–342

Ponomarev BP (1974) Methodological guidance on compiling agrometeorological forecasts of region-averaged yield of spring wheat in the steppe and forest-steppe zones of western Siberia. Hydrometeoizdat, Moscow, 174, 11

Ross YK (1966) An addition to the mathematical description of plant growth. Rep USSR Acad Sci 171 (N2):481–483

Ross YK (1967) A system of equations for quantitative growth of plants. In: Ross YK (ed) Phytoactinometric studies of vegetation cover. Tallin, Valgus, pp 64–88

Ross YK (1975) Radiation regime and architectonics of vegetation cover. Hydrometeoizdat, Leningrad, 342 pp

Ross YK, Bikhele ZN (1968) Calculation of photosynthesis of vegetation cover. In: Ross YK (ed) Photosynthesis and productivity of vegetation cover. Inst Phys Astron Estonian Acad Sci, Tartu, pp 75–110

Satyr' ZA (1973) Two-factor optimization model weather-yield. Collect Works IEM 3 (40): 109–114

Savvateev SP (1979) About a numerical model of water exchange in the system soil-plant-surface atmospheric layer. Collect Works State Hydrol Inst 257:96–101

Semikhatova OA (1980) Energy aspects of inegration of physiological processes in plants. Physiol Plant 27 (5):1005–1017

Semikhatova OA, Ivanova TI, Golovko TK (1979) Respiration for sustaining cell structure in arctic plants. Physiol Plant 26 (5):1093–1102

Sepp YV, Tooming HG (1982) Production process and actually possible yield of potato (dynamic model). Agric Biol 17 (1):89–97

Shchenina DN (1978) An addition to the methodology of prediction of district-averaged yield of spring wheat in the Kulundinsky zone of the south-east western Siberia. Collect Works West-Siberian Hydrometeorol Res Inst 32:3–11

Shlyachkova AA (1990) Problem of optimization of regime of agrocenosis phytomass collection. Collect Works AURIAM 26:97–103

Sirotenko OD (1969) Multidimensional regression analysis as a method of studying the relation of yield of agricultural crops and hydrometeorological factors. Meteorol Hydrol 12: 68–87

Sirotenko OD (1971) Component analysis in prognostic tasks of agrometeorology. Methodological guidelines. Hydrometeoizdat, Moscow, 53 pp

Sirotenko OD (1981) Mathematical modelling of the thermal and moisture regime and productivity of agroecosystems. Hydrometeoizdat, Leningrad, 167 pp

Sirotenko OD (1984) Methods to generalize the photosynthesis theory of productivity of agroecosystems. Rep VASHNIL 10:5–8

Sirotenko OD, Abashina EV (1985a) Mathematical model for growth and development of plants at the level of phytomers. Physiol Biochem Cult Plant 2:166–174

Sirotenko OD, Abashina EV (1985b) About selection of scenarios of meteorological conditions for prediction of yields by dynamic models WEATHER-YIELD. Collect Works AURIAM 10:61–73

Sirotenko OD, Boiko AP (1976) A dynamic model of agrocenosis. Collect Works IEM 8 (67): 12–36

Sirotenko OD, Mokievsky VM (1990) Parameterized model of water-heat regime of agroecosystem for calculation of gross evaporation in setting irrigation norms. Collect Works AURIAM 26:12–21

Sirotenko OD, Pavlova VN (1988) A stochastic model of climate for calculation of productivity of agroecosystems. Meteorol Hydrol 7:104–114

Sirotenko OD, Abashina EV, Pavlova VN (1985) A dynamic model WEATHER-YIELD for spring crops and its use for assessing agrometeorological conditions in the arid zone. Collect Works AURIAM 10:42–61

Sirotenko OD, Slavov NS, Valkov NG (1982) Water regime study of a dynamic maize model. Hydrol Meteorol XXXI (3):198, Sofia, pp 3–10

Sirotenko OD, Abashina EV, Shaakmedov SA (1987) Programming of yield using dynamic models. J Agric Sci 8 (369):55–59

Sirotenko OD, Velichko AA, Dolgy-Trach VA, Klimanov VA (1990) Global warming and the agroclimatic resources of the Russian plain. Izv USSR Acad Sci Ser Geogr 6:29–38

Sirotenko OD, Varcheva SE (1992) Use of climatological data in the optimization of long-term agricultural and environmental observations and experiments. Meteorol Hydrol 9:97–103

Svisyuk IV (1971) Prediction of region-averaged yield of spring barley and spring wheat. Meteorol Hydrol 7:80–86

Svisyuk IV (1991) Agrometeorological predictions, calculations, justifications. Hydrometeoizdat, Leningrad, 191 pp

Tarko AM, Sadullaev RI (1984) A mathematical model for growth and development of cotton with allowance for nitrogenous nutrition. Comput Cent USSR Acad Sci, Moscow 38 pp

Warrick RA, Gifford RM, Parry M (1989) Carbon dioxide, climate change and agriculture. In: Bolin B, Döös Bo R, Jäger J, Warrick R (eds) The greenhouse effect, climatic change, and ecosystems. Hydrometeoizdat, Leningrad, pp 432–512

Tooming H (1970) Mathematical description of net photosynthesis and adaptation processes in the photosynthetic apparatus of plant communities. In: Selik I (ed) Prediction and measurement of photosynthetic productivity. PUDOC, Wageningen, pp 103–114

Tooming HG (1976) Agrometeorological programming of crop yield. Socialistic land farming 22:1017–1020

Tooming HG (1977) Solar radiation and yield formation. Hydrometeoizdat, Leningrad, 199 pp

Tooming HG (1982) Method of reference yields. J Agric Sci 3 (306):89–94

Tooming HG (1984) Ecological principles of maximum productivity of crops. Hydrometeoizdat, Leningrad, 264 pp

Tsertsvadze SI, Meladze GG (1979) Prediction of republic-averaged yield of winter wheat. Collect Works ZakNIGMI 69 (75):90–94

Tupchiev VA, Nasyrov ZH, Gorbacher VA (eds) (1973) About asymptotics of solution of the diffusion coefficient for radial water flow to plant roots. Collect Works IEM 3 (40):50–56

Tyukalo TN (1980) A method for prediction of region-averaged yield of spring wheat in the Kemerovo region. Collect Works West-Siberian Hydrometeorol Res Inst 44:58–64

Ulanova ES (1963) A method for long-term prediction of agrometeorological conditions for formation of winter wheat yield. Meteorol Hydrol 11:12–30

Ulanova ES (1975) Agrometeorological conditions and yield of winter wheat. Hydrometeoizdat, Leningrad, 302 pp

Ulanova ES (1977) Methodological guidelines on compiling long-term forecasts of winter wheat yield in the main areas of cultivation of this crop. Hydrometeoizdat, Leningrad, 48 pp

Ulanova ES, Zabelin VN (1990) Methods of correlation and regression analysis in agrometeorology. Hydrometeoizdat, Leningrad, 207 pp

Ustinova OK, Abashina EV, Vol'vach VV (1990) A dynamic model WEATHER-YIELD-3 for potato as a basis for the system potato – pest (disease) – habitat. Collect Works AURIAM 25:77–83

Yusupov MC (1990) A mathematical model of soil – vegetation cover complex for cotton. Author's abstract of dissertation for Candidate of Physical and Mathematical Sciences. Comput Cent USSR Acad Sci, Moscow, 22 pp

Zainutdinov ZH (1978) An addition to the methodology of prediction of regional yield of winter wheat in Tadzhikistan. Collect Works SARNIGMI 56 (137):60–62

Zaslavsky BG, Oparina IV, Terleyev VV (1988) A dialogue system for generating a database of hydrophysical characteristies of soils. Rep Acad Agric Sci 11:40–42

Zilitinkevich SS (1970) Dynamics of atmospheric boundary layer. Hydrometeoizdat, Leningrad, 291 pp

[illegible]

Stomatal Density Responses to Global Environmental Change

D.J. BEERLING[1] and F.I. WOODWARD[1]

1 Introduction

The CO_2 concentration of the Earth's atmosphere is steadily increasing as a result of anthropogenic activities such as the combustion of fossil fuels and tropical deforestation which release stored carbon from terrestrial sinks. This accumulation of CO_2 in the atmosphere is the one fact in the global climate change debate upon which all scientists agree (Houghton et al. 1990), whereas the major uncertainties in the debate lie in predicting and observing the effects of this rise in CO_2 and other greenhouse gases on global climate.

As a result of past and future increases in the atmospheric CO_2 concentration (Houghton et al. 1990), there is considerable interest attached to understanding the physiological responses of vegetation to elevated CO_2 and temperature. These responses are important for predicting changes in the productivity of crop species (Adams et al. 1990) and natural vegetation types (Melillo et al. 1993).

Changes in the distribution and productivity of vegetation will affect mankind directly; but secondary, indirect effects are also possible. The processes of photosynthetic carbon fixation and regulation of water loss in higher plants are known to feedback and affect climate (Martin 1993), so that an understanding is required of the responses of these processes to changes in atmospheric CO_2 and climate if climate models are to accurately the effects of an increase in greenhouse gases on global forecast climate (Woodward and Smith 1994).

The experimental technique of CO_2 enrichment is now well established. A multitude of studies have been made to date employing this technique which have revealed many detailed facets of plant physiological responses to above-ambient CO_2 concentrations (see Strain and Cure 1985; Eamus and Jarvis 1989; Rozema et al. 1993). However, extrapolating the results from these studies to predict how natural vegetation may respond in the future is fraught with uncertainties. This is not only because many effects observed under high CO_2 concentrations may be artefacts of experimental design, especially nutrient limitation and pot size (Arp 1991; McConnaughay et al. 1993a,b), but also because of the difficulties in scaling up measurements made on single leaves or

[1] Department of Animal and Plant Sciences, The University of Sheffield, Sheffield S10 2TN, UK

individual plants to predict the response of canopies of mixed species at the regional scale and greater (Jarvis and McNaughton 1986; Ehleringer and Field 1993). Predictions are further confounded by ignorance of the interactive effects of elevated CO_2 and temperatures on plant–plant competition and plant–animal interactions.

An alternative source of information on the responses of natural vegetation to changes in atmospheric CO_2 concentration comes from the fossil record. The Quaternary Period, covering the last two million years, has been characterized by a repetitive cycle of warm temperate interglacials and cold glacial phases. Associated with these temperature fluctuations, at least for the last 200 000 years, were changes in atmospheric CO_2 concentration: during full-glacial periods, the CO_2 concentrations were ca. 180–220 ppmv (parts per million by volume) whilst during interglacials the CO_2 concentrations were 280–300 ppmv (Barnola et al. 1987; Jouzel et al. 1993). An appreciation of how vegetation responded to these past changes in atmospheric CO_2 concentration and temperature provides a basis for forecasting vegetation responses to future global climatic change. Quaternary palaeoecological studies may provide particularly appropriate insights into vegetation responses because they reveal the long-term acclimated responses of genotypes.

Plant ecophysiological studies have recently begun to encompass both experimental and palaeoecological approaches (Beerling and Woodward 1994a), particularly those investigating the response of stomatal density to global change (Beerling et al. 1993). The number and opening of stomata are crucial components of the diffusion pathways of CO_2 into the leaf during photosynthesis, and water vapour out of the leaf during transpiration. The ratio between these two fluxes is the transpiration ratio of leaves – a central feature of the optimization theory of stomatal control (Cowan 1977) in which the plant, through its control of stomatal aperture, maximizes its carbon gain for a given loss of water vapour. Pioneering studies by Wagner (1892) and Bonnier (1895) demonstrated that transplanting lowland species to high-altitude sites resulted in leaves with a higher stomatal density than lowland plants. However, because factors which can influence stomatal density, other than the partial pressure of CO_2, also change with altitude, these early studies did not link the observed stomatal density responses with changes in the concentration of atmospheric CO_2.

The recent work of Woodward (1987) refocused attention on stomatal density–CO_2 responses. Woodward (1987) showed that the stomatal density of several temperate tree species has decreased over the past 200 years in response to increases in the atmospheric CO_2 concentration from the pre-industrial levels to the present-day – a response confirmed experimentally by growing several species of tree seedlings at pre-industrial CO_2 concentrations. The documented decline in stomatal density of British trees provided evidence that man's recent perturbation of the atmospheric composition was having a direct effect on the morphology and development of vegetation. The work subsequently stimulated a renewed effort to understand stomatal density responses to CO_2 change and their consequences for plant physiology. As the responses of an increasing number of species begin to emerge from these studies it may prove possible to refine global models of photosynthesis and stomatal conductance which have been developed to improve our understanding of vegetation feedbacks on climate (Woodward and Smith 1994b).

To achieve this aim requires a better understanding of many aspects of the stomatal density response to CO_2 change. The genetic basis of stomatal density is largely unknown (Beerling and Chaloner 1991; 1993c; Jones 1992). In Sorghum, for example, stomatal density is controlled from a single locus (Suh et al. 1976), whereas in maize and wheat stomatal density appears to be under multigenic control (J. Gray, pers. comm.). Nothing is known about the genetic control of stomatal density responses to environmental change. Other areas of ignorance include whether stomatal density will decline further as atmospheric CO_2 concentration continues to rise or if there is a ceiling to the CO_2 response (Woodward 1987). With increased effort to understand stomatal density responses to global change come important issues for whole-plant physiology: how does stomatal density affect leaf stomatal conductance, photosynthesis, and the water economy of vegetation?

This review first briefly considers the records of past atmospheric CO_2 concentrations, as derived for the last 200 000 years from the ice core records, and the increases predicted to occur in the future by Houghton et al. (1992) and Wigley and Raper (1992) as a basis against which to compare past and future stomatal density responses. We then attempt to draw together the palaeoecological studies and experimental work in order to look for any clear patterns which may emerge among species between stomatal density and atmospheric CO_2. In the last section particular emphasis has been placed on the role of stomatal density studies in whole-plant physiology. We examine the effects of changes in stomatal density on stomatal conductance using the approach of Beerling and Woodward (1994a), which derives stomatal conductance from stomatal dimensions and equations of diffusion theory, and review the evidence for its subsequent effects on photosynthesis. The question of how stomatal density relates to water-use efficiency, either derived directly from instantaneous measurements of transpiration and photosynthesis, or from stable carbon isotope studies, is also considered.

2 Historical Records of Atmospheric CO_2 Concentration

2.1 Ice Core Records

Palaeo-data from ice cores provide an important way to understand the sensitivity of the Earth's climate to past changes in concentrations of radiatively active ("greenhouse") gases. Future changes in the concentration of these gases will trigger a multitude of feedbacks within the global climate system which are poorly understood (Houghton et al. 1990). Ice core records of past events give a direct integration of all the feedbacks from the different components of the climate system (oceans, terrestrial ecosystems, chemistry of the atmosphere and climate) and reveal a net global response, although interpretation poses difficulties (Raynaud et al. 1993).

The available ice core records cover the last 200 000 years and the last glacial–interglacial cycle in detail. The Siple Station core, from Antarctica, provides a record of atmospheric CO_2 with a high temporal resolution (Fig. 1), which documents the rapid rise over the last two centuries (Friedli et al. 1986). This short-term record provides strong evidence for the effects of human activities on the composition of the atmosphere and shows the close correspondence between atmospheric CO_2 measurements made at Mauna Loa, Hawaii, from 1958 onward (Boden et al. 1991) and the ice core measurements (Fig. 1).

Longer-term ice core records from the Byrd (Neftel et al. 1988) and Vostok sites in Antarctica (Barnola et al. 1987; Jouzel et al. 1993) demonstrate that atmospheric CO_2 concentrations during the last glacial maximum (LGM) were 25–30% lower than early Holocene levels (Fig. 2). A probable contributory source of this carbon is the oceans. Broecker and Peng (1986) proposed that CO_2 degassed from the oceans as the climate warmed; however, this mechanism can only account for a small part of the CO_2 increases, and a complete

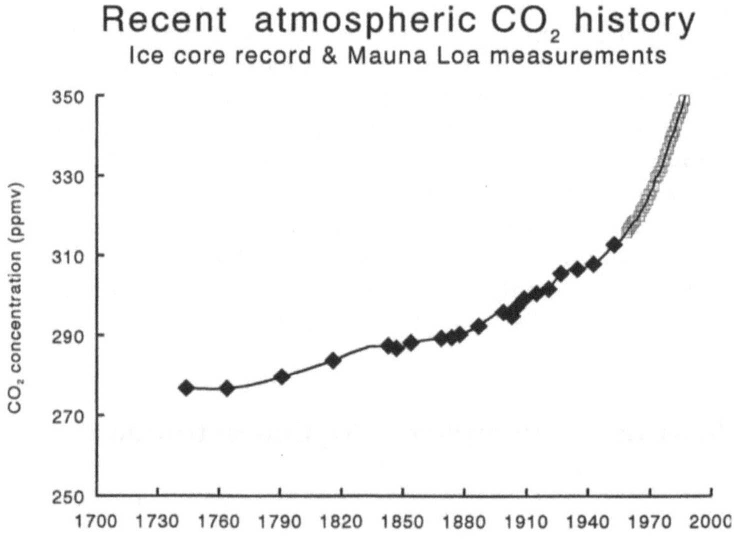

Fig. 1. Increases in atmospheric CO_2 concentration since 1750 as determined from the ice-core studies of Friedli et al. (1986) (*solid diamonds*) and the continuous monitoring record at Mauna Loa (*open squares*). (Boden et al. 1991)

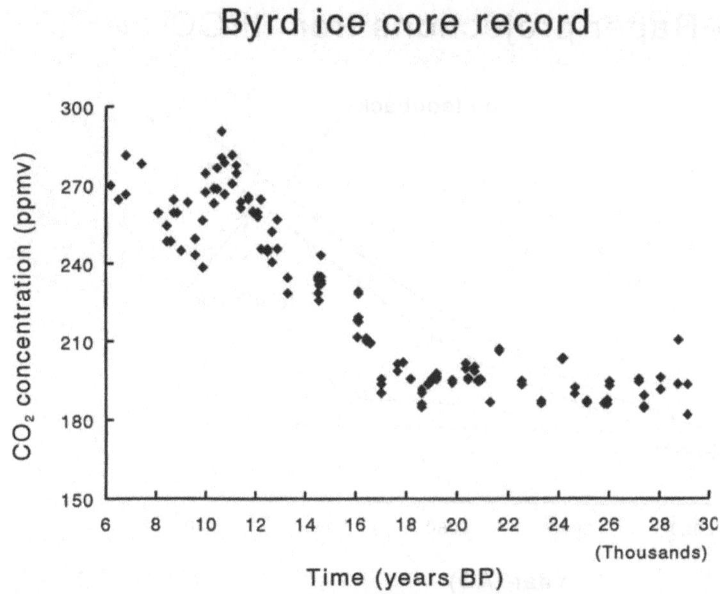

Fig. 2. Record of atmospheric CO_2 concentration from the Byrd ice core, Antarctica, on the timescale of calender years before present (B.P.). (After Neftel et al. 1988)

balance of the carbon budget for full-glacial–interglacial states remains elusive (Post 1993; Siegenthaler and Sarmiento 1993; Sundquit 1993).

3 Predictions of Future Increases in the Atmospheric CO_2 Concentration

The rate at which the concentration of atmospheric CO_2 is predicted to increase is dependent upon many factors which are largely unquantifiable. Therefore any predictions made must be based on assumptions of future patterns of fossil-fuel burning, socio-political factors and the role played by rising temperatures in oceanic CO_2 release over the past two centuries. The following section considers the best estimates for CO_2 increases and their reliability for use by plant scientists in simulating a future "greenhouse world" with CO_2 enrichment experiments.

3.1 Future Emission Scenarios

The Intergovernmental Panel on Climate Change (IPCC) has revised the initial (Houghton et al. 1990) set of emission scenarios for carbon in the light of new policies implemented or proposed for controlling CO_2 emissions and recent

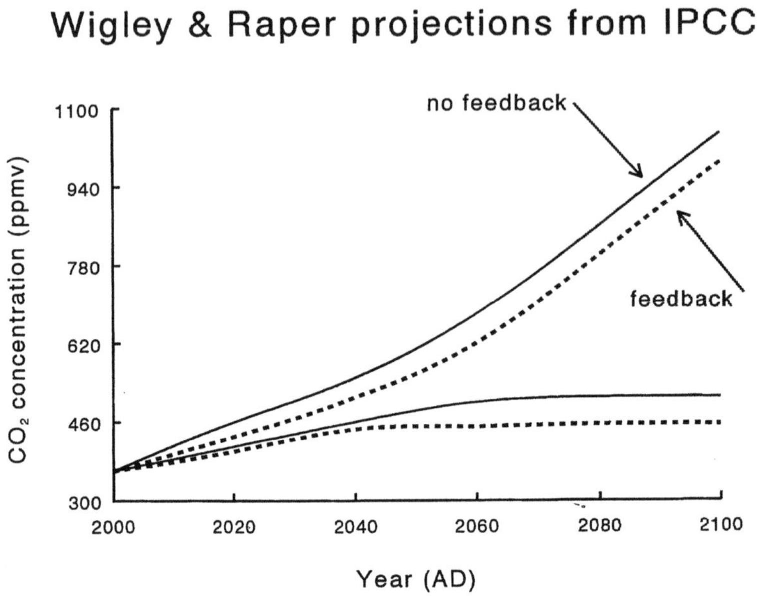

Fig. 3. The translation of the upper and lower IPCC scenarios into increases in atmospheric CO_2 concentration. Predictions exclude (*solid line*) and include (*dotted line*) feedback by vegetation acting as a terrestrial sink for carbon. Table 1 gives the underlying assumptions for these two scenarios. (After Wigley and Raper 1992)

political changes (Houghton et al. 1992). The new set of scenarios is important because it was not developed from a "top down" approach to achieve a prescribed set of CO_2 levels but instead was based on a wide range of assumptions based on socio-economic factors (Wigley and Raper 1992). Wigley and Raper (1992) converted the emission scenarios given in gigatonnes carbon per year (Gt C $year^{-1}$) to atmospheric CO_2 concentrations using a carbon-cycle model with and without feedback by vegetation. Feedback projections are made assuming that the "missing sink" (Enting and Mansbridge 1991) is explained solely by the CO_2 fertilization effect (i.e., enhanced biomass production under elevated CO_2 concentrations). Although contentious (Dai and Fung 1993; Houghton 1993), this approach sets the lower bound of the estimates and Wigley and Raper (1992) argue that modelling with other processes which are presently not accounted for would lead to CO_2 projections within the range of the feedback and no-feedback estimates. Associated temperature changes were derived from a coupled ocean general circulation carbon model (Maier-Reimer and Hasselmann 1987), assuming a number of future emission scenarios and a global climate sensitivity to a doubling of CO_2.

The predicted rise in atmospheric CO_2 concentration and temperature between 1990 and 2100 A.D. for the upper and lower IPCC scenarios with and without feedback in the model are given in Figs. 3 and 4 respectively. Table 1 gives the underlying assumptions for these two scenarios. In both cases, the feedback projections are lower than those made without feedback, showing the

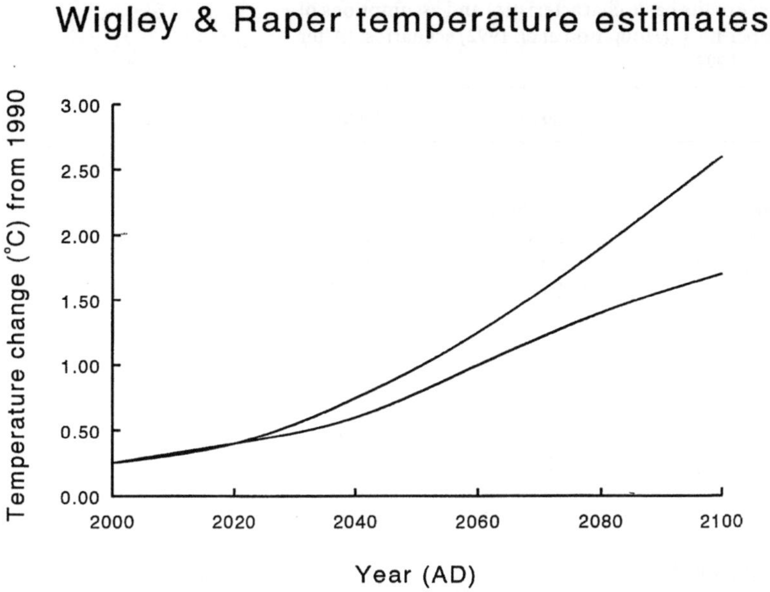

Fig. 4. Upper and lower temperature predictions from an increased warming as a result the accumulation of CO_2 in the atmosphere, assuming a climate sensitivity of 2.5 °C for a doubling of present CO_2 concentration. (After Wigley and Raper 1992)

importance of including vegetation as the "missing sink" for carbon. However, this does not include changes in the throughput of water from the soil through plants to the atmosphere – an important feature as water vapour is a strong "greenhouse gas". Overall, the modelling studies show that by 2100 vegetation will be exposed to CO_2 concentrations in the range of 460–1100 ppmv coupled with temperature increase of up to 3 °C, highlighting the need both for experimental treatments simulating these two increases on plant growth (Long 1991) and for the need for modelling studies to address possible plant responses (Beerling and Woodward 1994b). Whilst the independent effects of CO_2 concentration and temperature differ, the interactive effects of high CO_2 and temperature are likely to be of major importance for predicting plant growth and water-use efficiency at the canopy and regional scale (Eamus 1991; Long 1991; Gifford 1992; Beerling and Woodward 1994b).

The uncertainties of the predicted increases in atmospheric CO_2 concentration will remain until our quantitative understanding of the sources and sinks of the carbon cycle improves. Nevertheless, within the uncertainties of the socio-economic assumptions of the different IPCC 1992 scenarios, we can be reasonably confident that future increases will be within the range of current projections (Wigley and Raper 1992).

Greater uncertainty lies in predicting temperature changes. The climate model used by Wigley and Raper (1992) depends upon an estimate of climate sensitivity. Climate sensitivity is defined as the change in equilibrium surface temperature associated with a doubling of CO_2 concentration. Temperature

Table 1. Summary of the basic characteristics and assumptions of the lower and upper IPCC (Houghton et al. 1992) scenarios. (After Wigley and Raper 1992)

Parameter	Lower	Upper
Socio-economic growth		
Population growth	Medium-low	Medium
Economic growth	Low	High
Resource availability	Low	High
Policies	Existing	Existing
Emissions (2100)		
Fossil-fuel CO_2 (Gt C year^{-1})	4.8	35.9
Land-use CO_2 (Gt C year^{-1})	–0.2	–0.1
CH_4 (Tg CH_4 year^{-1})	546	1072
N_2O (Tg N year^{-1})	13.7	19.1
SO_x (Tg S year^{-1})	55	232
CFCs	Moderate	Low
Radiative forcing[a] (W m^{-2})		
CO_2 (NFB/FB)[b]	2.56/2.00	6.95/6.45
CH_4 plus stratospheric H_2O	0.25	0.98
N_2O	0.17	0.32
Halocarbons (NFB/FB)	0.40/0.36	0.23/0.40
Sulphate aerosol	0.2	−1.41
Total, best guess (includes feedback)	2.98	6.74

[a] Radiative forcing is defined as the trapping of thermal energy by infrared absorbing gases (8–12 µm waveband).
[b] Results for no-feedback carbon cycle model (NFB) and model with CO_2 fertilization feed back (FB).

projections (Fig. 4) were made assuming a climate sensitivity of 2.5 °C which has since been found to be comparable to the estimate derived from palaeoclimate reconstructions (Hoffert and Covey 1992). Superimposed on the anthropogenic component of future climate change will be the natural variability derived from external and internal forcing factors.

4 Stomatal Density Responses to Past Global Change

4.1 Stomatal Density Responses to Atmospheric CO_2 Over Centuries

Over the same period for which the Siple station ice core provides a detailed atmospheric CO_2 record, herbarium collections of leaves can provide corresponding information on the stomatal density of plants collected on particular dates. Correlations between stomatal density and increases in atmospheric CO_2 concentration can therefore be established. The emphasis at this stage is on correlation, since without CO_2-enrichment experiments the cause is not established. The importance of Woodward's (1987) work was that it established the causal link between CO_2 levels and stomatal density from independent experimental studies in controlled environments.

Three studies have investigated historical trends in stomatal density over the past two centuries based upon measurements of herbarium leaves (Woodward 1987; Peñuelas and Matamala 1990; Beerling and Chaloner 1993a), and in one study past stomatal densities were compared with those measured from present-day material (Körner 1988). In order to detect any such changes it is important that the environmental history of the leaf is accounted for (particularly with regard to solar radiation) and that counts are made from comparable areas of the leaf. Körner (1988) used historically reported stomatal density values of herbaceous alpine plants from the literature 7–12 decades ago and made comparisons against the present-day values from the same area around Innsbruck, Austria, and reported no overall changes. The results may have arisen through not paying sufficient attention to the environmental history of the leaves and because the leaves were not old enough for a significant trend to be detected. From 1918 to the present-day stomatal density is predicted to have changed by only ca. 10%: this small decrease appears to have been obscured by large variations in density due to environmental conditions (Woodward 1993).

Generally, the studies by Woodward (1987) and Beerling and Chaloner (1993a) of temperate British trees and Peñuelas and Matamala (1990) of Mediterranean trees show a significant ($P < 0.01$) decline in stomatal density over the past two centuries (Fig. 5). The Mediterranean trees, herbs, and shrubs appear to be less responsive to CO_2 increases than the British species, but this may reflect the different selected life forms since the British material represented only trees and it may be that shrubs and herbs respond differently. That all three studies show the same trend is remarkable considering the different species and geographical regions compared. Woodward (1987) measured the stomatal density of herbarium leaves collected in the south of England, Beerling and Chaloner (1993a) worked on herbaria leaves collected from Scotland and Durham, whilst Peñuelas and Matamala (1990) analysed herbarium leaves of Mediterranean species collected from the Catalonia region of Spain. The large amount of scatter in Fig. 5 may be due to year-to-year fluctuations in weather conditions. Potential problems of sun/shade effects were minimized by careful selection of leaves (in all three studies) from flowering shoots only, which are assumed to have developed under full irradiance. Several authors have suggested that a reduction in stomatal density in response to CO_2 increases is most marked pre-1900 (Peñuelas and Matamala 1990; Peñuelas and Azcón-Bieto 1992; Woodward 1993). Comparison with the ice-core data suggests that this interval was accompanied by a slow increase in atmospheric CO_2

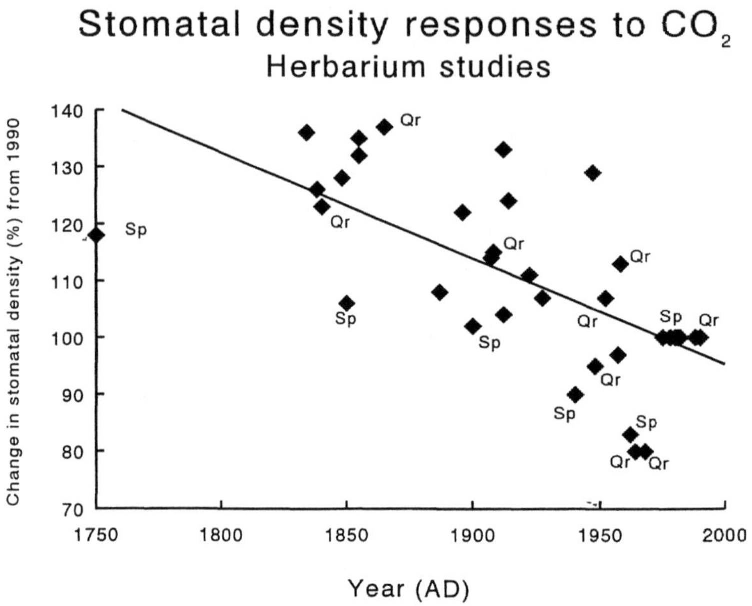

Fig. 5. Percentage change (calculated relative to the present-day values, taken as 100%) in stomatal density for herbarium-stored leaves from the data of Woodward (1987), Beerling and Chaloner (1993a) for *Quercus robur* (*Qr*) and Peñuelas Matamala (1990) Spanish species (*Sp*) (r = 0.66, $P < 0.01$)

concentration relative to the following century (Fig. 1). The early and apparently more rapid decline in stomatal density during this slow gradual increase in CO_2 concentrations seems surprising, but may represent an initial response of vegetation which was later moderated, possibly as a result of some degree of acclimation.

Drought is known to exert a marked effect on stomatal density (Tichá 1982) by reducing epidermal-cell expansion and causing the stomata to be packed closer together, i.e., increasing their density. In the semi-arid Mediterranean region, therefore, we might expect to observe the effects of water supply interacting with the effects of CO_2, although clearly the strength of this interaction would be dependent upon any increases in plant water-use efficiency in response to CO_2 increases of the past 200 years. To investigate this possible interaction historical trends in the stomatal density of herbarium specimens of the Mediterranean species *Quercus boissieri* reut var. *petiolaris* from between 1920 and 1980 were examined (G. Stanhill, pers. comm.). *Q. boissieri* grows at high altitude with a cool, humid climate and moderate rainfall. Plants growing at high altitude experience reduced partial pressures of CO_2 (Gale 1972) so that any increase in the concentration of atmospheric CO_2 may be of greater benefit to these species relative to their lowland counterparts (cf. tree ring studies of LaMarche et al. 1984; Graumlich 1991). In contrast to the trend shown by

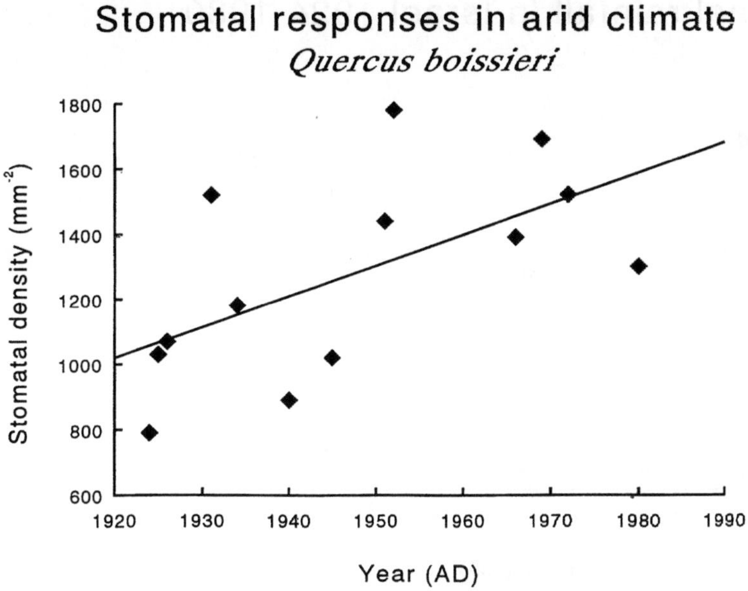

Fig. 6. Stomatal density of *Quercus boissieri* growing in the Mediterranean between 1920 and 1985, as determined from herbarium leaves. ($r = 0.53$, $P < 0.01$). (G. Stanhill, pers. comm.)

pooled species of different life histories (Fig. 5), Fig. 6 shows that the stomatal density of this *Quercus* spp., growing between 200 and 1000 m, appears to have increased by 0.6% a year over the last 85 years. Comparison of this trend with the patterns of seasonal rainfall over the same interval as the collection of the herbarium specimens (Fig. 7) show that rather than being associated with increased aridity of the region, water supply to the plants during the growing season generally increased. A similar trend of increasing stomatal density (+0.9% a year) was also shown by *Populus euphratica* growing in Israel over the same period at altitudes of 300 and 800 m. The stomatal density of the leaves of trees growing in these semi-arid regions is most likely to be determined by several environmental factors interacting simultaneously. Clearly, some other factor appears to be overriding the response expected from an increase in the CO_2 concentration. It may be, therefore, that for *Q. boissieri* and *P. euphratica*, woody species growing at different altitudes, an increase in the atmospheric CO_2 concentration, together with an increase in water supply, have resulted in these species increasing their stomatal density to maximise photosynthetic carbon fixation. Such a response, in terms of carbon gain, has been reported for mountain plants growing in CO_2-enriched atmospheres (Körner and Diemer 1994). An additional consideration is the effect of humidity changes resulting from annual changes in precipitation. If humidity has increased then stomatal density may also have increased, an effect reported for several agricultural species investigated under experimental

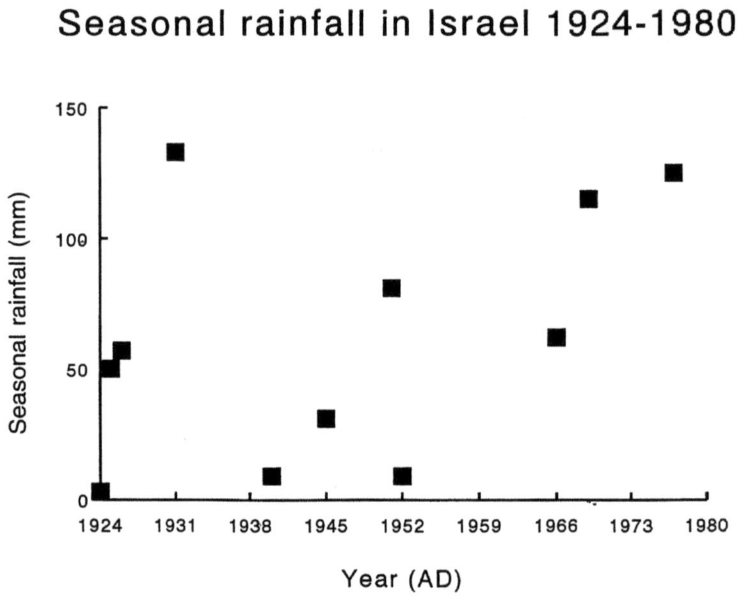

Fig. 7. Seasonal rainfall patterns in Israel 1924–1980, corresponding to the dates from which herbarium leaves were obtained

conditions (within the range 0.2–1.6 kPa) (Bakker 1991). Any possible decrease in transpiration rates due to the rise in atmospheric CO_2 concentration may also have been offset by the effects of relative humidity (rh) on stomatal conductance; increased rh tends to result in an increase in stomatal conductance (El-Sharkaway et al. 1985). In order to better understand the controls of stomatal density operating in the field on the vegetation of semi-arid regions, factorial experiments in controlled environments are desirable. The question as to whether it is possible to increase the stomatal density of vegetation from these regions by the manipulation of the environmental growth conditions is worthy of particular attention.

Beerling and Chaloner (1993b) also investigated the stomatal density responses of vegetation from arid regions. These authors measured the density of *Olea europaea*, a Mediterranean species whose distribution is tightly coupled with climate (Pollunin and Huxley 1974). The study also extended the time scale of the response by an order of magnitude by comparing values obtained from ancient leaf material preserved in Egyptian tombs with measurements from modern and herbarium Mediterranean leaves. Unlike *Q. boissieri* in Israel, the stomatal density of *O. europae* has decreased by ca. 50% over the last 3000 years of CO_2 increase. The older *O. europaea* leaves may have responded in this way because of the larger increase in atmospheric CO_2 experienced since 3300 years B.P.

The evolutionary response of stomatal density to changes in atmospheric CO_2 concentration is of interest for predicting how plants might respond to future global change. Present-day responses are dictated by the CO_2 concentrations under which plants evolved. The Vostok and Byrd ice-core records clearly demonstrate that plants have grown in CO_2 concentrations of less than 300 ppmv during the past 200 000 years. Long-term carbon-cycle modelling estimates that plants have probably spent at least the last 10 million years at atmospheric CO_2 concentrations at or below 300 ppmv (Berner 1990, 1992, 1993). This represents a considerable time interval for plants to evolve a stomatal density response. Predictions made for plants in future high CO_2 concentrations fail to incorporate any possibility of genetic change (Woodward 1988; Beerling and Chaloner 1993c). However, if responses of plants of the present-day could be compared over time scales greater than two centuries or three millennia then some idea of the plastic and genetic responses of plants to increasing CO_2 may be found. Quaternary palaeoecological studies enable an investigation of such responses.

4.2 Stomatal Density Responses Through Geological Time

Correlations of stomatal density with ice-core records of past CO_2 concentrations can be extended temporally by the analysis of cuticles isolated from fossil leaves. This approach also offers the potential for utilizing stomatal responses to CO_2 concentration to provide a proxy palaeo-CO_2 recorder. Palaeoecological data have two main advantages over the experimental studies (discussed in Sect. 5): firstly, they include the direct effects of temperature changes associated with the CO_2 changes, and secondly they provide a measure of how adapted or acclimated genotypes responded. Indeed, as Körner (1993) states:

> "Investigations of plants that have a history of enhanced or depleted CO_2 supply over geological periods can tell us whether carbon relations in plants, in relation to other metabolic processes, do exhibit homeostatic features – at least at the leaf scale. These living documents of past and ongoing natural CO_2 experiments will be the most profitable tools in teaching us long-term trends in adaptation."

Quaternary leaf macrofossils laid down over the last glacial cycle provide a measure of stomatal-density responses to the combined long-term increases in CO_2 concentration and temperature. The analysis of *Salix herbacea* leaf macrofossils mainly from deposits in the British Isles, and dating from the penultimate and last full-glacial interval through a range of ages spanning the last deglaciation, has shown that stomatal density appears to decline in response to increases in atmospheric CO_2 concentration (Beerling et al. 1993; Fig. 8). The trend shown by stomatal density was also mirrored by the stomatal index (the proportion of stomata to leaf epidermal cells), although stomatal index was only sensitive below CO_2 concentrations of 240 ppmv. The low CO_2 concentrations of both the penultimate and last glaciation resulted in the highest stomatal indices yet recorded for *S. herbacea* from the fossil record.

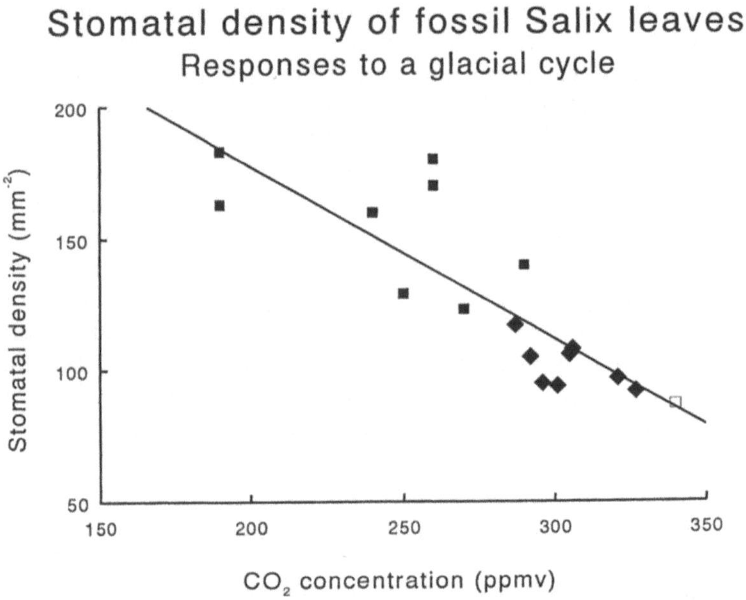

Fig. 8. The response of stomatal density of *Salix herbacea* to changes in atmospheric CO_2 concentration and temperature over the last glacial cycle. Measurements from fossil leaves (*solid squares*), herbarium leaves collected between 1845 and 1971 (*solid diamonds*) and fresh leaves from Bergen, western Norway (*open square*); ($r = 0.84$, $P < 0.001$). (After Beerling et al. 1993)

The time interval involved in this study is at least some 140000 years and so is sufficiently long to incorporate the possibility of a genetic response.

Further evidence of stomatal density responses to CO_2 dynamics on Quaternary time scales comes from the analysis of fossil *Pinus flexilis* needles preserved in packrat middens over the last 30000 years. Systematic measurement of the stomatal densities of needles from the central basin, USA showed a decline in response to CO_2 increases over the same period, as documented from the Byrd ice core (Van de Water et al. 1994). Furthermore, the abundance of packrat middens from the central basin area allowed these authors to eliminate possible temperature effects. Because of temperature increases since the last glaciation [ca. 30000 years B.P. (1950)] Van de Water et al. analysed needles from sites at progressively more northerly latitudes, so that observed changes in stomatal density could be attributed mainly to CO_2 changes rather than temperature increases. The evidence from the Quaternary studies of Beerling et al. (1993) and Van de Water et al. (1994) suggests that both gymnosperm and angiosperm species have responded consistently, in terms of stomatal density, to the atmospheric CO_2 dynamics of the last glacial cycle, despite the different climatic regions of each study.

Van der Burgh et al. (1993) have recently extended the temporal scale of stomatal responses back from the Quaternary into the Miocene and used the observed responses to estimate CO_2 concentrations of the time. Measurement of the stomatal index of Miocene fossil leaves of *Quercus petraea* showed that a high index corresponded to cooler climates and a low index corresponded to warmer climates. Climate conditions were inferred from the palynological record of the Lower Rhine Embayment. The results imply that high stomatal indices were associated with the low atmospheric CO_2 concentrations prevailing at the time (as a result of a reduced "greenhouse effect"). By constructing a calibration curve from herbarium leaves of *Q. petraea* (cf. Woodward 1987), Van der Burgh et al. (1993) estimated that atmospheric CO_2 concentrations in the late-Miocene to Pliocene epochs varied between 280 and 370 ppmv.

Fossil leaves from several other geological periods may also provide a palaeo-CO_2 signature for earlier phases of the Earth's history. In particular, long-term carbon-cycle models (Lasaga et al. 1985) and reconstructed patterns of CO_2 derived from fossil porphyrins (a derivative of chlorophyll) of phytoplankton from deep ocean cores (Freeman and Hayes 1992), both predict a large fall in atmospheric CO_2 concentration from the Cretaceous (K) to the Tertiary (T) (120–10 Ma). However, the predictions of the carbon model and the reconstructions differ by several orders of magnitude in the values of CO_2 itself, and neither include estimates of the effects of a K–T meteoric impact event on the atmospheric CO_2 concentration – which are likely to have been elevated by ca. ten times for periods of 10^4–10^5 years (O'Keefe and Ahrens 1989). The measurement of stomatal densities from fossil leaves of Cretaceous and Tertiary floras offers a powerful approach from which a CO_2 signal may be extracted (Beerling 1994) particularly across the K–T boundary where several well-preserved leaf floras have been described (Wolfe and Upchurch 1986).

The approach adopted for the study of Cretaceous–Tertiary fossil floras is currently being extended into the largest CO_2 excursions experienced since the evolution of land plants (J. McElwain and W.G. Chaloner, pers. comm.). These workers are focusing on Devonian and Carboniferous fossil floras from which cuticles can be isolated and stomatal density determined. Berner (1993) estimates atmospheric CO_2 concentrations ten times the pre-industrial concentration (ca. 280 ppmv) in the Devonian with a fall to two times the pre-industrial level in the Carboniferous. It follows therefore that observations from fossil material of different geological ages will provide the first attempt to verify predictions of the carbon-cycle models for the Paleozoic, Mesozoic and Cenozoic.

Interpreting stomatal density responses as a measure of ancient CO_2 changes obviously raises the question of how temperature may interact to determine density. Large CO_2 variations in the geological past are directly linked with warmer or cooler palaeoclimates through the greenhouse effect. Experimentally, the question of temperature effects on stomatal density has not been widely addressed, and what evidence there is remains inconclusive. Temperature may control the stomatal density of leaves firstly, and most directly, through leaf expansion rates (Huner et al. 1981) and secondly at the subcellular level by influencing stomatal complex formation (Beerling and Chaloner 1993a). The effects of temperature on leaf extension rates can be minimized by the selection of similarly sized leaves. Accounting for temperature effects at the subcellular level is more difficult.

We investigated the intraspecific response of stomatal density to temperature on seven different populations of the alpine dwarf shrub *Vaccinium myrtillus*. The results showed that five of the seven populations had slightly higher stomatal densities at the warmer temperatures of 22/12 °C day/night (compared to 15/5 °C; Fig. 9) although only the CG75 population showed a significant ($P < 0.05$) increase. Two populations responded in the opposite manner, with an decrease in density at warmer temperatures, in agreement with Huner et al. (1981). Higher densities at warmer temperatures are unlikely to be an effect of temperature on the rate of leaf expansion, since at higher temperatures leaf expansion increases and stomata are spaced further apart reducing density. These results provide some evidence that temperature can control stomatal development, but as yet it is difficult to reconcile the results from the few published studies available to provide quantification of temperature effects on stomatal density. For workers on ancient leaf material, especially, there is a need to experimentally characterize the effects of temperature on stomatal density for an improved interpretation of any possible CO_2 signal

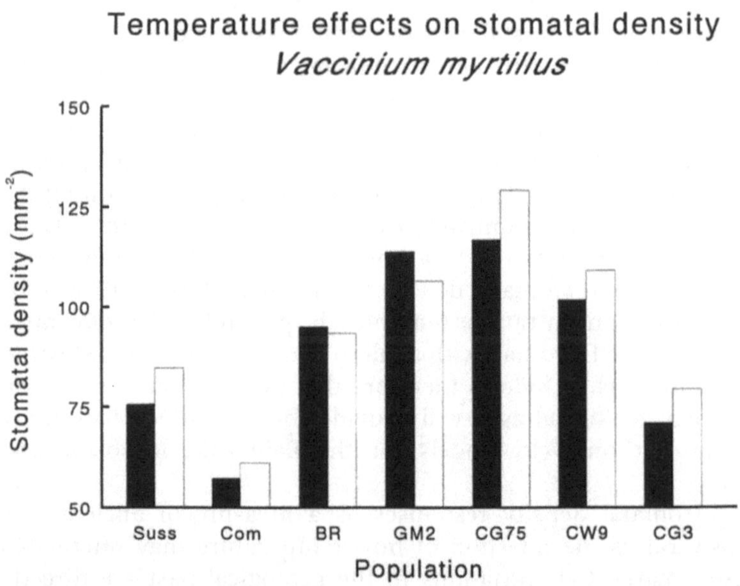

Fig. 9. Effects of temperature on the abaxial stomatal density of leaves of the dwarf shrub *Vaccinium myrtillus*. Population codes: *Suss* Sussex; *Com* Cairn O'Mount; *BR* Ballater Road, Scotland; *GM2* Glas Maol 650 m a.s.l.; *CG75* Cairngorm 750 m a.s.l.; *CW9* Cairnwell 900 m a.s.1.; *CG3* Cairngorm 1000 m a.s.. Population were grown at 15/5 °C day/night cycle (*solid*) and 22/12 °C day/night cycle (*open*) for 2 months under 13-h photoperiod. Values are the means of 10 observations from each of 3 leaves. Only the *CG75* population differed significantly ($P < 0.05$) between treatments (paired t-tests were performed on all populations, after $\log_{10}$ transformation of the original data using, UNISTAT 1993

from the fossil-leaf cuticles and this should be an important priority for future research.

Information regarding the response of stomatal density to atmospheric CO_2 concentration obtained from fossils, herbaria and leaves from archaeological sites increases our understanding of the history of plant responses to environmental change. Responses observed from the fossil record have three important implications for plant physiological responses to future increases in temperature and CO_2 concentration. These may be summarized as: (1) providing evidence that stomatal density responds to increases in CO_2 concentration in the same direction over long intervals of geological time, possibly by genetic adaptation; (2) the response is sustained over a wider CO_2 range than previously observed; and (3) plant water-use efficiency was likely to have been lower in the past when CO_2 levels were ca. 30% lower than at present (e.g., during full glacials; Fig. 2; considered in Sect. 6.2). From points (1) and (2) it appears more persuasive to extrapolate from short-term experiments representing a phenotypic response in order to forecast future stomatal responses (Sect. 5).

5 Stomatal Density Responses to Future Global Change

The response of stomatal density to future increases in atmospheric CO_2 concentration may be assessed from experimental studies in which plants are exposed to CO_2 enrichment using controlled environments. To some extent these experiments are unrealistic because plants are exposed in a single step to an abrupt doubling of CO_2 concentration, whereas the increase is likely to happen gradually over the next 100 years (Fig. 3; Houghton et al. 1992). Therefore several alternatives are currently being sought by plant ecologists to reconcile the temporal scale of CO_2 change in experiments with that most relevant in an ecological context. One alternative is to use areas of naturally high levels of CO_2 due to the presence of CO_2-rich springs (Woodward et al. 1991; Miglietta and Raschi 1993; Miglietta et al. 1993). At these localities, the release of CO_2 from the Earth's crust, as a result of outgassing from crustal magma and interactions between the groundwater chemistry and the underlying rocks, exposes vegetation close to the vents to above-ambient CO_2 concentrations. This vegetation will have had sufficient time to fully acclimate to these higher CO_2 concentrations and so any physiological responses detected may eventually be expected to occur in the future if the atmospheric CO_2 concentration continues to rise.

5.1 Stomatal Density Responses to CO_2 Enrichment Above Ambient

From data related to changes in stomatal density with altitude and studies of fossil and herbarium material (Au 1969; Körner et al. 1989; Woodward 1986,

1993; Beerling and Chaloner 1991, 1993a,b; Beerling et al. 1993), we may predict that in general the stomatal density of the majority of land plants will decrease in response to an increasing concentration of atmospheric CO_2 although this response would not be expected to continue indefinitely. Clearly, it is unlikely that the number of stomata on plant leaves will be reduced to zero at very high concentrations of atmospheric CO_2. Consequently, we might expect a plateau of stomatal response to rising CO_2 concentrations that may or may not be determined by the phenotypic or genotypic component.

Over the past two decades a number of experimental studies have addressed the responses of stomatal density to increasing CO_2 concentrations. Table 2 summarizes the results of these studies together with changes in gas-exchange and biomass responses. Of all the species so far investigated 50% showed a decline in stomatal density as atmospheric-CO_2 concentration increased above ambient (Table 2). This indicates that the plateau of response, previously thought to exist around 340 ppmv (Woodward 1987), appears to be much higher, at least for half of the species examined. Particularly significant is the observation that nine of the fourteen species showed a decrease in stomatal index, on the adaxial leaf surface, in response to increasing CO_2 concentrations (Table 2). Woodward (1988) demonstrated differences in stomatal-density sensitivity between vegetation types, and concluded that long-lived species, e.g. trees and shrubs, show little response to CO_2 concentrations above ambient, and suggested that the limit of plasticity for longer-lived plants may be the current atmospheric CO_2 concentration, although these life forms are sensitive to CO_2 concentrations below ambient. In contrast, annuals demonstrate a higher degree of plasticity at increased CO_2 concentrations. Variability in sensitivity of response may influence community structure in the future through possible changes in water-use efficiency and, if this occurs on a sufficiently large scale, feedback, affecting climate via evapotranspiration rates (Martin 1993).

Sites with naturally elevated concentrations of atmospheric CO_2 could provide a test for the long-term applicability of short-term experiments and the trends observed from fossil and herbarium leaves. Sites of naturally elevated CO_2 which are not significantly contaminated by other gases are uncommon, nevertheless several suitable gas vents have been located recently in Florida (Woodward et al. 1991) and Italy (Miglietta and Raschi 1993; Miglietta et al. 1993). At all these sites, the plants growing nearest the vents are exposed to the highest concentrations of atmospheric CO_2 with a decreasing CO_2 gradient moving away from the vent (Miglietta and Raschi 1993). In Florida, CO_2 concentrations close to the vent were ca. 400–500 ppmv. Populations of the herbaceous annual *Boehmeria cylindrica* growing near the vent showed a small reduction in stomatal density relative to populations growing under ambient CO_2 concentrations (F.I. Woodward, unpubl.). More recently, several studies have been made on vegetation growing at the Solfatara spring, Viterbo, Italy where CO_2 concentrations are extremely high (ca. 7000 ppmv) near the vent with a gradient to present concentrations (350 ppmv) within 50 m of the vent.

Table 2. Stomatal density and stomatal index responses observed from C3 plants grown in controlled environment experiemnts to increases in CO_2 concentrations above ambient

Species	CO_2 concentration (μmol mol^{-1})	Duration of exposure (d)	Stomatal density[a]		Stomatal index		Gas-exchange responses		Authors
			ab	ad	ab	ad	A[b]	g_s[c]	
Amaranthus retroflexus	525, 700	28	–	–	–	–	+	–	Woodward and Bazzaz (1988)
Ambrosia artemisiifolia	525, 700	28	–	–	–	–	+	–	
Castanea sativa	700	730	nd	–	nd	–	nd	–	Mousseau and Enoch (1989)
Glycine max	520, 718, 910	21	–	–	–	–	nd	nd	Thomas and Harvey (1983)
Liquidambar styraciflua	520, 718, 910	21	–	na	–	na	nd	nd	Thomas and Harvey (1983)
Lolium perenne	680	35	0	0	nd	nd	nd	nd	Ryle and Stanley (1992)
Lycopersicum esculentum	650–3200	30	–	–	–	–	nd	nd	Madsen (1973)
Maranthes corymbosa	350, 700	210	–	na	nd	nd	+	–	Eamus et al. (1993)
Ochroma lagopus	675	60	0	0	nd	nd	–	–	Oberbauer et al. (1985)
Oryza sativa	500, 600, 990	104	0	0	nd	nd	nd	nd	Rowland-Bamford et al. (1990)
Pelargonium x *hortorum*	1000	55	–	–	nd	nd	+	nd	Kelly et al. (1990)
Pentaclethra macroloba	675	123	–	0	nd	nd	–	–	Oberbauer et al. (1985)
Phaseolus vulgaris	1200	nd	–	0	nd	nd	nd	nd	O'Leary and Knecht (1981)

Table 2. *Continued*

Species	CO_2 concentration (μmol mol^{-1})	Duration of exposure (d)	Stomatal density[a]		Stomatal index		Gas-exchange responses		Authors
			ab	ad	ab	ad	A[b]	g_s[c]	
Phaseolus vulgaris	350, 700								
high nutrient		28	0	0	0	0	nd	nd	Radoglou and Jarvis (1992)
low nitrient		28	0	0	0	0	nd	nd	
Populus euramericana	660	15	+	+	+	+	–	–	Gaudillère and Mousseau (1989)
Populus trichocarpa (4 clones)	700	92	0	0	0	0	nd	–	Ragdolou and Jarvis (1990)
Quercus pubescens at naturally elevated CO_2 site, central Italy	1000	>20 yrs	0	na	0	nd	nd	nd	Miglietta and Raschi (1993)
Setaria faberii	525, 700	28	–	–	–	–	+	–	Woodward and Bazzaz (1988)
Sinapis alba	350, 700	60							
low wind speed (0.3 m s^{-1})			0	–	nd	nd	nd	nd	Retuerto and Woodward (1993)
high wind speed (3.7 m s^{-1})			0	+	nd	nd	nd	nd	
Vaccinium myrtillus	450	21	+	–	–	–	+	–	Woodward (1986, 1987)
Vivia faba	350, 700	45	+	+	+	+	nd	nd	Radoglou and Jarvis (1993)
Zea mays	520, 718, 910	21	–	–	–	–	nd	nd	Thomas and Harvey (1983)

Key to responses used throughout the table: +, increased response to CO_2 enrichment; – decreased response to CO_2 enrichment; 0, no responses to CO_2 enrichment; nd, no data; na, not applicable.

[a] ab and ad = abaxial (lower) and adaxial (upper) leaf surfaces respectively.

[b] Photosynthetic rate.

[c] Stomatal conductance.

At this site individuals of *Quercus ilex* were grown in pots across the CO_2 gradient (in order to standardize soil nutrient availability). The results indicated a consistent decline in density with increasing CO_2 concentration (E. Paoletti, pers. comm.). A similar result has also been reported for *Scirpus lacustris* (I. Bettarini, pers. comm.) and *Phragmites australis* (Van Gardingen et al. 1993) growing naturally at the same site. Other examples of plants producing leaves with a lower stomatal density at naturally elevated CO_2 springs relative to those growing at ambient CO_2 concentrations have also been reported for *S. herbacea* and *Polygonum viviparum* growing at Icelandic CO_2 springs (B. Sveinbjornsson, pers. comm.). The studies on plant-leaf morphology from naturally elevated CO_2 sites are encouraging in that the responses seen from studies of fossil and herbarium leaves, together with the controlled environment work, appear consistent with those of vegetation exposed to naturally elevated CO_2 concentrations at these sites.

5.2 An Index of Sensitivity of Stomatal Responses to CO_2 Change

In an attempt to standardize the available data and to understand better differences between species, we calculated an index of sensitivity to CO_2 change, defined as:

$$\text{Index of sensitivity (IS)} = \frac{(sd_e - sd_a)/sd_a}{(Ca_e - Ca_a)/Ca_a} \times 100, \qquad (1)$$

where sd_e is the stomatal density (mm^{-2}) at an elevated CO_2 concentration (Ca_{ppmv}) and sd_a is the stomatal density at ambient CO_2 concentration (Ca_a). A negative IS value indicates a reduction in stomatal density with increasing CO_2 concentration and a positive value an increase in stomatal density with increased CO_2. The index of sensitivity was calculated for 109 species of angiosperms from around the world (Appendix), in some cases values were reported for the same species more than once and these were included in the analysis.

Frequency distributions of sensitivity index, constructed from the entire data set extracted from the literature (Appendix), show that the majority of observations have negative IS values when the stomatal responses are considered for both leaf surfaces (Fig. 10a) or when each leaf surface response is considered separately (Fig. 10b–c). All three distributions differed significantly ($P < 0.001$) from the expected normal distribution (Table 3). However, despite a major proportion of observations being negative the most extreme values were positive, indicating an increase in density with increasing CO_2, and resulting in a positive skewness statistic, even for abaxial IS values (Table 3). Note that extreme values have not been plotted on Fig. 10a–c in order to maintain a common x-axis, the full range for each data set is given in Table 3. Marked differences were apparent in the responses of the abaxial and adaxial leaf surfaces; the adaxial surface showing much greater sensitivity. Similar results have been reported for plants of *Vaccinium myrtillus* grown in a controlled environment under different CO_2 concentrations (Woodward 1986). The large scatter of each of the frequency distributions may readily be explained by the different sources of information used for their construction which includes data from historical and altitudinal studies, in which plants are responding to

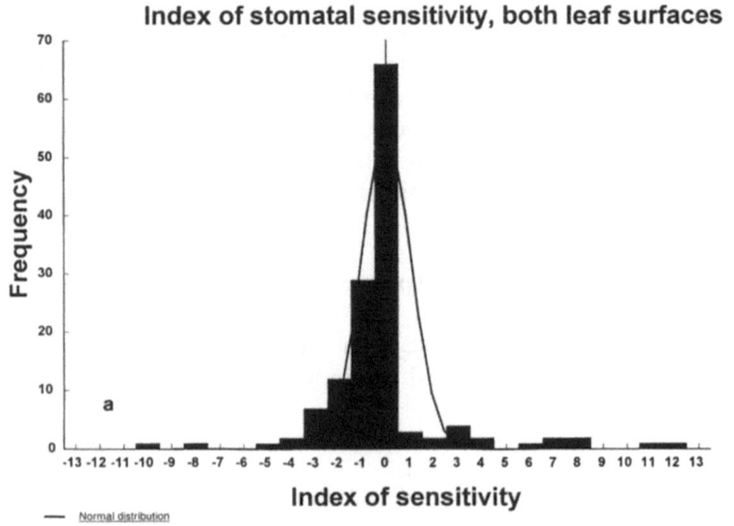

a

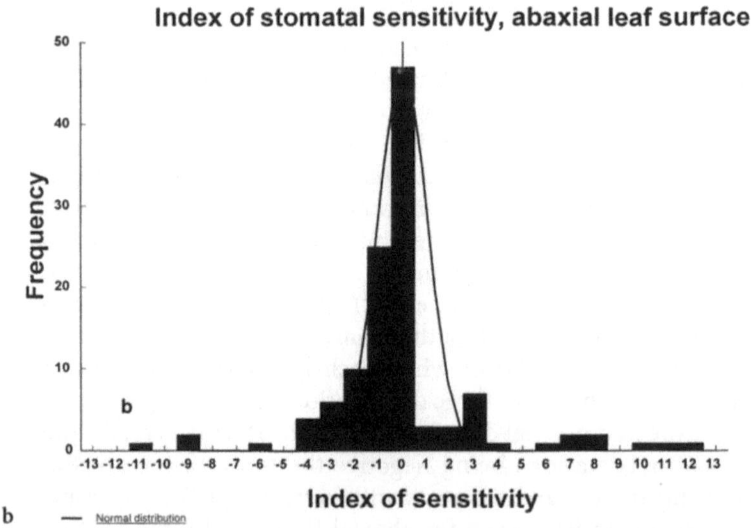

b

Fig. 10a–c. Frequency histograms of the stomatal index of sensitivity (IS) to CO_2 change for **a** both leaf surfaces, **b** abaxial surface and **c** adaxial surface. *Solid curves* indicate the fitted normal distribution centered around zero. The histograms were constructed from a standardized stomatal-density response to a unit change in atmospheric CO_2 concentration using Eq. (1). Statistical analysis of the data for each distribution is given in Table 3

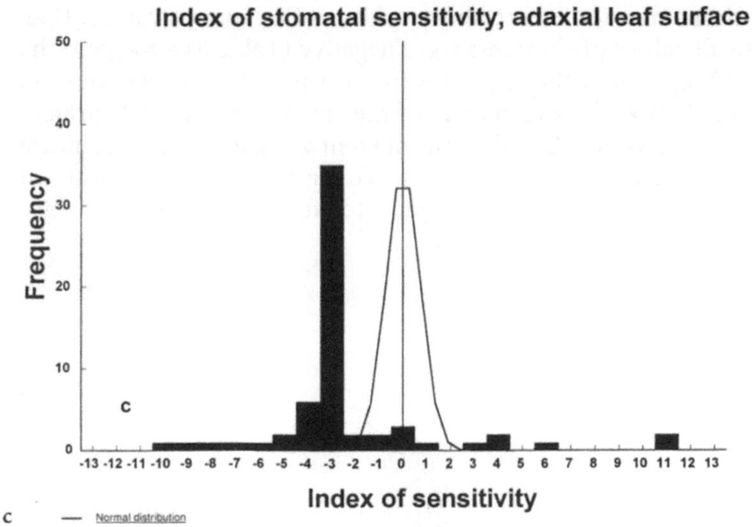

Fig. 10a–c. *Continued*

Table 3. Statistical summary and analysis of the index of stomatal sensitivity frequency distributions

Distribution	Mean	Range (min-max)	Chi-Square statistic[a]	*P*	Skewness[b] (±1 SE)	*P*
Fig. 10a	0.48	−10.2–42.7	335.9	0.001	5.35 ± 0.2	0.001
Fig. 10b	0.64	−10.8–42.7	201.5	0.001	3.75 ± 0.2	0.001
Fig. 10c	1.3	−14.5–28.5	2206.9	0.001	1.81 ± 0.3	0.001
Fig. 11a	−1.7	−1.32–0.22	136.6	0.001	−1.87 ± 0.3	0.001
Fig. 11b	−0.12	−1.32–0.22	199.2	0.001	−2.10 ± 0.3	0.001
Fig. 11c	−0.20	−1.03–0.20	15.91	0.05	−1.08 ± 0.4	0.001

[a] Observed frequencies were compared with expected frequencies, calculated from the fitted normal distribution, using the chi-square statistic to test the null hypothesis that the observed frequency distribution did not differ significantly from the normal distribution (Sokal and Rohlf 1981). All statistical analyses were performed using the UNISTAT (1993) statistical package.
[b] Skewness statistic was used to test the null hypothesis that there is no skewness, according to the test given by Sokal and Rohlf (1981, pp. 174–175).

lower-than-present CO_2 concentrations, growth chamber experiments and results from CO_2-rich spring sites responding to above present CO_2 concentrations. Data from altitudinal studies in particular introduce complicating environmental gradients, especially the decreasing partial pressure of oxygen and temperature and the increasing precipitation with altitude. In order to examine the more direct effects of CO_2 on IS values we have reconstructed frequency histograms for data derived only from controlled environment experiments. The results indicate a much narrower range of IS values, which may be expected for stomatal responses to above-ambient CO_2 concentrations (Fig. 11c). In this case all values of skewness were negative (Table 3) and again the majority of the observations themselves were also negative. There were no marked differences between the responses of the abaxial and adaxial surfaces (Fig. 11b,c). It seems therefore, based on the current available data, that future increases in CO_2 concentration are likely to reduce the stomatal density of vegetation but to a lesser extent than that reported for the CO_2 increase since the pre-industrial era (Woodward 1987).

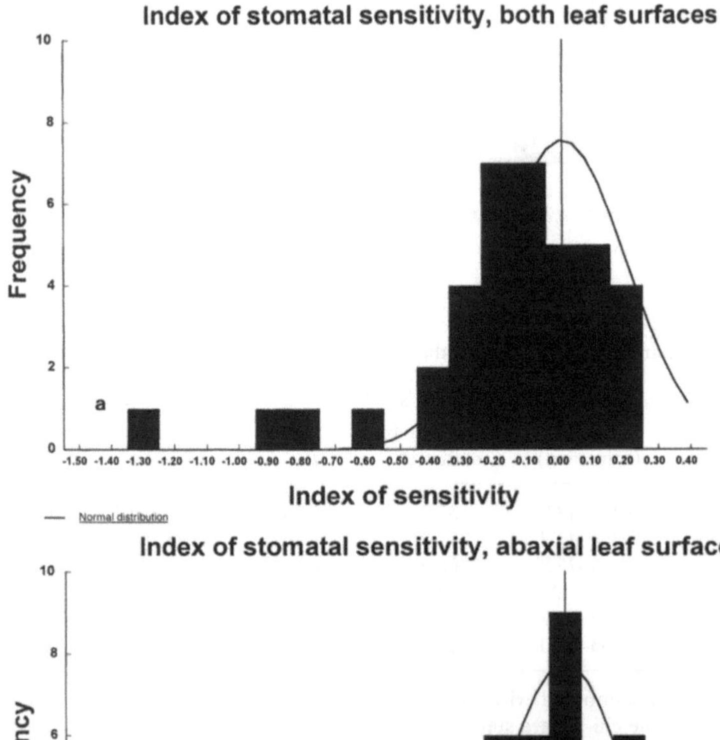

a

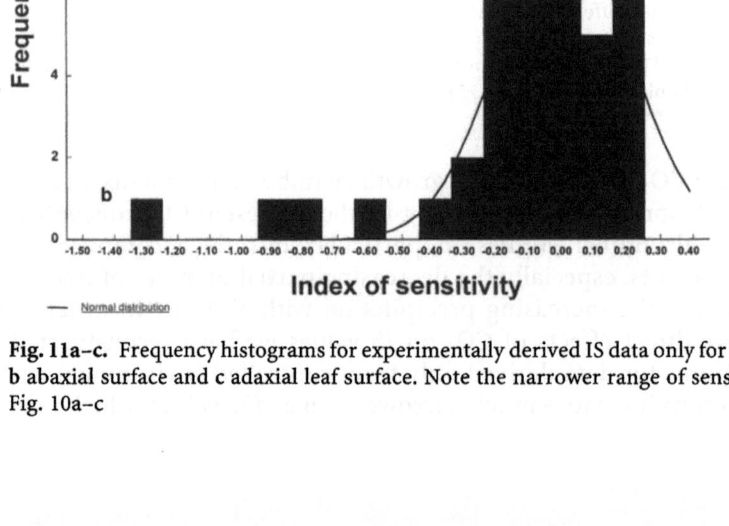

b

Fig. 11a–c. Frequency histograms for experimentally derived IS data only for **a** both leaf surfaces, **b** abaxial surface and **c** adaxial leaf surface. Note the narrower range of sensitivity compared to Fig. 10a–c

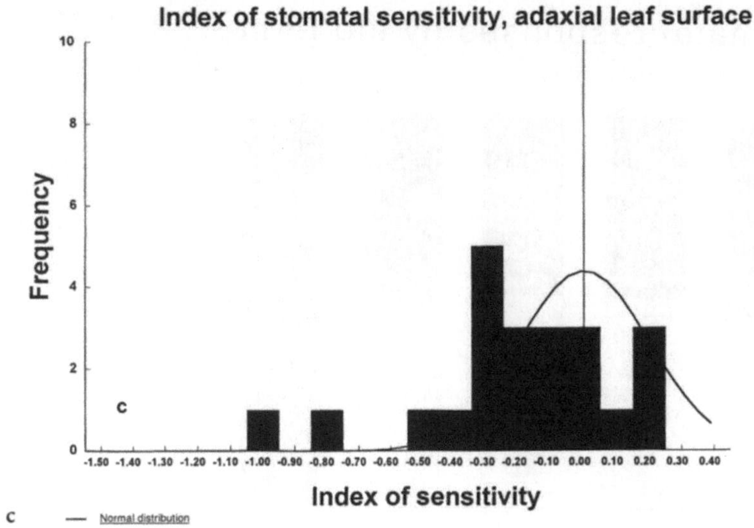

Fig. 11a–c. *Continued*

For comparisons of the stomatal responses between vegetation types with different life histories the data set was divided into five categories based on the strength of the density–CO_2 response: (1) strongly negative (≤–4.5), (2) weakly negative (–0.5 to –4.49), (3) little or no response (0.49 to –0.49), (4) weakly positive (0.5 to 4.49) and (5) strongly positive (≥4.5). The distribution of vegetation types within these categories (Fig. 12) gives some indication of which life form may be the most likely to respond in future and the direction of change. Interestingly, crop species (both C3 ad C4) consistently showed no response, in terms of stomatal density, to increases in CO_2 concentration. Only about 40% of tree species studied responded negatively. Shrubs, perennial herbs and annuals all show a weakly negative sensitivity which dominates the overall responses. Perennial herbs often have a positive response, perhaps due to the preponderance of alpine species in this category. However, a cautious interpretation of Fig. 12 is required since the database is limited in the range of habitats and species covered, alpine constituting by far the largest group. Nevertheless, with the exception of crop species, stomatal responses to CO_2 enrichment are not consistent with a particular life form, indicating species-specific responses (Fig. 12).

We further analysed the data set to test for consistency of stomatal responses depending upon the climate in which the species grew. The results, classified according to the Holdridge life zones (Holdridge 1967) show a more consistent pattern (Fig. 13). Species comprising tropical rainforest (little or no response), dry tropical forest, savannah and cool temperate forest (all weakly negative) all showed consistent responses (Fig. 13), with all other biomes dominated by a single type of response comprising over 60% of all species

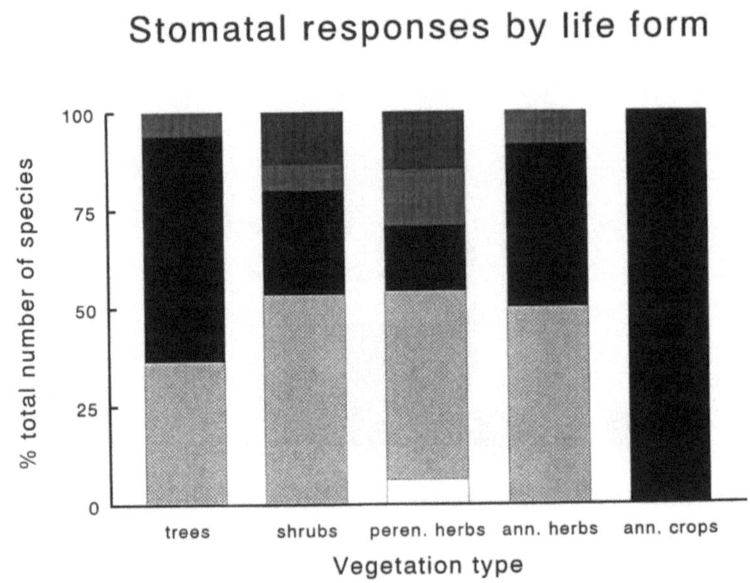

Fig. 12. Classification of standardized stomatal density responses into life-history types. Stomatal sensitivity codes: strongly negative (*unshaded*), weakly negative (*cross-hatched*), little/no response (*black*), weakly positive (*stippled*) and strongly positive (*dark-stippled*)

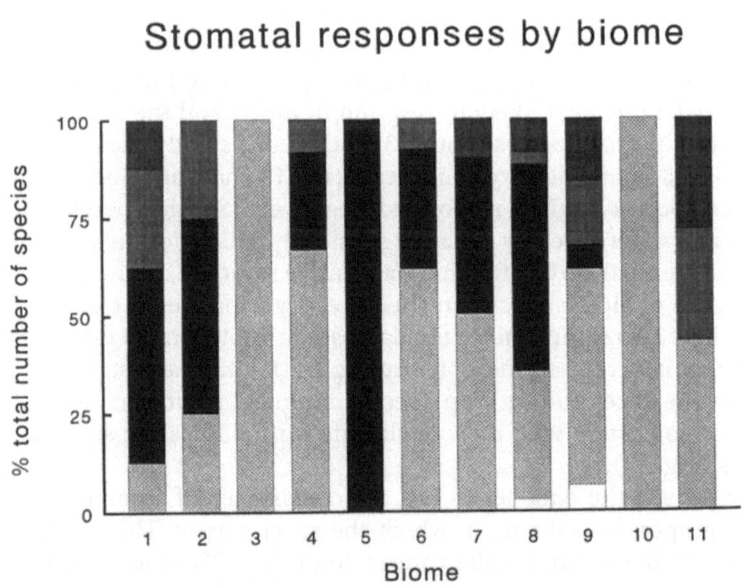

Fig. 13. Classification of standardized stomatal density responses into the global vegetation types of Holdridge (1967). Key of responsiveness as for Fig. 12. Biome codes: *1* tropical rainforest; *2* subtropical forest; *3* dry tropical forest; *4* chaparral; *5* savannah; *6* steppe; *7* warm temperate forest; *8* cool temperate forest; *9* boreal forest; *10* cool forest; *11* tundra

examined. Although ideally a larger database is required with greater coverage of species from all vegetation types in order to draw more robust conclusions, the most common category of response is weakly negative implying some decrease in density for an increase in atmospheric CO_2 concentration. No data were available for plants from hot or cool deserts.

6 The Interrelationship Between Whole Plant Physiology and Stomatal Density

Stomatal density has long been reported either as part of a comparative exercise or as an anatomical response, the significance of which was not always appreciated. Relatively little attention has focused on how stomatal density effects the gas-exchange characteristics of vegetation and its water economy, since the early work of Wilson (1972, 1975) and Wilson and Cooper (1969) investigated stomatal diffusion resistances in different lines of *Lolium perenne*. This final section seeks to attempt a synthesis and exploration of how changes in stomatal density might mediate changes in whole-plant gas exchange and water economy.

6.1 Gas Exchange

From simple diffusion theory, changes in the stomatal density of leaves would be expected to have a major effect on stomatal conductance (g_s) to water vapour (Penman and Schofield 1951; Wilson 1975) unless stomatal width tends to adjust to compensate for different densities (Wilson 1972; Jones 1977). In general an increase in stomatal density without any changes in pore width or length would be expected to reduce stomatal resistance (or increase stomatal conductance) (Jones 1992). This hypothesis has been developed in two ways. Firstly, Eq. (2) is used to calculate how g_s responds to changes in stomatal density, pore length and width. Secondly, measurements of the gas exchange of natural vegetation with different stomatal densities reported in the literature are reviewed in order to test the model predictions. The relevant literature covers comparative studies of plants growing under ambient and enriched CO_2 concentrations and studies on mountain plants originating from different altitudes grown in controlled environments. Mountain plants have tended to evolve genetically based differences in stomatal density, perhaps as a consequence of growing at reduced partial pressures of CO_2 (Woodward and Bazzaz 1988; Friend and Woodward 1990).

Stomatal conductance was calculated using the formula originally developed by Parlange and Waggoner (1970) and subsequently modified by Van Gardingen et al. (1989) to give stomatal conductance in units of $mol\,m^{-2}\,s^{-1}$, and then by Beerling and Woodward (1994a) to incorporate a temperature effect on water vapour diffusion:

$$g_s = \frac{nD'P_a}{RT\sum_{i=1}^{k}\left[\frac{d}{\pi a_i b_i} + \frac{Ln(4a_i/b_i)}{\pi a_i}\right] \Big/ k}, \tag{2}$$

where a_i = pore length (m)
b_i = width of the stomatal pore (m)
D' = temperature dependence of diffusion coefficient of water vapour in air corrected for temperature ($m^2 s^{-1}$), according to Eq. (3) below.
d = depth of pathway for gas exchange (m)
g_s = stomatal conductance ($mol\,m^{-2} s^{-1}$)
k = number of stomatal pores measured
n = stomatal density (m^{-2})
P_a = atmospheric pressure (101 325 Pa)
R = universal gas constant (8.314 $J\,mol^{-1} K^{-1}$)
T = mean leaf and air temperature (K)

Although the direct effects of temperature are included in the above model in the denominator, there is also the additional effect of temperature on the diffusivity of water vapour. Jones (1992) provides an equation to describe this effect. Therefore, Eq. (2) was modified according to Weyers and Meidner (1990) to reduce the free air water vapour diffusion by 0.1 to take account of wall collisions in narrow channels, resulting in the function:

$$D' = \left(D(T/293.2)^{1.75}\right)0.9, \tag{3}$$

where D = coefficient of self-diffusion of water vapour/air at 25 °C ($m^2 s^{-1}$) and T = temperature of the air (K).

Sensitivity analysis of the temperature component revealed that with a fixed diffusion pathway, pore length and width the relationship between g_s and stomatal density was not greatly affected by temperature. With a stomatal density of 100 mm^{-2}, g_s increased by 10% with an increase in temperature from 5 to 35 °C; similarly at a density of 1000 mm^{-2}, g_s increased by 8% over the same temperature range.

The model has been used to assess the effects of increasing stomatal density on g_s by assuming different fixed pore widths and (1) a fixed pore length and (2) a decreasing pore length. Both changes have been observed in natural vegetation. For example, Beerling and Woodward (1994a) reported no relationship between stomatal density and pore length in the arctic alpfine *S. herbacea* whilst other workers have found that pore length decreases as stomatal density increases in Poplar, barley and in species of Appalachian hardwoods (Miskin and Rasmusson 1970; Carpenter and Smith 1975; Jones 1977; Pallardy and Kozlowski 1979).

Figure 14a shows that, for a fixed pore length, g_s increases linearly with increasing density; this response is maintained regardless of the width of the

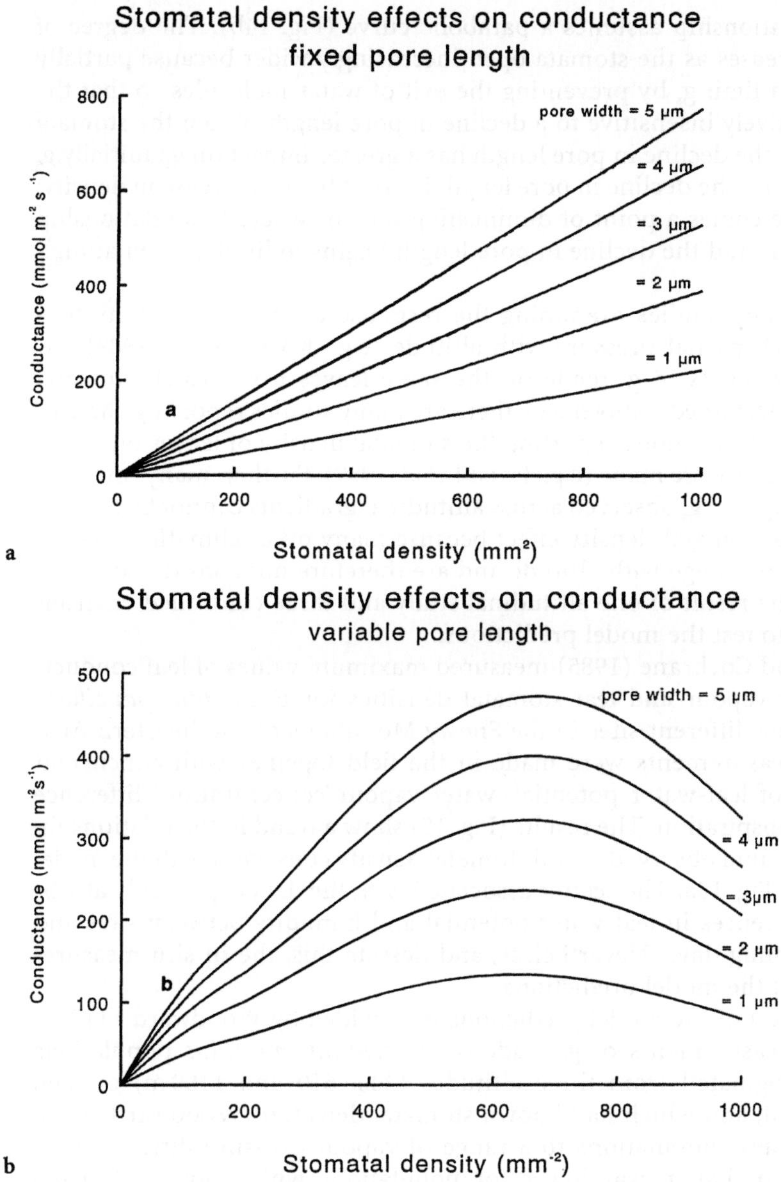

Fig. 14a,b. Modelled responses of stomatal conductance to changes in stomatal density. **a** – No change in pore length and **b** – a decrease in pore length from 25–5 µm. Each set of lines represents an increase in stomatal pore width from 1–5 µm

stomatal pore. In practice it is unlikely that a single species will exhibit such a large range of densities, but the computations nevertheless illustrate that an increase in stomatal density results in leaves with a higher g_s. Where the increase in density is associated with a decrease in pore length (from 25 to

5 µm), the relationship assumes a parabolic curve (Fig. 14b). The degree of curvature increases as the stomata open increasingly wider because partially closed stomata limit g_s by preventing the exit of water molecules so that the system is relatively insensitive to a decline in pore length. When the stomata are fully open, the decline in pore length has a greater impact on g_s; initially g_s increases because the decline in pore length is offset by the increase in density; however, there comes a point of diminishing returns where this relationship no longer holds and the decline in pore length begins to limit g_s even at high densities.

Despite many studies examining the response of stomatal density to a decrease in CO_2 partial pressure with altitude (e.g., Körner et al. 1989) few report measurements of g_s made on the same leaves from which stomatal density was determined. Conversely, there are many studies reporting changes in g_s with altitude but none reporting the stomatal density of leaves on which the measurements were made (e.g., Friend et al. 1989). Further, many interpretations of changes in g_s observed across altitudinal gradients cannot be directly attributed to a stomatal density effect because many other climatic variables affecting g_s also change with altitude and are therefore not considered. Here, we consider the result of one altitudinal study and three controlled environment studies to test the model predictions.

Körner and Cochrane (1985) measured maximum values of leaf conductance to water vapour and leaf stomatal densities for *Eucalyptus pauciflora* growing at four different sites in the Snowy Mountains of southeastern Australia. The measurements were made in the field together with concurrent observations of leaf-water potential, water-vapour concentration difference and rate of transpiration. The results (Fig. 15) show a trend in the relationship between maximum observed g_s and stomatal density consistent with the model predictions in Fig. 14a. The scatter associated with the data is probably attributable to differences in leaf-water potential and humidity between sites and between sampling days. Nevertheless, and despite this, the in situ measurements support the model predictions.

Another test of the model predictions is provided by Woodward's (1986) analysis of measurements of g_s made on *Vaccinium myrtillus* populations which had originated from three altitudes (200, 610 and 1100 m) on Ben More, Scotland, and which had known stomatal densities (Woodward 1986). Exposure of these populations to a range of vapour pressure differences of the air showed that g_s was lower for populations with a lower stomatal density (which occurs with decreasing altitude of origin) (Fig. 16a). This effect can clearly be seen in Fig. 16b (where a lower ab:ad stomatal-density ratio represents lower overall densities) which represents "slices" through Fig. 16a at three different VPDs. Woodward and Bazzaz (1988) reported a strong linear relationship between g_s and stomatal density for four species, one tree species (*Acer pseudoplatanus*) and three herbaceous species (*Rhamnus catharticus*, *Geum urbanum* and *Rumex crispus*), grown under two different partial pressures of CO_2 (34 and 22.5–26 Pa) (Fig. 17a). The benefit of including data

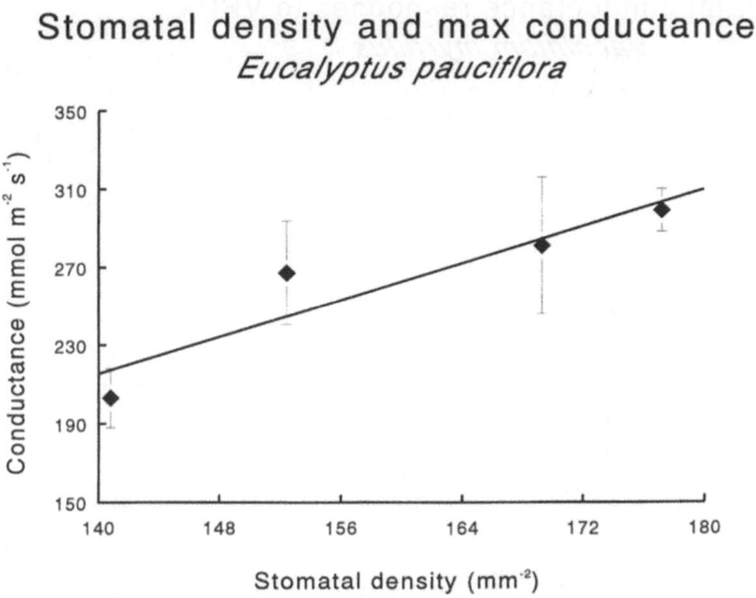

Fig. 15. Relationship between maximum reported values of leaf conductance to water vapour and increasing stomatal density for four populations of *Eucalyptus pauciflora* growing in the Snowy Mountains, southeastern Australia; mean ± 1 SE, r = 0.927, $P < 0.001$. (After Körner and Cochrane 1985)

from leaves which developed at different partial pressures of CO_2 is that it allows the relationship to be extended over a greater range of stomatal densities over which stomatal conductance remains linear (Fig. 17). Körner et al. (1983) showed a similar relationship for three species of *Ranunculus* originating from different altitudes on Mt. Wilhem, New Guinea, when grown under controlled environment conditions to standardize for temperature and humidity. A significant ($P < 0.001$) correlation between stomatal density and stomatal conductance has also been reported for three perennial grasses native to the southern prairies of Texas (Anderson and Briske 1990). These controlled environment studies, and the data of Körner and Cochrane (1985), show that the response of g_s to increasing stomatal density is in general agreement with the model predictions (Fig. 14a).

While there is limited support for the model predictions of changes in stomatal conductance with stomatal density, there are fewer investigations on the effects of stomatal density on photosynthetic rates (Heichel 1971; Miskin et al. 1972). An early investigation reported no significant relationship between photosynthetic rates and stomatal density for different genotypes of *Lolium perenne* (Wilson and Cooper 1969). Since then Woodward and Bazzaz (1988) have reported some data for the four species considered previously (Fig. 18). The relationship between stomatal density and photosynthesis shows lower rates of photosynthesis at the lower CO_2 concentrations; this effect is

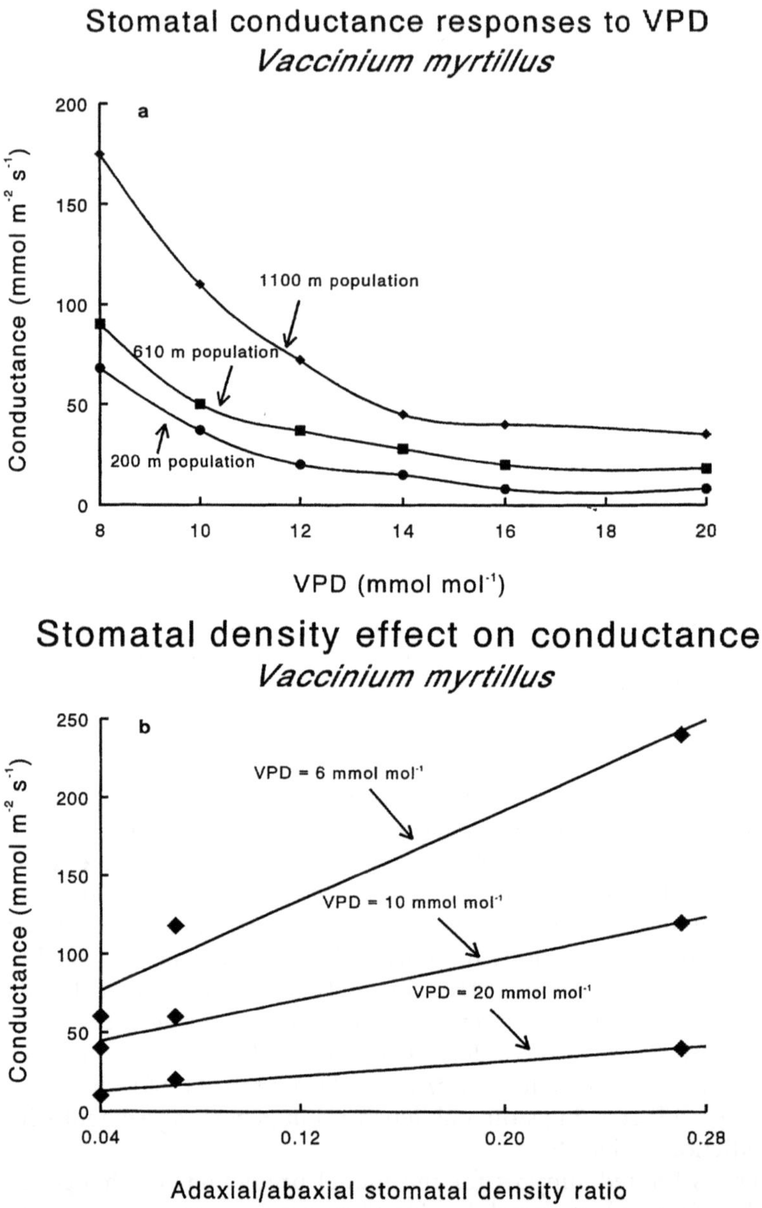

Fig. 16a,b. The response of stomatal conductance to **a** increases in the water-vapour pressure deficit *VPD* for three populations of *Vaccinium myrtillus* and **b** the relationship between stomatal conductance and stomatal density. (After Woodward 1986)

particularly marked at low saturation vapour pressure deficits (Woodward 1987). At low partial pressures of CO_2 the relationship appears to be asymptotic with a maximum photosynthetic rate reached at quite low densities (150 mm^{-2}) (Fig. 18) possibly because even with a greater number of stomata

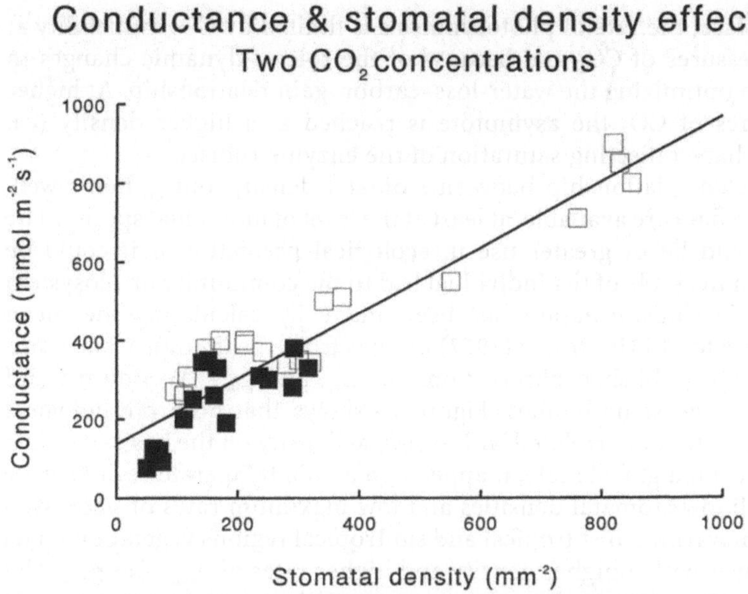

Fig. 17. The effects of stomatal density on stomatal conductance of four species (*Acer pseudoplatanus, Rhamnus catharticus, Geum urbanum, Rumex crispus*) at saturated leaf vapour pressure differences grown at 22.5–26 Pa CO_2 (*open squares*) and at 35 Pa CO_2 (*solid squares*); ($r = 0.935$, $P < 0.001$). (After Woodward and Bazzaz 1988)

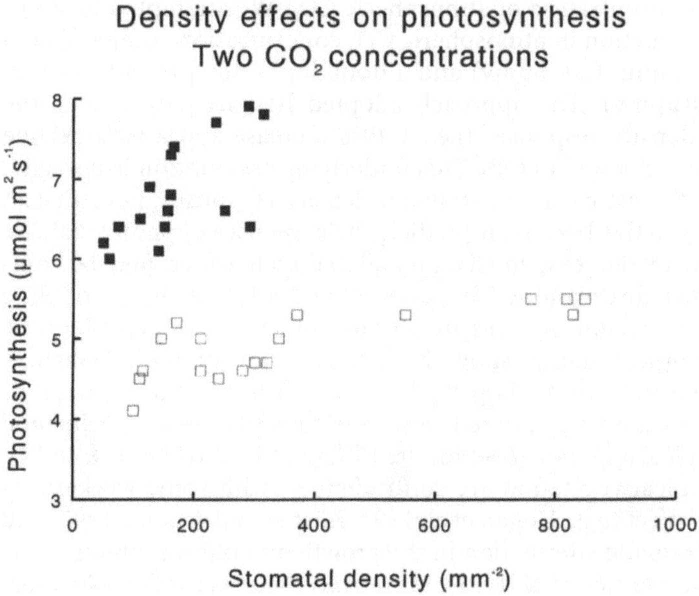

Fig. 18. Relationship between photosynthesis and stomatal density (species and *symbols* as in Fig. 17)

on the leaf surface, the rate of photosynthesis is limited by CO_2 availability at low partial pressures of CO_2 and because of the role of dynamic changes in stomatal width optimizing the water-loss–carbon-gain relationship. At higher partial pressures of CO_2 the asymptote is reached at a higher density (ca. 300 mm^{-2}), perhaps reflecting saturation of the enzyme rubisco.

The predicted relationship between stomatal density and g_s holds well, where sufficient data are available, at least at the level of individual species. The correlation would be of greater use in ecological prediction if it could be translated from the scale of the individual leaf to the community or ecosystem level. This scale transformation has been made by calculating the mean stomatal density for 15 Holdridge (1967) biomes from the literature and comparing them with published observations of A_{max} and g_{max} (Woodward and Smith 1994) for the same biomes. Figure 19 shows that both physiological processes are positively correlated with stomatal density on the basis of biome type. Therefore, on a global scale, it appears that cool temperate regions have vegetation with low stomatal densities and low maximum rates of photosynthesis, whilst in warm, moist tropical and subtropical regions vegetation tends to produce leaves with a higher density and higher rates of A_{max} and g_{max}. The pattern in Fig. 19b suggests that where water and low temperatures are not limiting factors, plants produce leaves with more stomata in order to maximize biomass production in warmer climates. This implies a possible causal link between the typical stomatal density evolved by a vegetation type and its climate.

The value of Fig. 19b is that it enables predictions of A_{max} to be made for each biome given an assumed change in stomatal density to either an increase or decrease in the concentration of atmospheric CO_2. We attempted to simulate the effect of a reduction in atmospheric CO_2 concentration comparable to the last glacial maximum (200 ppmv) and a doubling of the present-day CO_2 concentration (700 ppmv). The approach adopted has been to assume the relevant stomatal density responses (i.e., a 40% increase and a 50% decline respectively) to these changes in CO_2. This underlying assumption is unrealistic given the individualist nature of stomatal-density responses, nevertheless the approach is instructive because it predicts the responses of photosynthetic rates to an increase or decrease in CO_2 on a global scale which may be compared with published observations. The results (Fig. 20a) show that a doubling in CO_2 concentration would increase the photosynthetic rates of all biomes. Tropical and subtropical biomes show the greatest photosynthetic enhancement whilst cool tundra regions show the least enhancement. The predictions for these two biomes can be compared with recent results from experimental trials for moist tropical and moist/wet tundra biomes. The direct effects of CO_2 enrichment on tropical vegetation are contradictory, with some workers reporting little or no effect (e.g., Hogan et al. 1991; Körner and Arnone 1992) and others reporting dramatic stimulation in the growth and photosynthetic rates of tropical vegetation (Ziska et al. 1991). Moreover, the effects of elevated CO_2 on the stomatal density of tropical species also appear species-specific. After 3

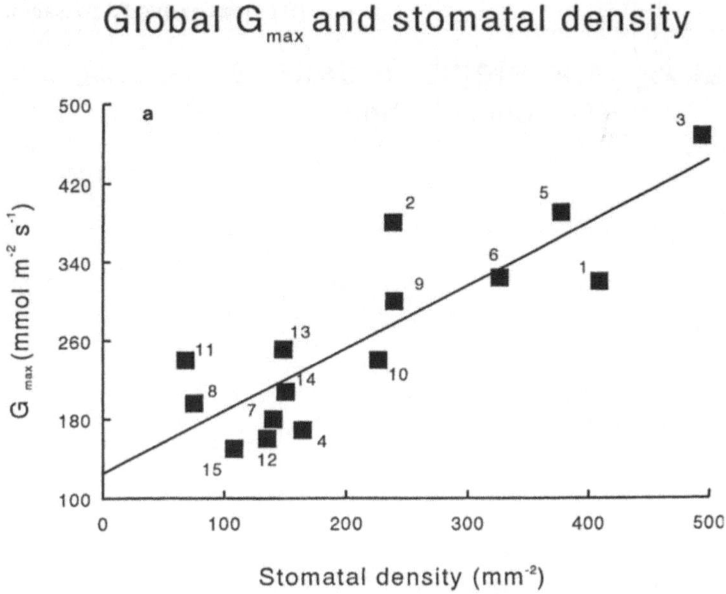

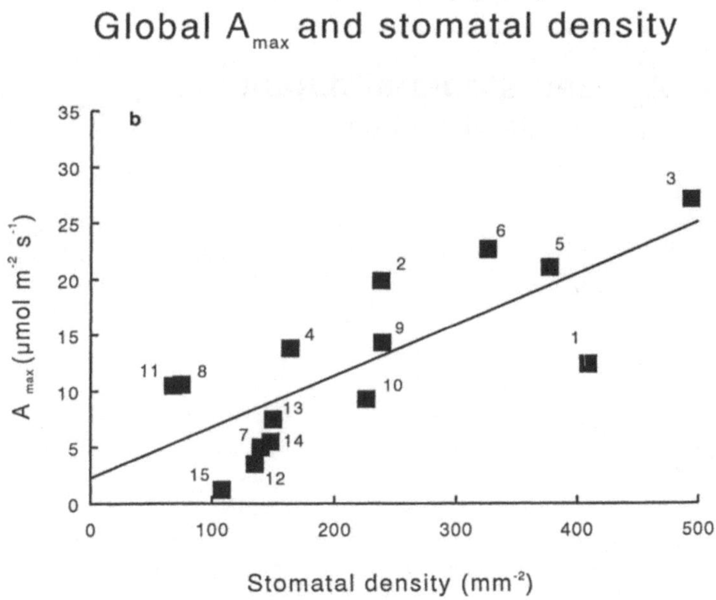

Fig. 19a,b. The relationship between mean stomatal density and **a** maximum stomatal conductance (g_{max}) and **b** maximum photosynthetic rates (A_{max}) for 15 different Holdridge (1967) biomes. Biomes were: *1* tropical wet forest; *2* subtropical wet forest; *3* subtropical moist forest; *4* subtropical desert; *5* warm temperate moist forest; *6* warm temperate dry forest; *7* warm temperate rainforest; *8* cool temperate wet forest; *9* cool temperate moist forest; *10* boreal wet forest; *11* boreal moist forest; *12* polar rain tundra; *13* polar wet tundra; *14* polar moist tundra; *15* polar dry tundra. A_{max} and g_{max} for each biome were taken from Woodward and Smith (1994a) and represent the maximum observed values reported in the literature. Stomatal-density data were compiled from: Au (1969), Beerling and Chaloner (1993a,b), Beerling et al. (1993), Bolhar-Nordenkampf and Draxler (1993), Bongers and Popma (1990), Carpenter and Smith (1975), Gindel (1969), Körner (1988), Körner and Cochrane (1985), Körner et al. (1986), Körner et al. (1989), Jones (1992), Mousseau and Enoch (1989), Oberbauer and Strain (1986), Oberbauer et al. (1985), Peñuelas and Matamala (1990), Weyers and Meidner (1990), Wilmer (1983), Woodward (1987) and Woodward and Bazzaz (1988). Regression coefficients: **a** g_{max} ($r = 0.856$, $P < 0.01$); **b** A_{max} ($r = 0.773$, $P < 0.01$)

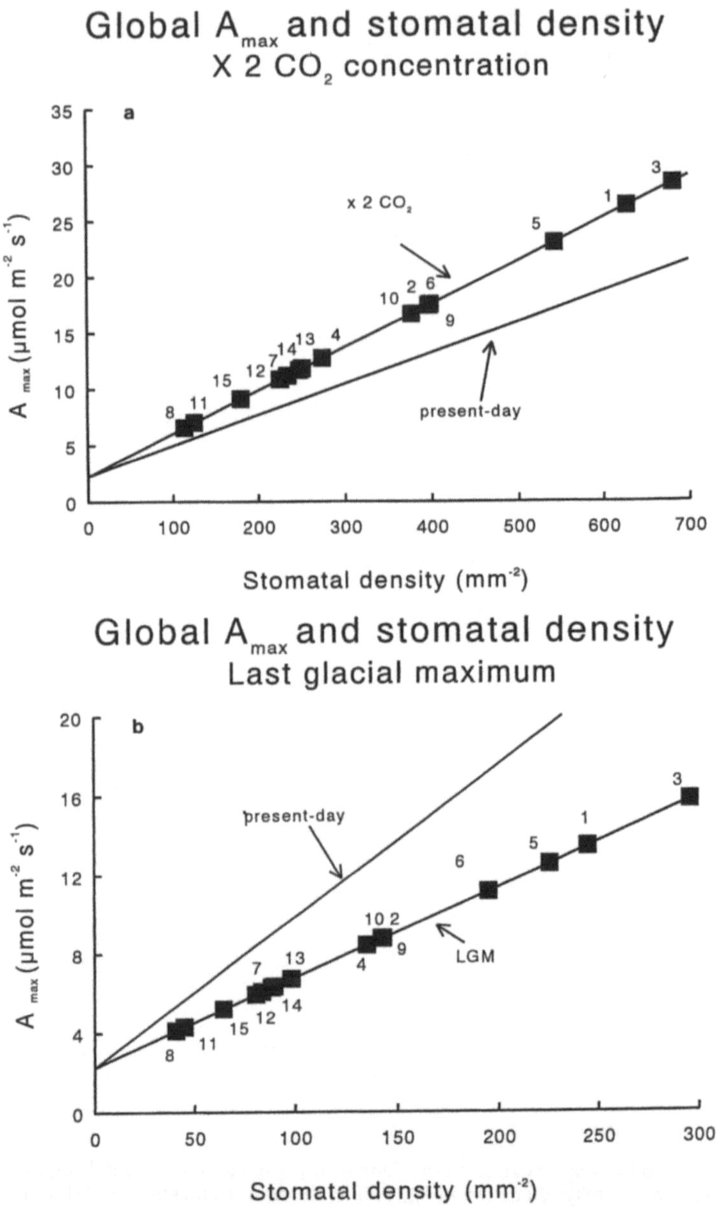

Fig. 20a,b. Predicted changes in A_{max} for each biome for **a** 2-fold CO_2 increase and **b** the last glacial maximum; based on forecast changes in stomatal density. Biome codes correspond to those in Fig. 17

months exposure to elevated CO_2 in an artificial tropical ecosystem, Körner and Arnone (1992) reported on significant differences in comparison to the control vegetation for the fifteen tropical species grown. Oberbauer et al. (1985), on the other hand, reported that of two species from the Atlantic lowland rainforest in Costa Rica, *Ochroma lagopus*, a fast-growing early successional and *Pentaclethera macroloba*, a slow-growing late successional, only *P. macroloba* responded to CO_2 enrichment with a decline in stomatal density. In tundra regions, no effects of increased CO_2 and temperature on net gas exchange responses have been reported even after exposure for 2 years (Tissue and Oechel 1987; Ziska et al. 1991), in accordance with the predictions of Fig. 20a.

Figure 20b predicts that during the last glacial maximum the photosynthetic rates of all biomes were lower. Several studies growing plants across a glacial-to-present atmospheric CO_2 gradient confirm this suggestion. Polley et al. (1992, 1993a,b) reported a decrease in photosynthetic rates of several C3 crop species as the atmospheric CO_2 concentration decreased from 350 to 150 ppmv. A different line of evidence also supports this result. Beerling and Woodward (1994a) inferred lower rates of photosynthesis from stomatal measurements made on late-Quaternary fossil leaves and Eqs. (2) and (3). It is clearly unlikely that lower photosynthetic rates of plants growing during the last glacial maximum would have been brought about solely by an increase in the stomatal density of the leaves. Photosynthetic rates are determined firstly by substrate availability and then by the ability of the plant to optimize carbon gain with respect to water loss. Lower photosynthetic rates of plants growing in a pre-industrial CO_2 concentration of ca. 270 ppmv have also been reported (Woodward 1987; Overdieck 1989). The stomatal density of *Acer pseudoplatanus* certainly contributed to the effect reported by Woodward (1987); however, it is unknown whether the reduced carbon exchange rate Overdieck (1989) reported for the shrub *Lyonia mariana* was attributable to changes in stomatal density and/or stomatal conductance.

The correlation between biome stomatal density and maximum photosynthetic rates provides a means of estimating, albeit rather simply, the proportional change in photosynthetic rates for different areas of the globe under different concentrations of atmospheric CO_2. Presently, the predicted changes in photosynthetic rates under both higher and lower CO_2 concentrations are supported by experimental studies. The approach may be further refined as more data become available by incorporating differential stomatal density responses of species from different biomes rather than assuming all species respond in the same manner.

The results of the modelling studies and the measurements on living plants together demonstrate that stomatal density clearly has a major impact on stomatal conductance. Therefore, in the past, when CO_2 levels were lower, g_s was higher but photosynthetic rates were lower. Overall, the effects of CO_2 on these two gas-exchange functions combine to reduce the water-use efficiency of vegetation. If the atmospheric CO_2 concentration continues to increase then a reduction in stomatal density is a likely response (see Sect. 5) associated with a decrease in g_s (this Section). The latter occurs through a combination of (1) CO_2 causing partial closure of the stomata pores (Mansfield et al. 1990) and (2) because of the physical relationship between stomatal density and g_s.

6.2 Water-Use Efficiency

Increased water-use efficiency is a ubiquitous response shown by plants when grown under elevated concentrations of atmospheric CO_2 (Morison 1985; Mott 1990). Formally, the water-use efficiency (WUE) of vegetation is defined as the number of units of CO_2 fixed per unit of water lost. Experimental data (e.g., Jones 1977; Woodward 1987) indicate that leaves with different stomatal densities appear to function with different WUE under both present and elevated CO_2 concentrations. On this basis, therefore, historical patterns of stomatal density, as determined from herbarium and fossil leaves, may be used to infer parallel long-term trends in the water-use efficiency of the plants (Woodward 1987; Beerling et al. 1993). These inferred changes in WUE can, in theory, then be tested independently by measuring the stable carbon isotope discrimination of the material.

The isotope discrimination term, for the stable ^{13}C and ^{12}C isotopes of carbon, proposed by Farquhar et al. (1982) relates to the leaf gas-exchange parameters as follows:

$$\Delta^{13}C = a + (b - a) p_i / p_a , \qquad (4)$$

where $\Delta^{13}C$ is the net isotopic discrimination against ^{13}C, relative to a standard, between air and plant; a is the diffusive discrimination in air (4.4‰) and b is the discrimination due to the enzyme rubisco (27‰); and p_i and p_a are the internal and external CO_2 partial pressures respectively (Farquhar et al. 1982). The theoretical inverse relationship between WUE and $\Delta^{13}C$ has been confirmed in a number of studies with container-grown crop plants, including wheat (Farquhar and Richards 1984), peanut (Hubick et al. 1986) and sunflower (Virgona et al. 1990), and more recently in a wide range of temperate field-grown crops under irrigated and droughted conditions (Knight et al. 1994). Non-crop species growing in natural conditions also appear to exhibit the same relationship between WUE and $\Delta^{13}C$, in such diverse habitats as grasslands (Smedley et al. 1991), cold deserts (Toft et al. 1989), tropical forests (Ehleringer et al. 1987) and deserts (Ehleringer 1993).

As yet there have been no experimental studies of plants grown under elevated CO_2 concentrations which have attempted to couple WUEs derived from gas-exchange measurements with an inferred WUE from measured carbon-isotope discrimination. Several difficulties may be anticipated in such experiments because the relationship between WUE and discrimination requires all other factors to be equal if it is to represent stomatal control and overall plant WUE. However, many other environmental factors also influence discrimination in leaves (Farquhar and Lloyd 1993) which are unlikely to be taken into account.

Experimental testing of this aspect of carbon isotope work is required; meanwhile a historical approach may be employed. We have compared the WUE trends in herbarium collections of leaves using two surrogate indicators of WUE ((Δ)^{13}C and stomatal density) obtained from the same leaves. Leaves of the past two centuries will have developed under the rapidly increasing CO_2 concentrations of this interval and may be expected to show improved WUE as a result. The improvement in WUE ought to be detectable in both carbon-isotope discrimination and stomatal-density responses. We have tested this hypothesis by measuring leaf discrimination and stomatal density for leaves of six temperate tree species collected over the past 200 years. The results show, rather surprisingly, that leaves possessing the highest stomatal density show the lowest discrimination (Fig. 21). This result runs counter to the theoretical prediction that leaves with more stomata should show a greater discrimination as a result of discrimination being dominated by the C-fixing enzyme rubisco (27‰; Farquhar et al. 1989) rather than stomatal diffusion resistance. However, discrimination, besides being related to the p_i/p_a ratio, is also proportional to atmospheric CO_2 concentration. Therefore, we have attempted to allow for the direct effects of CO_2 on $\Delta^{13}C$ by relating observed $\Delta^{13}C$ values to atmospheric CO_2 concentrations prevailing at the time of leaf collection, obtained from ice-core studies, to give corrected values of $\Delta^{13}C$. This linear relationship explained 99.75% of the variance, and clearly dominated the relationship in Fig. 21. Nevertheless, the difference between this corrected value and the observed value represents the $\Delta^{13}C$ changes due mainly to WUE. Recalculated in this way, our results show that

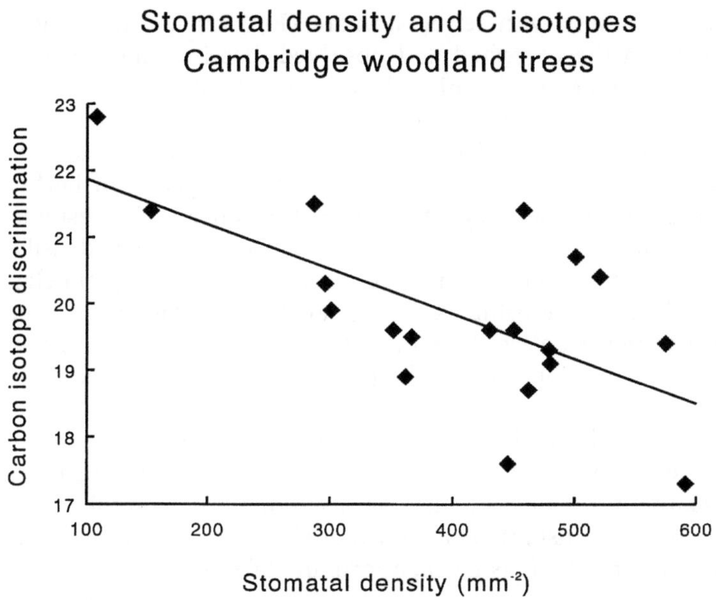

Fig. 21. The relationship between carbon-isotope discrimination and the stomatal density of the leaves of six species of temperate British trees (*Carpinus betulus*, *Fagus sylvatica*, *Populus nigra*, *Quercus petraea*, *Q. robur* and *Tilia cordata*; $r = 0.65$, $P < 0.05$)

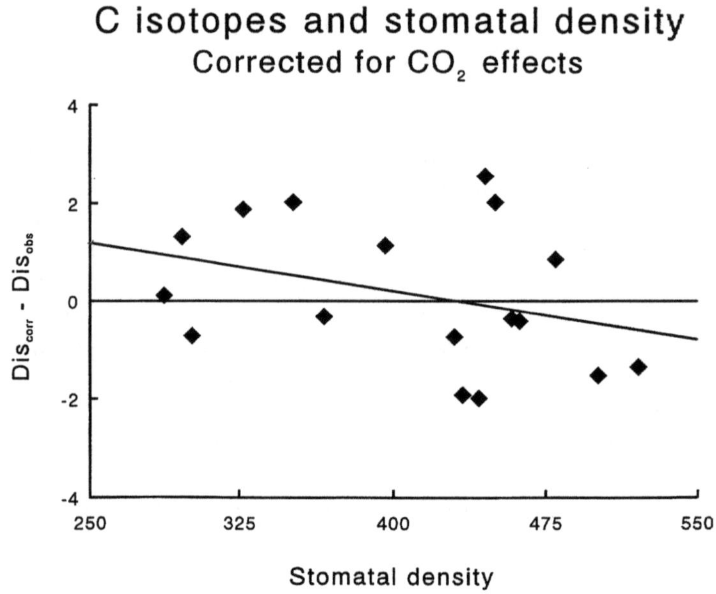

Fig. 22. The relationship between the difference in carbon isotope discrimination, corrected for CO_2 effects, and observed values and stomatal density. Data from the same species as Fig. 21 ($r = 0.33$, $P < 0.05$)

in fact discrimination is less with fewer stomata (Fig. 22; i.e., the larger the discrepancy between the corrected and observed discrimination values the greater the discrimination). The historical approach of using carbon-isotope discrimination to infer long-term changes in WUE has also been applied by Peñuelas and Azcón-Bieto (1992). These authors reported a small (ca. 1‰) mean decline in $\Delta^{13}C$ between 1750 and 1988 for 12 species of herbs, shrubs and trees from the Catalonia region of the Mediterranean. This result is in keeping with what might have been expected from the stomatal-density responses of the vegetation from this region, i.e., an increased WUE (Peñuelas and Matamala 1990), but a relationship which appears to contradict that in Fig. 21. The tabulated data nevertheless show that 30% of the species had leaves with an increased discrimination value at the present-day compared with the 1750s.

Another approach to determining WUE from carbon-isotope measurements utilises an integrated model of stomatal conductance and assimilation rates. In this model, the $\Delta^{13}C$ values of a leaf are used directly to calculate the p_i/p_a ratio from Eq. (4). The instantaneous ratio of CO_2 assimilation rate of a leaf, A, to its transpiration rate, E, is given, approximately, by:

$$\frac{A}{E} = \frac{p_a(1 - p_i/p_a)}{1.6\nu}, \tag{5}$$

where ν is the leaf water vapour pressure deficit (Farquhar et al. 1982, 1989). The factor of 1.6 is the ratio of diffusivities of water vapour and CO_2 in air. By allowing for respiratory loss of carbon at night WUE can be defined as:

$$\mathrm{WUE} = \frac{(1-\varnothing)p_a(1-p_i/p_a)}{1.6\nu}, \quad (6)$$

where ø is the proportion of fixed carbon respired by leaves at night and other parts of plants throughout the day. The proportion of ø is difficult to measure experimentally and ν is unknown historically. An adequate relative estimate of leaf WUE (Farquhar et al. 1982; Farquhar and Richards 1984; Hubick and Farquhar 1989), which avoids these complications is provided by:

$$\mathrm{WUE} = \frac{(p_a - p_i)}{1.6\nu}, \quad (7)$$

where ν′ is taken as the water vapour deficit of the air measured at a meteorological station. $\delta^{13}C$ analyses of plant material enable an estimation of p_i/p_a to be made [from Eq. (7)], which together with known values of p_a (from the ice-core record) and ν′ enable WUE of plants to be estimated historically.

WUE can then be estimated directly from Eq. (7) together with historical measurements of the vapour pressure deficit of the air which precludes calculation of leaf $\Delta^{13}C$. Woodward (1993) adopted this approach obtaining a historical sequence of VPD data from UK meteorological stations, p_i from leaf $\delta^{13}C$ composition and p_a from ice-core studies. The reported trend, determined by regression analysis, indicates an increase of 4 mmol CO_2/mol H_2O per century (Woodward 1993). This evidence is independent of the stomatal-density work and indicates the same pattern of improved WUE over the last two centuries. The increase is about 28% over the last century and is in keeping with experimental observations on the influence of increasing CO_2 on plant WUE (Eamus and Jarvis 1989). Araus and Buxó (1993) attempted to use Eq. (7) but without historical measurements of air VPD. However, despite the increase in WUE obtained using this approach, the authors assumed that the ν′ of the air remained constant over the last 7000 years, which is unlikely to be correct (Houghton et al. 1990).

This section has considered evidence for historical patterns in plant WUE from the stable carbon-isotope composition of herbarium and fossil leaves which can be compared with the inferred WUE changes from the stomatal-density record. Both sources of information on plant physiological processes converge to provide a similar conclusion, that changes in the atmospheric concentration of CO_2 have influenced plant functioning to an extent that it is detectable in terms of leaf morphology and integrated measures of gas-exchange processes, represented by the leaf carbon-isotope composition.

7 Conclusions

Stomata are easy to see either on leaf surfaces, or if necessary, on surface replicas. In the late nineteenth and early twentieth centuries a number of scientists realised the significance of these pores to the physiology and ecology of plants and made extensive surveys of stomatal density for different species (Köhne 1889; Salisbury 1927; Wood 1934; Metcalfe and Chalk 1950). Intriguing patterns emerged between stomatal density, the occurrence of amphistomaty and habitat. Little headway was made on the theoretical aspects of stomatal distribution and ecology, until relationships between amphistomaty and photosynthetic physiology (Longstreth et al. 1980), water availability (Parkhurst 1978; Mott et al. 1982) and irradiance (Mott et al. 1982) were investigated. This work encouraged an agricultural interest in breeding plants for low stomatal density with expectations of a decreased transpiration rate and increased water-use efficiency. Although the objective of reducing stomatal density was achieved, in some cases, transpiration also increased (Jones 1977) because the pore length of individual stomata increased in parallel with reduced density.

This review of the responses of stomatal density to CO_2, temperature and water relations and the relationships with the geographical distribution, indicates that there are quite distinct patterns of response. Over long time intervals, water-use efficiency does increase as stomatal density decreases (Fig. 21) and plants in the most photosynthetically active vegetation types tend to posses the highest stomatal densities (Fig. 17b). Similarly, across a wide range of species, stomatal density has declined over the last two centuries (Fig. 5), probably with little if any change in pore length. However, where the environment is extreme, then stomatal density has increased over the last 100 years (Fig. 6), a feature which may be explained by a gradual increase in water supply to the growing plant. It is clear that short-term artificial selection for stomatal density, as in agricultural crop-breeding programmes, does not mimic the long-term development and selective forces acting on stomatal density in plants in natural habitats. Indeed, it is particularly intriguing that crop species show the least response of stomatal density to CO_2 of all species (Fig. 10).

This review and that of Tichá (1982) lead to an impression that although stomatal density varies markedly between and within species, nevertheless the responses might best be considered as levels in a hierarchy. In the hierarchy, responses to water supply might be expected to override all others, while responses to irradiance might be next in importance with the response to CO_2 operating at an even lower level. Although this view can provide a framework for interpreting responses to changing environments, the final resolution of plant responses to environmental change will only be achieved when the actual response mechanisms are defined.

Appendix Species used for calculation of the index of sensitivity of stomatal responses to CO_2 change

Species	Author	Species	Author
Acer pseudoplatanus	Woodward and Bazzaz (1988)	*Olea europaea*	Beerling and Chaloner (1993b)
Achillea moschata	Körner (1988)	*Oryza sativa*	Rowland-Bamford et al. (1990)
Alnus glutinosa	Peñuelas and Matamala (1990)	*Oxyria digyna*	Au (1969); Körner (1988)
Amaranthus caudatus	Peñuelas and Matamala (1990)	*O. digyna*	Körner et al. (1989)
A. retroflexus	Woodward and Bazzaz (1988)	*Papaver alpinum*	Peñuelas and Matamala (1990)
Ambrosia artemisifolia	Woodward and Bazzaz (1988)	*Pedicularis kerneri*	Körner (1988)
Anthyllis vulneraria	Körner et al. (1989)	*Pelargonium x hortorum*	Kelly et al. (1990)
Arabidopsis thaliana	V. Putland (unpublished)	*Pentaclethra macroloba*	Oberbauer et al. (1985)
Arabis alpina	Körner (1988)	*Phaseolus vulgaris*	O'Leary and Knecht (1981)
Arctostaphylos uva-ursi	Körner et al. (1989)	*P. vulgaris*	Radoglou and Jarvis (1992)
Artemisia genipi	Körner (1988)	*Pinus pinea*	Peñuelas and Matamala (1990)
Bellidiastrum michelii	Körner et al. (1989)	*P. uncinata*	Peñuelas and Matamala (1990)
Betula nana	Beerling (1993)	*Pistacia lentiscus*	Peñuelas and Matamala (1990)
B. pendula	Lee (1993)	*Pittosporum pullifoluim*	Körner et al. (1983)
Betula pendula	Peñuelas and Matamala (1990)	*Plantago major*	Broadmeadow and Griffiths (1993)
Biscutella laevigata	Körner et al. (1989)	*Polygala amara*	Körner et al. (1989)
Boehmeria cylindrica	Woodward and Beerling (1996)	*Polygonum viviparum*	Körner (1989)
Buxus sempervirens	Peñuelas and Matamala (1990)	*Populus alba*	Lee (1993)
Campanula barbata	Körner (1988)	*P. euramericana*	Gaudillere and Mousseau (1989)
Cardamine resedifolia	Körner (1988)	*Populus x Beau*	Lee (1993)
Castanea sativa	Mousseau and Enoch (1989)	*Populus x Beau*	Radoglou and Jarvis (1993)
Cerastium uniflorum	Körner (1988)	*Populus x Col R*	Lee (1993)
Ceratonia siliqua	Peñuelas and Matamala (1990)	*Populus x Col R*	Radoglou and Jarvis (1993)
Coleus blumei	V. Putland (unpublished)	*Populus x Ras*	Radoglou and Jarvis (1993)
Cynodon dactylon	Peñuelas and Matamala 1990	*Populus x Robusta*	Lee (1993)
Erigeron uniflorus	Körner (1988)	*Populus x Robusta*	Radoglou and Jarvis (1993)
Eucalyptus pavciflora	Körner and Cochrane (1985)	*Potentilla frigida*	Körner (1988)
Fagus sylvatica	Körner et al. (1989)	*Primula auricula*	Körner et al. (1989)

Species	Author	Species	Author
Gentiana alpina	Penuelas and Matamala (1990)	*Quercus pubescens*	Miglietta and Raschi (1993)
G. verna	Körner (1988)	*Q. robur*	Beerling and Chaloner (1993a)
G. verna	Peñuelas and Matamala (1990)	*Q. robur*	Woodward and Bazzaz (1988)
Geum montanum	Körner et al. (1989)	*Ranunculus glacialis*	Körner (1988)
G. reptans	Körner et al. (1989)	*R. grahamii*	Körner et al. (1986)
G. rivale	Körner et al. (1989)	*R. lappaceus*	Körner et al. (1986)
G. urbanum	Woodward and Bazzaz (1988)	*Rhamnus catharticus*	Woodward and Bazzaz (1988)
Globularia cordifolia	Körner et al. (1989)	*Rhododendron hirsutum*	Körner et al. (1989)
G. nudicaulis	Körner et al. (1989)	*Rumex acetosella*	Körner et al. (1986)
Glycine max	Thomas and Harvey (1983)	*R. alpinum*	Körner et al. (1989)
Griselina littoralis	Körner et al. (1986)	*R. crispus*	Woodward and Bazzaz (1988)
Gaultheria antipoda	Körner et al. (1986)	*R. scutatus*	Körner et al. (1989)
G. depressa	Körner et al. (1986)	*Salix herbacea*	Beerling et al. (1993a)
G. mundula	Körner et al. (1983)	*Saxifraga moschata*	Körner (1988)
Helleborus foetidus	Peñuelas and Matamala (1990)	*Schizachyrium scoparium*	Malone et al. (1993)
Hutchinsia alpina	Körner (1988)	*Scirpus olneyi*	Drake (1992)
Hypochaeris radicata	Körner et al. (1986)	*Setaria faberii*	Woodward and Bazzaz (1988)
Juniperus communis	Peñuelas and Matamala (1990)	*Sibbaldia procumbens*	Körner (1988)
Leontodon hispidus	Körner et al. (1989)	*Silene acaulis*	Körner et al. (1989)
Linaria alpina	Körner et al. (1989)	*Sinapis alba*	Retuerto and Woodward (1993)
L. alpina	Körner (1988)	*Sorbus aucuparia*	Körner et al. (1989)
Liquidambar styraciflua	Thomas and Harvey (1983)	*Styphelia suaveoleus*	Körner et al. (1983)
Lobelia teleki	Körner et al. (1989)	*Triticum aestivum "yaqui"*	Malone et al. (1993)
Lolium perenne	Ryle and Stanley (1992)	*Triticum aestivum "seri"*	Malone et al. (1993)
Lycopersicum esculentum	Madsen (1973)	*Tropaeolum major*	V. Putland (unpub.)
Maranthes corymbosa	Eamus et al. (1993)	*Tussilago farfara*	Körner et al. (1989)
Nardus stricta	Woodward and Bazazz (1988)	*Vaccinium ambylandrum*	Körner et al. (1983)
Nothofagus menziesii	Körner et al. (1986)	*V. myrtillus*	Körner et al. (1989)
Ochroma lagopus	Oberbauer et al. (1985)	*V. myrtillus*	Woodward (1986)
		Vicia faba	Radoglou Jarvis (1993)
		Zea mays	Thomas and Harvey (1983)

Acknowledgements. We thank the Herbarium of the Hebrew University for supplying the leaf material of *Quercus boissieri* and *Populus euphratica* and Dr. G. Stanhill for providing the stomatal-density and meteorological data, Dr. J. Peñuelas for providing the stomatal-density data for each of the 14 species studied in the Mediterranean, Dr. A. Fuchs for the Byrd ice-core data and Dr. P.L. Mitchell, Professor W.G. Chaloner and Dr. G. Stanhill for constructive comments on an earlier draft of the manuscript.

References

Adams RM, Rosenzweig C, Peart RM, Ritchie JT, McCarl BA, Glyer JD, Curry RB, Jones JW, Boote KJ, Allen LH (1990) Global climate change and US agriculture. Nature 345:219–223

Anderson VJ, Briske DD (1990) Stomatal distribution and conductance of three perennial grasses native to the southern true prairie of Texas. Am Midl Nat 123:152–159

Araus JL, Buxó R (1993) Changes in carbon isotope discrimination in grain cereals from the north-western Mediterranean basin during the past seven millennia. Aust J Plant Physiol 20:117–128

Arp WJ (1991) Effects of source-sink relations on photosynthetic acclimation to elevated CO_2. Plant Cell Environ 14:869–875

Au S (1969) Internal leaf surface and stomatal abundance in arctic and alpine populations of *Oxyria digyna*. Ecology 50:131–135

Bakker JC (1991) Effects of humidity on stomatal density and its relation to leaf conductance. Sci Hortic (Amst) 48:205–212

Barnola JM, Raynaud D, Korotkevich YS, Lorius C (1987) Vostok ice core provides 160 000-years record of atmospheric CO_2. Nature 329:408–414

Beerling DJ (1993) Changes in the stomatal density of *Betula nana* L. leaves in response to increases in the atmospheric CO_2 concentration since the late-glacial. Spec Pap Palaeobot 49:181–187

Beerling DJ (1994) Palaeo-ecophysiological studies on Cretaceous and Tertiary fossil floras. In: Boulter MC, Fisher HC (eds) Cenozoic plants and climates of the Arctic. Springer, Berlin Heidelberg New York, pp 23–33

Beerling DJ, Chaloner WG (1991) Stomatal density as an indicator of atmospheric CO_2 concentration. Holocene 2:71–78

Beerling DJ, Chaloner WG (1993a) The impact of atmospheric CO_2 concentration and temperature change in stomatal density: observations from *Quercus robur* lammas leaves. Ann Bot 71:231–235

Beerling DJ, Chaloner WG (1993b) Stomatal density responses of Egyptian *Olea europaea* L. leaves to CO_2 change since 1327 BC. Ann Bot 71:431–435

Beerling DJ, Chaloner WG (1993c) Evolutionary responses of stomatal density to global CO_2 change. Biol J Linn Soc 48:343–353

Beerling DJ, Woodward FI (1994a) Ecophysiological responses of plants to global environmental change since the last glacial maximum. New Phytol 125:641–648

Beerling DJ, Woodward FI (1994b) The climate change experiment (CLIMEX): phenology and gas exchange responses of boreal vegetation to global change. Global Ecol Biogeog Lett 4:17–26

Beerling DJ, Chaloner WG, Huntley B, Pearson JA, Tooley MJ (1993) Stomatal density responds to the glacial cycle of environmental change. Proc R Soc Lond B251:133–138

Berner RA (1990) Atmospheric carbon dioxide levels over Phanerozoic time. Science 254:1382–1386

Berner RA (1992) Palaeo-CO_2 and climate. Nature 358:114

Berner RA (1993) Paleozoic atmospheric CO_2: importance of solar radiation and plant evolution. Science 261:68–70

Boden TA, Sepanski RJ, Stoss F (eds) (1991) Trends '91: a compendium of data on global change. Carbon dioxide information analysis centre, Oak Ridge Tennessee

Bolhar-Nordenkampf HR, Draxler G (1993) Functional leaf anatomy. In: Hall DO, Scurlock JMO, Bolhar-Nordenkampf HR, Leegood RC, Long SP (eds) Photosynthesis and production in a changing environment: a field and laboratory manual. Chapman & Hall, London, pp 91–112

Bongers F, Popma J (1990) Leaf characteristics of the tropical rain forest flora of Los Tuxtlas, Mexico. Bot Gaz 155:354–365

Bonnier G (1895) Recherches expérimentales sur l'adaptation des plantes au climat alpin. Annls Sci Nat (Bot) 20:217–360

Broadmeadow MSJ, Griffiths H (1993) Genotypic variation in *Plantago major* in response to elevated CO_2. J Exp Bot [Suppl] 44:15

Broeker WS, Peng TH (1986) Glacial to interglacial changes in the operation of the global carbon cycle. Radiocarbon 28:309–327

Carpenter SB, Smith ND (1975) Stomatal distribution and size in southern Appalachian hardwoods. Can J Bot 53:1153–1156

Cowan IR (1977) Stomatal behaviour and environment. Adv Bot Res 4:117–228

Dai A, Fung IY (1993) Can climate variability contribute to the "missing" CO_2 sink? Global Biogeochem Cycles 7:599–609

Drake BG (1992) A field study of the effects of elevated CO_2 on ecosystem processes in a Chesapeake Bay Wetland. Aust J Bot 40:579–595

Eamus D (1991) The interaction of rising CO_2 and temperatures with water use efficiency. Plant Cell Env iron 14:843–852

Eamus D, Jarvis PG (1989) The direct effects of increase in the global atmospheric CO_2 concentration on natural and commercial temperate trees and forests. Adv Ecol Res 19:1–55

Eamus D, Berryman CA, Duff GA (1993) Assimilation, stomatal conductance, specific leaf area and chlorophyll responses to elevated CO_2 of *Maranthes corybosa*, in a tropical monsoon rain forest species. Aust J Plant Physiol 20:741–755

Ehleringer JR (1993) Carbon and water relations in desert plants: an isotopic perspective. In: Ehleringer JR, Hall AE, Farquhar GD (eds) Stable isotopes and plant carbon-water relations. Academic Press, San Diego, pp 155–172

Ehleringer JR, Field CB (eds) (1993) Scaling physiological processes leaf to globe. Academic Press, San Diego

Ehleringer JR, Lin ZF, Field CB, Sun GC, Kuo CY (1987) Leaf carbon isotope discrimination of plants from a subtropical monsoon forest. Oecologia 72:109–114

El-Sharkaway MA, Cock JH, Hernandez ADP (1985) Stomatal response to air humidity and its relation to stomatal density in a wide range of warm climate species. Photosyn Res 7:137–149

Enting IG, Mansbridge JV (1991) Latitudinal distribution of sources and sinks of CO_2: results from an inversion study. Tellus 43B:156–170

Farquhar GD, Lloyd J (1993) Carbon and oxygen isotope effects in the exchange of carbon dioxide between terrestrial plants and the atmosphere. In: Ehleringer JR, Hall AE, Farquhar GD (eds) Stable isotopes and plant carbon-water relations. Academic Press, San Diego, pp 47–70

Farquhar GD, Richards RA (1984) Isotopic composition of plant carbon correlates with water use efficiency of wheat genotypes. Aust J Plant Physiol 11:539–552

Farquhar GD, O'Leary MH, Berry JA (1982) On the relationship between carbon isotope discrimination and the intercellular carbon dioxide concentration of leaves. Aust J Plant Physiol 9:121–137

Farquhar GD, Hubick KT, Condon AG, Richards RA (1989) Carbon isotope fractionation and plant water-use efficiency. In: Rundel PW, Ehleringer JR, Nagy KA (eds) Stable carbon isotopes in ecological research. Springer, Berlin Heidelberg New York, pp 21–40

Freeman KH, Hayes JM (1992) Fractionation of carbon isotopes by phytoplankton and estimates of ancient CO_2 levels. Global Biogeochem Cycles 6:185–198

Friedli H, Lötscher H, Oeschger U, Siegenthaler U, Stauffer B (1986) Ice core record of the $^{13}C/^{12}C$ ratio of atmospheric CO_2 in the past two centuries. Nature 324:237–238
Friend AD, Woodward FI (1990) Evolutionary and ecophysiological responses of mountain plants to the growing season environment. Adv Ecol Res 20:59–124
Friend AD, Woodward FI, Switsur VR (1989) Field measurements of photosynthesis, stomatal conductance, leaf nitrogen and $d^{13}C$ along altitudinal gradients in Scotland. Funct Ecol 3:117–122
Gale J (1972) Availability of carbon dioxide for photosynthesis at high altitudes. Ecology 53:494–497
Gaudillère JP, Mousseau M (1989) Short term effect of CO_2 on leaf development and gas exchange of young poplars (*Populus euramericana* cv I 214). Oecol Plant 10:95–105
Gifford RM (1992) Interaction of carbon dioxide with growth limiting environmental factors in vegetation productivity: implications for the global carbon cycle. Adv Bioclimatol 1:25–58
Gindel I (1969) Stomatal number and size as related to soil moisture in tree xerophytes in Israel. Ecology 50:263–267
Graumlich LJ (1991) Subalpine tree growth, climate, and increasing CO_2: an assessment of recent growth trends. Ecology 72:1–11
Heichel GH (1971) Stomatal movements, frequencies, and resistances in two maize varieties differing in photosynthetic capacities. J Exp Bot 22:644–649
Holdridge LR (1967) Life zone ecology. Tropical Science Center Costa Rica
Hoffert MI, Covey C (1992) Deriving global climate sensitivity from palaeoclimate reconstructions. Nature 360:573–576
Hogan KP, Smith AP, Ziska LH (1991) Potential effects of elevated CO_2 and changes in temperature on tropical plants. Plant Cell Environ 14:763–778
Houghton RA (1993) Is carbon accumulating in the northern temperate zone? Global Biogeochem Cycles 7:611–617
Houghton JL, Jenkins GJ, Ephraums JJ (1990) Climate change: the IPCC scientific assessment. Cambridge University Press, Cambridge
Houghton JL, Callander BA, Varney SK (1992) Climate change 1992: the supplementary report to the IPCC scientific assessment. Cambridge University Press, Cambridge
Hubick KT, Farquhar GD (1989) Carbon isotope discrimination and the ratio of carbon gained to water lost in barley cultivars. Plant Cell Environ 12:795–804
Hubick KT, Farquhar GD, Shorter R (1986) Correlation between water-use efficiency and carbon isotope discrimination in diverse peanut (*Arachis*) germplasam. Aust J Plant Physiol 13:803–816
Huner NPA, Palta JP, Li PH, Carter JV (1981) Anatomical changes in leaves of puma rye in response to growth at cold-hardening temperatures. Bot Gaz 142:55–62
Jarvis PG, McNaughton KG (1986) Stomatal control of transpiration: scaling up from leaf to region. Adv Ecol Res 15:1–49
Jones HG (1977) Transpiration in barley lines with differing stomatal frequencies. J Exp Bot 28:162–168
Jones HG (1992) Plants and microclimate. Cambridge University Press, Cambridge
Jouzel J, Barkov NI, Barnola JM, Bender M, Cappellaz J, Genthon C, Kotlyakov VM, Lipenkov V, Lorius C, Petit JR, Raynaud D, Raisbeck G, Ritz C, Sowers T, Stievenard M, Yiou F, Yiou P (1993) Extending the Vostok ice-core record of palaeoclimate to the penultimate glacial period. Nature 364:407–412
Kelly DW, Hicklenton PR, Reekie EG (1990) Photosynthetic response of geranium to elevated CO_2 as affected by leaf age and time of CO_2 exposure. Can J Bot 69:2482–2488
Knight JD, Livingston NJ, Van Kessel C (1994) Carbon isotope discrimination and water use efficiency of six crops grown under wet and dryland conditions. Plant Cell Env 17:173–179
Köhne E (1889) Über das vorkommen von papillen und oberseitigen spaltöffnungen auf Blattern von Laubholzgewachsen. Mitt Dtsch Den Gesell 2:53–73
Körner C (1988) Does global increase in CO_2 alter stomatal density? Flora 181:253–257

Körner C (1993) CO_2 fertilization: the great uncertainty in future vegetation development. In: Solomon AM, Shugart HH (eds) Vegetation dynamics and global change. Chapman & Hall, London, pp 53–70
Körner C, Arnone J (1992) Responses to elevated carbon dioxide in artificial tropical ecosystems. Science 257:1672–1675
Körner C, Cochrane PM (1985) Stomatal responses and water relations of *Eucalyptus pauciflora* in summer along an elevational gradient. Oeocolgia 66:443–455
Körner C, Diemer M (1994) Evidence that plants from high altitudes retain their greater photosynthetic efficiency under elevated CO_2. Funct Ecol 8:58–68
Körner C, Allison A, Hilscher H (1983) Altitudinal variation of leaf diffusive conductance and leaf anatomy in heliophytes of montane New Guinea and their interrelation with microclimate. Flora 174:91–135
Körner C, Bannister P, Mark AF (1986) Altitudinal variation in stomatal conductance, nitrogen content and leaf anatomy in different plant life forms in New Zealand. Oecologia 69:577–588
Körner C, Neumayer M, Menendez-Riedl SP, Smeets-Scheel A (1989) Functional morphology of mountain plants. Flora 182:353–383
LaMarche VC, Graybill DA, Fritts HC, Rose MR (1984) Increasing atmospheric carbon dioxide: tree ring evidence for growth enhancement in natural vegetation. Science 225:1019–1021
Lasaga AC, Berner RA, Garrels RM (1985) An improved geochemical model of atmospheric CO_2 fluctuations over the past 100 million years. Geophys Monogr 32:397–411
Lee HSJ (1993) Biomass growth: *Betula pendula*. In: An invesitgation into the impact of elevated CO_2 upon the response of European forests. EPOC-0013 Final Rep
Long SP (1991) Modification of the response of photosynthetic productivity to rising temperature by atmospheric CO_2 concentrations: has its importance been underestimated? Plant Cell Environ 14:729–739
Longstreth DJ, Hartsock TL, Nobel PS (1980) Mesophyll cell properties for some C3 and C4 species with high photosynthetic rates. Physiol Plant 48:497–498
Lorius C, Jouzel J, Raynaud D (1992) The ice core record: past archives of the climate and signpost to the future. Philos Trans R Soc Lond B338:227–234
Maier-Reimer E, Hasselmann K (1987) Transport and storage of CO_2 in the ocean – an inorganic ocean circulation transport model. Clim Dynam 2:63–90
Madsen E (1973) Effect of CO_2-concentration on the morphological, histological and cytological changes in tomato plants. Acta Agric Scand 23:241–246
Malone SR, Mayeux HS, Johnson HB, Polley HW (1993) Stomatal density and aperture length in four plant species grown across a subambient CO_2 gradient. Am J Bot 80:1413–1418
Mansfield TA, Hetherington AM, Atkinson CJ (1990) Some current aspects of stomatal physiology. Annu Rev Plant Physiol Mol Biol 41:55–75
Martin P (1993) Vegetation responses and feedbacks to climate: a review of models and processes. Clim Dynam 8:201–210
McConnaughay KDM, Berntson GM, Bazzaz FA (1993a) Plant responses to carbon dioxide. Nature 361:24
McConnaughay KDM, Berntson GM, Bazzaz FA (1993b) Limitations to CO_2-induced enhancement in pot studies. Oecologia 94:550–557
Melillo JM, McGuire AD, Kicklighter DW, Moore B, Vorosmarty CJ, Schloss AL (1993) Global climate change and terrestrial net primary productivity. Nature 363:234–240
Metcalfe CR, Chalk L (1950) Anatomy of dicotyledons. Clarendon Press, Oxford
Miglietta F, Raschi A (1993) Studying the effect of elevated CO_2 in the open in a naturally enriched environment in Central Italy. Vegetatio 104/105:391–400
Miglietta F, Raschi A, Bettarini I, Resti R, Selvi F (1993) Natural CO_2 springs in Italy: a resource for examining long-term response of vegtation to rising atmospheric CO_2 concentrations. Plant Cell Environ 16:873–878
Miskin KE, Rasmusson DC (1970) Frequency and distribution of stomata in barely. Crop Sci 10:575–578

Miskin KE, Rasmusson DC, Moss DN (1972) Inheritance and physiological effects of stomatal frequency in barley. Crop Sci 12:780–783

Morison JIL (1985) Sensitivity of stomata and water use efficiency to high CO_2. Plant Cell Environ 8:467–474

Mott KA (1990) Sensing of CO_2 by plants. Plant Cell Environ 13:731–737

Mott KA, Gibson AC, O'Leary JW (1982) The adaptive significance of amphistomatic leaves. Plant Cell Environ 5:455–460

Mousseau M, Enoch HZ (1989) Carbon dioxide enrichment reduces shoot growth in sweet chestnut seedlings (*Castanea sativa* Mill). Plant Cell Environ 12:927–934

Neftel A, Oeschger H, Staffelbach T, Stauffer B (1988) CO_2 record in the Byrd ice core 50 000–5000 years BP. Nature 331:609–611

Oberbauer SF, Strain BR (1986) Effects of canopy position and irradiance on the leaf physiology and morphology of *Pentaclethra macroloba* (Mimosaceae). Am J Bot 73:409–416

Oberbauer SF, Strain BR, Fetcher N (1985) Effect of CO_2-enrichment on seedling physiology and growth of tropical tree species. Physiol Plant 65:352–356

O'Keefe JD, Ahrens TJ (1989) Impact production of CO_2 by the Cretaceous/Tertiary extinction bolide and the resultant heating of the Earth. Nature 338:247–249

O'Leary JW, Knecht GN (1981) Elevated CO_2 concentration increases stomate numbers in *Phaseolus vulgatris* leaves. Bot Gaz 142:438–441

Overdieck D (1989) The effects of preindustrial and predicted future atmospheric CO_2 concentration on *Lyonia mariana* L.D. Don. Funct Ecol 3:569–576

Pallardy SG, Kozlowski TT (1979) Frequency and length of stomata of 21 *Populus* clones. Can J Bot 57:2519–2523

Parkhurst DF (1978) The adaptive significance of stomatal occurrence on one or both surfaces of leaves. J Ecol 66:367–383

Parlange JY, Waggoner PE (1970) Stomatal dimensions and resistance to diffusion. Plant Physiol 46:337–342

Penman HL, Schofield RK (1951) Some physical aspects of assimilation and transpiration. Symp Soc Exp Bot 5:115–129

Peñuelas J, Azcón-Bieto J (1992) Changes in leaf $\Delta^{13}C$ of herbarium plant species during the last 3 centuries of CO_2 increase. Plant Cell Environ 15:485–489

Peñuelas J, Matamala R (1990) Changes in N and S leaf content, stomatal density and specific leaf area of 14 plant species durng the last three centuries of CO_2 increase. J Exp Bot 41:1119–1124

Polley HW, Johnson HB, Mayeux H (1992) Growth and gas exchange of oats (*Avena sativa*) and wild mustard (*Brassica kaber*) at subambient CO_2 concentrations. Int J Plant Sci 15:453–461

Polley HW, Johnson HB, Marino BD, Mayeux HS (1993a) Increase in C3 plant water use efficiency and biomass over glacial to present CO_2 concentrations. Nature 361:61–64

Polley HW, Johnson HB, Mayeux HS, Malone SR (1993b) Physiology and growth of wheat across a subambient carbon dioxide gradient. Ann Bot 71:347–356

Pollunin O, Huxley A (1974) Flowers of the Mediterranean. Chatto and Windus, London

Post WM (1993) Uncertainties in the terrestrial carbon cycle. In: Soloman AM, Shugart HH (eds) Vegetation dynamics and global change. Chapman & Hall, London, pp 116–132

Radoglou KM, Jarvis PG (1990) Effects of CO_2 enrichment on four Poplar clones. II. Leaf surface properties. Ann Bot 65:627–632

Radoglou KM, Jarvis PG (1992) The effects of CO_2 enrichment and nutrient supply on growth, morphology and anatomy of *Phaseolus vulgaris* L. seedlings. Ann Bot 70:245–256

Radoglou KM, Jarvis PG (1993) Effects of atmospheric CO_2 enrichment on early growth of *Vicia faba*, a plant with large cotyledons. Plant Cell Environ 16:93–98

Raynaud D, Jouzel J, Barnola JM, Chappellaz J, Delmas RJ, Lorius C (1993) The ice core record of greenhouse gases. Science 259:926–934

Retuerto R, Woodward FI (1993) The influences of increased CO_2 and water supply on growth, biomass allocation and water use efficiency of *Sinapis alba* L. grown under different wind speeds. Oecologia 94:415–427

Rowland-Bamford AJ, Nordenbrock C, Baker JT, Bowes G, Hartwell Allen L (1990) Changes in stomatal density in rice grown under various CO_2 regimes with natural solar irradiance. Environ Exp Bot 30:175–180

Rozema J, van de Geijn, Cambridge ML (eds) (1993) CO_2 and the biosphere. Kluwer, Dordrecht

Ryle GJA, Stanley J (1992) Effect of elevated CO_2 on stomatal size and distribution in perennial ryegrass. Ann Bot 69:563–565

Salisbury EJ (1927) On the causes and ecological significance of stomatal frequency, with special reference to the woodland flora. Philos Trans R Soc Lond B216:1–65

Siegenthaler U, Sarmiento JL (1993) Atmospheric carbon dioxide and the ocean. Nature 365:119–125

Smedley MP, Dawson TE, Comstock JP, Donovan LA, Sherrill DE, Cook CS, Ehleringer JR (1991) Seasonal carbon isotope discrimination in a grassland community. Oecologia 85:314–320

Sokal RR, Rohlf FJ (1981) Biometry. Freeman, New York

Strain BR, Cure JD (eds) (1985) Direct efects of increasing carbon dioxide on vegetation. DOE/ER-0238, US Dep Energy, Washington, DC

Suh HW, Dayton AD, Casady AJ, Liang GH (1976) Diallel cross analysis of stomatal density and leaf blade area in grain sorghum. *Sorghum bicolor*. Can J Genet Cytol 18:679–686

Sundquist ET (1993) The global carbon budget. Science 259:937–941

Thomas JF, Harvey CN (1983) Leaf anatomy of four species grown under continuous CO_2 enrichment. Bot Gaz 144:303–309

Tichá I (1982) Photosynthetic characteristics during ontogenesis of leaves: 7 stomatal density and sizes. Photosynthetica 16:375–417

Tissue DT, Oechel WC (1987) Responses of *Eriophorum vaginatum* to elevated CO_2 and temperature in the Alaskan tussock tundra. Ecology 68:401–410

Toft NL, Anderson JE, Nowak RS (1989) Water use efficiency and carbon isotope composition of plants in a cold desert environment. Oecologia 80:11–18

UNISTAT (1993) UNISTAT Statistical package. Users guide version 1.2 Unistat, London

Van der Burgh J, Visscher H, Dilcher DL, Kürschner WM (1993) Paleoatmospheric signatures in Neogene fossil leaves. Science 260:1788–1790

Van de Water PK, Leavitt SW, Betancourt JL (1994) Trends in stomatal density and $^{13}C/^{12}C$ of *Pinus flexilis* needles during the last glacial-interglacial cycle. Science 264:239–243

Van Gardingen PR, Jeffree CE, Grace J (1989) Variation in stomatal aperture in leaves of *Avena fatua* L. observed by low-temperature electron microscopy. Plant Cell Environ 12:887–898

Van Gardingen PR, Grace J, Jeffree CE (1993) Long-term effects of CO_2 increase on a Mediterranean ecosystem. J Exp Bot (Suppl) 44:13

Virgona JM, Hubick KT, Rawson HM, Farquhar GD, Downes RW (1990) Genotypic variation in transpiration efficiency, carbon-isotope discrimination and carbon allocation during early growth in sunflower. Aust J Plant Physiol 17:207–214

Wagner A (1892) Zur Kenntniss des Blattbaues der Alpenflanzen und dessen biologischer Bedeutung. Sber Akad Wiss Wien 100:487–547

Weyers J, Meidner H (1990) Methods in stomatal research. Longman, Essex

Wigley TML, Raper SCB (1992) Implications for climate and sea level of revised IPCC emissions scenarios. Nature 357:293–300

Wilmer CM (1983) Stomata. Longman, London

Wilson D (1972) Effects of selection for stomatal length and frequency on theoretical stomatal resistance to diffusion in *Lolium perenne*. New Phytol 71:811–817

Wilson D (1975) Leaf growth, stomatal diffusion resistances and photosynthesis during droughting of *Lolium perenne* populations selected for contrasting stomatal length and frequency. Ann Appl Biol 79:67–82

Wilson D, Cooper JP (1969) Effect of light intensity and CO_2 on apparent photosynthesis and its relationship with leaf anatomy in genotypes of *Lolium perenne* L. New Phytol 68:627–644

Wolfe JA, Upchurch GR (1986) Vegetation, climatic and floral changes at the Cretaceous-Tertiary boundary. Nature 324:148–152

Wood JG (1934) The physiology of xerophytism in Australian plants. J Ecol 22:69–85
Woodward FI (1986) Ecophysiological studies on the shrub *Vaccinium myrtillus* L. taken from a wide altitudinal range. Oecologia 70:580–586
Woodward FI (1987) Stomatal numbers are sensitive to increases in CO_2 from pre-industrial levels. Nature 327:617–618
Woodward FI (1988) The responses of stomata to changes in atmospheric levels of CO_2. Plants Today 1:132–135
Woodward FI (1993) Plant responses to past concentrations of CO_2. Vegetatio 104/105:145–155
Woodward FI, Bazzaz FA (1988) The responses of stomatal density to CO_2 partial pressure. J Exp Bot 39:1771–1781
Woodward FI, Beerling DJ (1993) Plant-CO_2 responses in the long term: plants from CO_2 springs in Florida and tombs in Egypt. In: High CO_2 springs and their use in biological research. Cambridge University Press, Cambridge, (in press)
Woodward FI, Smith TM (1994a) Predictions and measurements of the maximum photosynthetic rate, A_{max}, at the global scale. In: Schulze ED, Caldwell MM (eds) Ecophysiology of photosynthesis. Springer, Berlin Heidelberg New York, pp 491–509
Woodward FI, Smith TM (1994b) Global photosynthesis and stomatal conductance: modelling the controls by soil and climate. Adv Bot Res 20:1–41
Woodward FI, Thompson GB, McKee IF (1991) The effects of elevated concentrations of carbon dioxide on individual plants, populations, communities and ecosystems. Ann Bot (Suppl 1) 67:23–38
Ziska LH, Hogan KP, Smith AP, Drake BG (1991) Growth and photosynthetic responses of nine tropical species with long-term exposure to elevated carbon dioxide. Oecologia 86:383–389

Sensory and Behavioural Responses of Terrestrial Invertebrates to Biogenic Carbon Dioxide Gradients

G. Stange[1]

1 Introduction

As CO_2 is both a major end product of metabolism in heterotrophic organisms and the predominant carbon source in autotrophic organisms, it is not surprising that variations in its concentration have effects on animals. The effects are highly diverse, particularly if concentrations are above values normally encountered in the habitat, and they can occur on a considerable number of levels (Fuzeau-Braesch and Nicolas 1981; Nicolas and Sillans 1989). The scope of the present review is limited to a single effect, namely the action of CO_2 on identified peripheral sensory organs specialized to its detection, and the resulting coordinated behavioural responses at concentrations that occur naturally in the habitats of the organisms under consideration.

At first sight, it might appear that CO_2 is too ubiquitous to be relevant as a sensory signal. Indeed, the ability to measure its concentration in the atmosphere appears to be absent in land vertebrates, although its concentration inside the body is sensed and participates in the control of breathing. This might be seen as a consequence of the fact that the noses of vertebrates are exposed to exhaled air in which the concentration of CO_2 exceeds the atmospheric background by several orders of magnitude.

For invertebrates, the situation is different. Over the last 30 years, investigators of the physiology of this taxon have repeatedly, and often unexpectedly, identified sensory structures which respond to CO_2, as well as a wide range of behavioural responses. The bulk of the observations has been made in insects, but exceptions make it necessary to expand the scope of this article beyond that class.

The CO_2 sense is involved in the detection of airborne molecules and is, therefore, part of the sensory modality of olfaction. However, it has emerged that CO_2-sensitive sensory receptor structures are usually specialized for the detection of that stimulus, justifying their treatment as a separate group. In addition, other olfactory signals can sometimes be present at very low concen-

[1] Research School of Biological Sciences, Institute of Advanced Studies, Australian National University, P.O. Box 475, Canberra A. C. T. 2601, Australia and Abteilung Kaissling, Max-Planck-Institut für Verhaltensphysiologie. 82319 Seewiesen Germany

trations, making their detection an all-or-nothing task that can involve the sensing of a few molecules. In comparison, the mean atmospheric CO_2 background is in the order of 10^{19} molecules l^{-1}, implying that the sensing task involves the measurement of concentration differences or contrasts. To a stationary detector of small size, concentration differences can only occur in the dimension of time (temporal gradients); however, spatial gradients can be measured if the detector is either moved or if several detectors are present.

Sources of gradients include localized pockets of high CO_2 levels in confined environments such as cavities in the soil, burrows or hives; transient high-concentration plumes of the breath of large mammals; and the soil–atmosphere interface as an extended source, as well as the smaller and less-localized contrasts resulting from the respiration or CO_2 assimilation of plants, ranging in scale from the immediate vicinity of leaves to the planetary boundary layer.

The material presented here is organized in order of decreasing magnitude of the gradients involved and hence in order of increasing demand on the signal detection capability of the sensory systems. In each case, information on the magnitude and structure of the input signals is provided, together with a description of the properties of the detectors, if that information is available. In addition, the behavioural responses are described. Where possible, an attempt is also made to put the CO_2 sense in perspective with other senses. That section is preceded by an synopsis of general principles of the structure and function of the sensory organs.

2 General Structure and Function of the Detectors

In arthropods, sensory organs conveying modalities other than vision are composed of sensilla; they occur on a wide variety of body appendages. In insects, general olfactory sensilla occur predominantly on the antennae; CO_2-sensitive sensilla have been located in a wide range of positions, including the antennae (honeybees, flies, termites), labial palps (adult Lepidoptera), maxillary palps (larvae of beetles and Lepidoptera, adult mosquitoes), head capsule (centipedes) and tarsi of the forelegs (ticks).

The basic design of a sensillum is shown in Fig. 1. One or more sensory neurons (also referred to as receptor neurons or receptors) are located underneath a protrusion of the cuticle, the sensory hair. In the case of the olfactory sensillum shown here, the hair cuticle contains many fine pores which allow access of the stimulus molecules to the interior. A distal extension of the cell body of each sensory neuron, the dendrite, protrudes into the hair. The dendrite contains the sites of primary transduction of the stimulus. The primary transduction process consists of two steps; firstly, an interaction of the stimulus with protein molecules in the cell membrane which leads to the release of a chemical signal, the second messenger. Molecules of that messenger then react with a different type of membrane-bound protein molecules, the

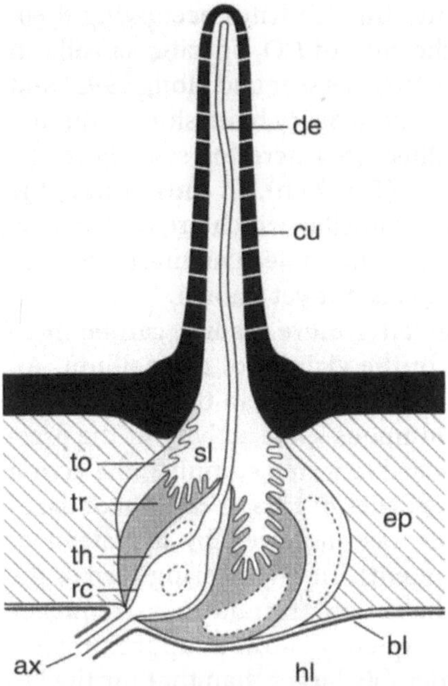

Fig. 1. General schematic of an olfactory sensillum in the body wall of an arthropod. The receptor neuron cell body (*rc*), with the distal dendrite (*de*) and the proximal axon (*ax*), is enveloped by three accessory cells, the tormogen (*to*), trichogen (*tr*) and thecogen (*th*) cells. Folded membranes of those cells face the sensillum lymph space (*sl*). The sensillum forms part of and is surrounded by epithelium (*ep*). The external boundary is formed by the cuticle with wall pores (*cu*), the internal boundary towards the hemolymph space (*hl*) by the basal lamina (*bl*). (After Keil and Steinbrecht 1984)

ion channels. Activation of the ion channels leads to a change in the conductivity of the dendrite membrane. Due to a voltage difference between the inside of the dendrite and the surrounding sensillum lymph space, maintained by ion transport through the membranes of the accessory cells, the change in conductivity leads to a change in the current across the membrane and to a change in the transmembrane potential in the cell body of the receptor neuron, the receptor potential. The proximal extension of the receptor neuron, the axon, contains a third type of protein, the voltage-sensitive channels. Their activation triggers the occurrence of brief depolarizations of the axon membrane, the action potentials or spikes, which conduct the sensory message to higher-order neurons in the central nervous system. The message is encoded in the spike rate. As the transduction mechanism consists of several cascaded processes, a considerable gain is possible; for the case of pheromone-sensitive receptor neurons a single stimulus molecule is sufficient to evoke an action potential. In the case of general olfactory neurons and of CO_2-receptor neurons, a sustained train of action potentials often occurs under stimulus-free conditions; in those cases, stimuli can either increase the spike rate (excitatory response) or decrease it (inhibitory response).

In general olfactory sensilla, the hairs and the dendrites occupying them are often of elongated, tubular shape; for the case of CO_2-specific sensilla in stable flies (Lewis 1971), mosquitoes (McIver 1972; Desser and Hong 1992) and moths (Stange et al. 1995) it has emerged that the sensory hairs show a characteristic thickening of their distal parts, as illustrated here for sensilla in the labial palp of the moth *Cactoblastis cactorum* (Fig. 2a,b). In those cases, the dendrite is not tubular, but rather spread into lamellae which are folded (Fig. 2c), leading to a large increase in surface area. The molecular mechanism of primary transduction in CO_2-receptor neurons is not yet known.

Both receptor potentials and spikes, the latter more readily, can be measured by the insertion of a microelectrode in the vicinity of a sensillum. An example of such recordings, for the case of a CO_2 receptor in the honeybee, is presented in Fig. 3. The receptor exhibits continuous spike activity in the presence of a background of 300 ppm; when exposed to a step stimulus, the action potential rate increases, after a latency in the order of tens of milliseconds, to an initial peak (phasic component), in this case followed by a plateau of maintained increased activity (tonic component). In engineering terms, the combination of latency and phasic-tonic behaviour results in a frequency response which ranges from DC to a high-frequency cutoff at approximately 10 Hz; the gain for high frequencies is considerably larger than that for the DC component, resulting in strong emphasis on the signalling of changes.

The concentration transfer function for large signals (usually measured as step responses to changes from zero to a given concentration) is logarithmic over a range of concentrations followed by saturation, in common with most other sensory modalities. This results in a dynamic range that covers several orders of magnitude.

When a continuous stimulus background is presented, the action potential response settles from an initial phasic peak to a much lower plateau value; subsequently, small changes around that background are encoded with considerable gain. An example of responses to small changes in CO_2 concentration, superimposed on a background of 350 ppm, is presented in Fig. 4.

3 CO_2 Microclimate, Behavioural Effects of CO_2 and Properties of CO_2-Receptor Neurons in Different Habitats

3.1 Soil and Logs

The soil is a carbon dioxide source as a result of the metabolic activity of microorganisms and plant roots; the latter form localized sources of CO_2 and are of particular interest as they are food sources for phytophagous animals. The predominant sink is the soil surface; as dilution in the open atmosphere is dominated by relatively fast, turbulent mixing, the bulk of the diffusion resistance leading to gradient buildups is within the soil itself, where molecular

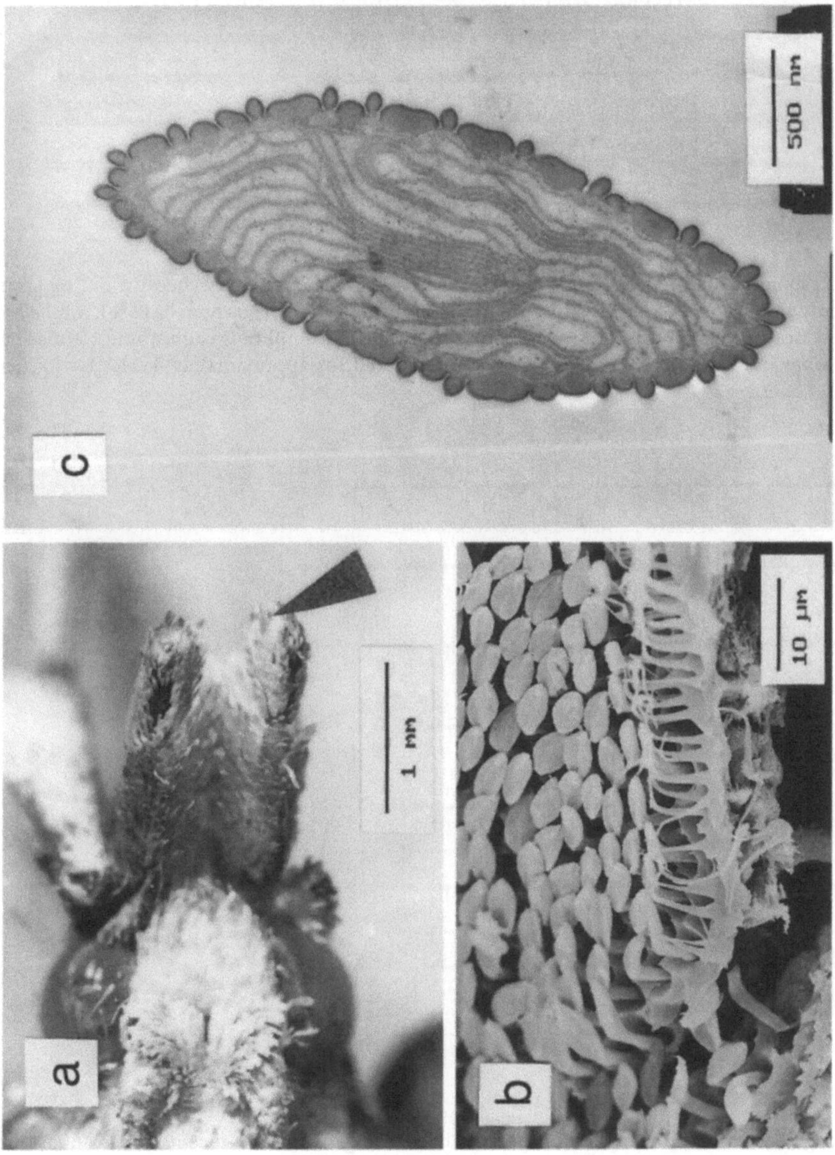

Fig. 2. a Head of a moth, female of *Cactoblastis cactorum*, showing the labial palps. A fine pore at the tip (*arrow*) forms the opening of a cavity in the terminal segment. **b** View of internal wall of cavity, showing a layer of stalked, bulbous sensilla. **c** Cross section through bulbous part of sensillum, showing folded lamellae of the sensory dendrite

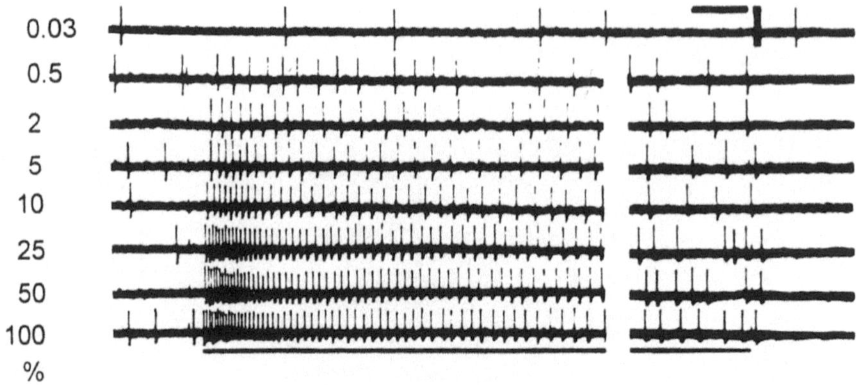

Fig. 3. Responses of a CO_2-receptor neuron on the antenna of a honeybee drone, to large increments of CO_2 concentrations superimposed on a background of 300 ppm (0.03%). *Bar below* indicates duration of stimulus. *Bar top right* indicates time and amplitude calibration: 100 ms and 0.4 mV. *Gaps* indicate where recordings were interrupted for approximately 4 s. (After Lacher 1964)

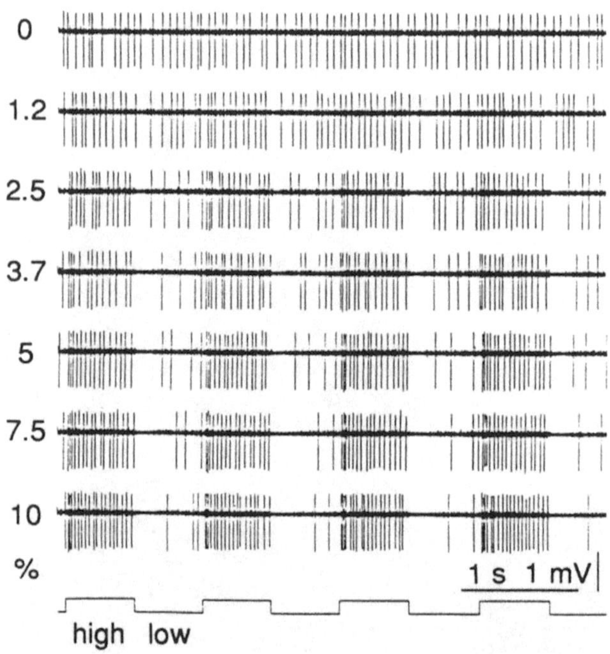

Fig. 4. Segments of continuing action potential recordings from CO_2-receptor neurons in the labial palp organ of the moth *Helicoverpa armigera*. Responses to square wave modulation around a background of 350 ppm; modulation of the response becomes obvious at a contrast of 2.5% (9 ppm). (Stange 1992)

diffusion is the determining factor. As soil consists of solid particles interspersed with water and with air, the diffusion resistance depends on the proportion of those components. The diffusion constant of CO_2 in air is $0.16\,cm^2\,s^{-1}$ in the gas phase and $1.6 \times 10^{-5}\,cm^2\,s^{-1}$ in the liquid phase; for the calculation of gradients in soil, a value of $6.5 \times 10^{-2}\,cm^2 s^{-1}$ has been used (Dusenbery 1987). Treating a typical root of 0.37 mm diameter as a line source with a rate of CO_2 production per area of $1.6 \times 10^{-6}\,ml\,cm^{-2}\,s^{-1}$, it was calculated that the relative gradient (relative to background of 1) in the immediate vicinity of the root is $5 \times 10^{-3}\,cm^{-1}$. On the other hand, the minimal detectable gradient for an ideal CO_2 receptor was given as $2 \times 10^{-6}\,cm^{-1}$; this level occurs at a distance of 90 cm from a single root fibre.

3.1.1 Nematodes

Early measurements (Klingler 1963) showed that plant-parasitic nematodes (*Ditylenchis dipsaci*) can detect artificial CO_2 gradients. They are attracted towards a source if the absolute gradient is > 0.1% $CO_2\,cm^{-1}$ on a background of 0.35% CO_2, or a relative gradient of $0.3\,cm^{-1}$.

More recently, Pline and Dusenbery (1987) have observed in the nematode *Meloidogyne incognita* that a sudden change in CO_2 concentration of 0.01% on a background of 0.03 or of 0.05% on a background of 1% evoked an increase in the rate of locomotion. The behavioural threshold for net movement along continuous spatial gradients was $10^{-2}\,cm^{-1}$. Thus, the sensitivity of the behavioural response, determined under experimental conditions unlikely to fully reflect the behavioural capabilities of the animals under natural conditions, is only marginally smaller than what is required to detect a root at close distance.

Another possible function of the CO_2 response could consist of leading the nematode to the optimal depth in the soil profile (Pline and Dusenbery 1987). According to Boynthon (1941), mean concentrations in the soil of an apple orchard at 30, 60 and 90 cm depth are 2.2, 3.6 and 5.0% CO_2. The question arises as to how the much smaller root-evoked gradients can remain detectable against a vertical gradient of that magnitude.

It has often been suggested that CO_2 gradients by themselves are too unspecific for host detection to be successful and that they must be accompanied by other cues, for instance chemicals specific to the host plant. McCallum and Dusenbery (1992) addressed this question, using the behaviour of nematodes as a bioassay at the output of a gas chromatograph to search for behaviourally active volatile substances in the vapour from the roots of host tomato plants. Apart from CO_2, no such substances were found.

3.1.2 Insect Larvae, Collembola

Responses to CO_2 gradients have been observed in the larvae of beetles with habitats in the soil or in decaying logs. Feeding larvae of *Orthosoma brunneum*, which mine in logs feeding on fungi, gather in places containing more than the atmospheric background of 0.03% CO_2, while non-feeding larvae do not respond (Paim and Beckel 1964b). This response coincides with the observation that the cores of such logs and mines of insects contain uniform CO_2 concentrations of up to 15% (Paim and Beckel 1963). Wood containing less than 2% CO_2 was not infested.

In contrast to what has been reported for nematodes, effective chemical stimuli other than CO_2 have been reported for beetle larvae: Bjostad and Hibbard (1992) demonstrated that corn rootworms, the larvae of the beetle *Diabrotica virgifera virgifera* which feed on maize roots, are attracted to a maize-specific semiochemical, MBOA, provided that a CO_2 gradient is also present.

The location of the CO_2-sensitive sensory organs in beetle larvae was determined by amputation of various combinations of receptor organs, followed by behavioural experiments. Klingler (1966) found that, in *Otiorrhynchus sulcatus*, the responses are mediated primarily by the terminal segments of the maxillary palps, while the antennae and labial palps play a minor role, if any. White et al. (1974) found for *Orthosoma brunneum* that the receptors are situated on the distal segments of the maxillary palps and the labial palps, the former being of primary importance.

Sensitivity to CO_2 gradients, sometimes in conjunction with other chemical stimuli, has also been reported for carrot-fly larvae (Jones and Coaker 1977), grass grubs (Galbreath 1988) and fungivorous collembola (Bengtsson et al. 1991). The roles of CO_2 ("primary plant metabolite") and other volatile chemicals ("secondary plant metabolites") in host orientation of phytophagous soil animals were reviewed by Jones and Coaker (1978).

3.1.3 Adult Beetles

The presence of elevated CO_2 levels also affects oviposition site selection by adult females of beetles with soil-inhabiting larvae. Paim and Beckel (1964a) reported for *Orthosoma brunneum* that females oviposited most frequently in places where the CO_2 concentration was highest. In *Diabrotica virgifera virgifera* (Lance 1992), females oviposit preferentially at sites where humidity is elevated and where bacterial odours are present; CO_2 at a concentration of 5% significantly increased oviposition if the flow rate was 40 ml h^{-1} or more.

3.1.4 Centipedes

CO_2-sensitive receptor neurons have not yet been characterized electrophysiologically in any of the species listed so far, but a detailed characterization has been performed in another soil inhabitant, the Japanese house centipede *Thereuonema hilgendorfi* (Yamana et al. 1986; Yamana and Toh 1987, 1990), which is carnivorous. The neurons are located in the organ of

Tömösvary, or temporal organ, which is located on the head in the vicinity of the compound eyes. The organ occurs throughout the Chilopoda but also in other soil inhabitants (Collembola, Symphyla, Protura). In recordings of action potentials, after adaptation to CO_2-free air, the response to CO_2 is inhibitory: the action potential rate decreases in proportion to the logarithm of CO_2 concentration over a range from 0.001 to 0.1% for the phasic component and from 0.05 to 5.0% for the tonic component. Above these ranges, the responses saturate. It was also found that slight increases and decreases of CO_2 concentration from a given adapting level result in large changes in the action potential rate. For instance, the rate decreases by a factor of two when concentration is changed from 0.1 to 0.13%. Weak responses were also observed on exposure to fatty acids, amines and changes in humidity, but might be attributable to secondary effects on the ionic composition of the aqueous medium surrounding the receptor neurons. Changes in temperature do not change spontaneous activity. Thus, it appears that the centipede temporal organ is specialized to the detection of CO_2, with emphasis on the detection of rapid temporal gradients. The detection of prey and the regulation of respiration were suggested as functional roles.

3.2 Hives and Nests of Social Insects

Anderson and Ultsch (1987) measured the concentrations of O_2 and CO_2 in several arthropod microhabitats including the burrows of spiders and a tiger beetle larva, mounds of ants and stumps inhabited by termites as well as decaying logs. An O_2 deficit and increased CO_2 occurred in all microhabitats; CO_2 levels were 1.1% in spider burrows, ant mounds and under decaying logs. In the burrows of tiger beetle larvae, within decaying logs and inside stumps inhabited by termites the CO_2 concentration rose as high as 6–7%. Part of that rise can be atrributed to the metabolism of the inhabitants, the remainder to diffusional exchanges with the soil.

3.2.1 Ants

The behaviour of individuals of social insects is strongly influenced by chemical stimuli released by conspecifics. Groups of workers separated from their nests tend to aggregate rapidly in clusters. The effective stimulus can be a species-specific pheromone, as shown for the termite *Kalotermis flavicollis* (Verron 1963). In ants, this behaviour can also be evoked by CO_2: when away from the nest and in close quarters, fire ant workers of the genus *Solenopsis* attempt to dig through porous barriers put up to separate them from other members of their colony (Wilson 1962, 1971). The same effect can be induced by the presence of tubes containing concentrations of CO_2 that are higher than ambient but considerably lower than those than cause anesthesia (Hangartner 1969).

Burkhardt (1991) measured the locomotor activity of workers of the ant *Pheidole pallidula* in artificial nests that were continuously flushed with 0, 0.1, 0.25 and 0.5% CO_2. Locomotor activity increases with concentration and is significantly different between all concentrations tested. Foraging ants returning to nests move more rapidly with increasing CO_2 concentration in the nest. It was suggested that the CO_2 concentration, under natural conditions, is a measure of colony size. The observations imply that ants can detect the absolute level of CO_2.

3.2.2 Termites

Termites perish quickly when exposed to open air: they are adapted to living their entire existence in the closed environment of the nest and the tunnels that connect it to sources of food and water (Lüscher 1961). As a consequence, the exchange of water, O_2, CO_2 and heat must take place by diffusion through the wall of the nest. A medium-sized colony of the African species *Macrotermes* contains two million individuals that need 140 l of O_2 per day and produce the corresponding amount of CO_2. Nevertheless, the CO_2 concentration inside the hive does not exceed 2.7%. The task of ventilation is not accomplished by the activity of individuals, such as in the case of bees (see below), but rather by the design of the nest which consists of an interior chamber which is heated by metabolic heat. Warm air rising in the centre circulates through an elaborate system of ducts close to the porous nest walls and returns to the bottom of the nest, exchanging gases on the way.

High CO_2 concentrations do not adversely affect termites (Lee and Wood 1971) and the structure of the nest walls cannot be easily modified to adjust to fluctuations in gas concentration; therefore, sensory organs monitoring CO_2 within the context of the control of ventilation do not seem to be advantageous. However, a CO_2-sensitive receptor neuron has been identified in a sensillum on the antenna of *Schedorhinotermes lamanianus* workers (Kaib et al. 1993), albeit with a different putative functional role. Each sensillum bears two odour-sensitive neurons, with tubular unlamellated dendrites which respond with excitation to chemicals such as 1-hexanol. One of the neurons is inhibited by CO_2; at the concentrations prevailing in the nest the inhibition is complete and the neuron is insensitive to odour stimulation. Therefore, the responses of the two neurons are context specific: both will respond to an odour stimulus outside the nest or at the periphery, but only one responds inside the nest. A simple comparison of excitation level could inform the individual about its location.

3.2.3 Fig Pollinators

Figs of the species *Ficus religiosa* depend for pollination on symbiontic wasps, *Blastophaga quadraticeps* (Galil et al. 1973). Those insects are solitary but the insides of figs contain hollow spaces loosely comparable to hives or nests. Female wasps penetrate the cavity inside an immature fig, which is initially female, and simultaneously pollinate the flowers and oviposit; a new generation of wasps develops in galls inside the fig, which simultaneously enters its male phase. During the early male phase, the internal atmosphere contains 10% CO_2, 10% O_2 and some ethylene. The presence of a high level of CO_2 causes the young female wasps to stay inactive until the males mate with them. Subsequently, the wall of the fig is perforated by the males leading to equilibration with the outside atmosphere. As a result, females become active, leave their galls, load pollen pockets with pollen and emerge from the fig. The decrease in internal CO_2 also triggers metabolic changes in the fig which finally lead to ripening of seeds. It remains unproven whether the observed behavioural changes are evoked via sensory organs or via a less direct effect of CO_2.

3.2.4 Honeybees

The highly-developed social structure of honeybees (*Apis mellifera*) has enabled them to succeed in occupying habitats ranging from the tropics to the subarctic by maintaining perennial colonies in the protection of the hive. A major prerequisite for this is the tight control of the hive climate, including adjustments of ventilation, humidity and temperature. Active control of the temperature in the hive (Seeley 1985) allows survival in winter; during the period of brood rearing, from early spring to late autumn, the temperature in the brood nest is stabilized at between 32 and 35 °C, optimal for the development of brood. The penalty for this stability, as in warm-blooded mammals and birds, is the continuous generation of heat via the burning of glucose and the accompanying release of CO_2. In resting bees heat production is about $20\,W\,kg^{-1}$, but by fast, repetitive contractions of the flight muscles it can rise to $500\,W\,kg^{-1}$. As a consequence, a hive with 50000 bees (5 kg) produces at least $240\,ml\,min^{-1}$ of CO_2, which has to be removed from the hive. Atmospheric exchange in the hive is facilitated by fanning: under the appropriate conditions, workers sit on the honeycomb or in the nest entrance and beat their wings, driving air out of the nest.

When a current of CO_2 was injected into a hive (Seeley 1974), at a rate which increases CO_2 concentration from an initial level of 0.4 to 8%, over a period of 12 min, the number of fanning bees increased, with a delay, by a factor of 35; eventually, 40% of the colony population participated in fanning. After cessation of the injection, CO_2 level and fanning activity returned to normal within 5 min. The response is not caused by oxygen depletion as replacing the hive atmosphere with pure nitrogen does not evoke fanning, but rather leads to immobilization.

In the absence of experimental interference (Seeley 1974), the CO_2 concentration in the hive fluctuates and the fanning activity almost exactly duplicates the changes, although the two phenomena are slightly out of phase, with the fanning activity trailing by a few minutes. Over an observation period of 52 h, the mean CO_2 concentration in a large hive was 0.44%, and in a small hive 0.78%.

In honeybees, the CO_2-receptor neurons are located on the antennae (Lacher 1964). The antennae of workers are covered with approximately 12 000 sensilla of various types, containing approximately 100 000 sensory neurons. The majority is sensitive to olfactory stimuli such as floral odours or several types of pheromones involved in social communication. Two morphologically distinct sensillum types, which are clustered together in small groups (sensilla coeloconica and sensilla ampullacea), number 236 per antenna, with the sensilla ampullacea forming the majority; the number of sensory neurons in each sensillum ampullaceum is one. In recordings from their vicinity, a cell type was observed which responds to CO_2. A wide range of other stimuli, mechanical and olfactory, was ineffective. Responses to a concentration series, on a background of 0.03%, are presented in Fig. 3. It was noticed that in comparison to other olfactory-receptor neurons, the sensitivity to differences is extraordinarily large: a concentration step from 0.5 to 1% CO_2 evoked a change in action potential rate, in the phasic peak, by 80 impulses s^{-1}.

3.3 Haematophagous Arthropods in the Open Atmosphere

The localization of the source of an olfactory stimulus in the open atmosphere requires strategies that are quite distinct from those useable in confined spaces. Most importantly, there is usually a net horizontal air movement, implying that an insect which has detected an odour from a distant source will have to employ means of taking an upwind course in order to reach the source. A resting insect can detect wind direction with mechanoreceptive sensory organs, and upwind direction could be maintained by flying in short hops; insects in continuous flight, however, cannot use mechanoreceptive reactions to steer upwind because they are supported by the moving air. Instead, information about ground velocity must be derived from visual inputs (David 1986).

Furthermore, the structure of gradients in the open atmosphere (Murlis et al. 1992) is quite different from that in confined spaces: continuous and monotonic gradients exist only under calm conditions. In the presence of wind, two processes control the dispersion of molecules released by a point source. The first is molecular diffusion, which in the atmosphere becomes a small-scale phenomenon, whereas large-range dispersion is dominated by turbulent diffusion, to a varying extent depending on conditions. At night, under calm weather conditions, the atmosphere can be stable and layered. During daytime, solar irradiance leads to differential heating of the ground, leading to the development of large convection cells (Desjardins et al. 1982, 1989). These vortices or eddies are continuously broken up into ever smaller cells, down to a minimum size characterized by a parameter known as the Kolmogoroff length, which is in the order of 1 cm.

If an odour source releasing material into the wind is smaller than the Kolmogoroff scale, odour molecules initially form a filament that expands by molecular diffusion until its diameter reaches the size of the smallest eddies, at which point turbulent diffusion becomes dominant. The rate of molecular diffusion depends on properties of the molecules, that of turbulent diffusion does not. As the odour plume develops further, filaments become stretched and tangled by small eddies; larger eddies cause fluctuations in the overall direction of the plume. Thus, an olfactory receptor that is moving relative to the plume will be exposed to an irregular sequence of brief stimulus pulses (Murlis and Jones 1981; Kramer 1986).

In warm-blooded animals, the requirement to maintain an elevated body temperature results in a high rate of metabolism, leading to the release of large amounts of CO_2. In breath, the concentration of CO_2 is 5% and the rate of release is 25 ml min^{-1} for a chicken, 250 ml min^{-1} for a human and 2500 ml min^{-1} for a horse or cow (Kline et al. 1991). The initial concentration exceeds the atmospheric background by two orders of magnitude and will therefore require considerable dilution until it becomes indistinct from the background. In consequence, a plume of increased CO_2 is a giveaway cue for the presence of a large animal in the upwind direction. At the source, a plume of breath of a large mammal is larger than the Kolmogoroff distance, particularly as additional turbulence is created by movement of the plume relative to the surrounding air and by convection resulting from temperature differences. Therefore, the initial filamentous structure does not exist at the outset. However, it would appear that the pulsed structure at larger distances is likely to prevail; it could be enhanced additionally as breathing is periodical in the first place.

To date, the volume resolution (a few millilitres) of existing technical detectors for CO_2 is many orders of magnitude less than that of arthropod CO_2-receptor neurons (a few nanolitres). Therefore, we cannot measure the actual concentration that is encountered by an insect when a plume of air from a distant host passes by, and it is possible that a concentration close to that at the source prevails in small pockets of air over a considerable distance. In the case of the orientation of moths to pheromones, the efficiency of mate finding is primarily steered by the frequency with which odour pulses would contact the receptor neurons rather than by concentration, provided that the latter was above a threshold (Mafra-Neto and Cardé 1994). For the case of CO_2, where a large background is present, the threshold of detectability must ultimately be determined by the contrast sensitivity of the receptor neurons and/or interference from other sources.

The mechanisms of host finding in haematophagous arthropods have attracted considerable attention, predominantly because many species are vectors of infectious diseases with enormous medical and economical consequences and less because of their nuisance value. For example, malaria and yellow fever are transmitted by mosquitoes, trypanosomiasis by tsetse flies and virus encephalitis by ticks. Given the above, it is not surprising that the ability to detect CO_2 appears to be ubiquitous in haematophages and that it forms a major cue in the search for a host.

There are several distinguishable steps in host finding, mediated by a range of sensory inputs. The first step is activation, whereby a stimulus, usually chemical, triggers activity in a resting haematophage. The next step is orientation behaviour and locomotion towards the host, which, in flying insects, may follow the classical downwind model described above, involving the tracking of the chemical stimulus together with the maintenance of upwind flight via visual cues. At closer approach, the visual contours of the host and other close-range sensory stimuli such as temperature and humidity may become important. Finally, after landing on a putative host, mechanical and contact chemical inputs of the correct combination are prerequisites for commencement of feeding.

3.3.1 Mosquitoes and Flies

Wright and Kellogg (1962) suggested that yellow fever mosquitoes, *Aedes aegypti*, are stimulated to fly in search of a warm-blooded host by an alteration in the ambient CO_2 level and that they are guided to the appropriate surface partly by colour, but mainly by warm, moist convection currents rising from it. Nelson (1965) observed that both males and females of *Culicoides variipennis* were attracted in large numbers by traps baited with dry ice; Snow (1970) found in field experiments that human breath from which CO_2 was removed attracted significantly lower numbers of mosquitoes than the untreated control. Omer (1979) found in wind-tunnel experiments that females of *Anopheles arabiensis* and *Culex pipiens fatigans* responded to changing concentrations of CO_2 with upwind flight, but not to a constant concentration of 0.5% CO_2. Odour from a human hand was a stronger stimulus than CO_2.

Gillies (1980) reviewed the literature of the time on the role of CO_2 in host finding by mosquitoes and came to the following conclusions. Firstly, CO_2 acts both as an activator and an attractant. Secondly, its orientating effect appears to be comparable to that of odour factors, with the important difference that a CO_2 stimulus is only effective if it is pulsed. Under natural conditions this would normally be the case due to the irregular and filamentous structure of the plume. Concerning the threshold, it was suggested that the crucial factor might be a change in concentration and not the absolute concentration. Thirdly, in terms of attractant effect, CO_2 alone is less powerful than the combined effect of all host factors; one of the most striking effects of CO_2 is the modification or augmentation of the effects of other stimuli.

Field experiments on the effect of CO_2 emission rates on the range over which mosquitoes were attracted were presented. At an emission rate of

1 l min^{-1}, the maximum range across which mosquitoes were attracted was between 26 and 36 m.

McIver and McElligott (1989) also tested the effect of release rates on the range of attraction by CO_2 on some southwestern Ontario mosquito species; a release rate of 1 l min^{-1} is attractive over a range of 3–7 m; at 4 l min^{-1} the range is 7–11 m. The question whether the discrepancy with earlier results is due to different meteorological conditions or to species differences was discussed.

In recent reviews (Bowen 1991; Takken 1991; Davis and Bowen 1994), further in-depth information on the role of other sensory modalities, in addition to CO_2, is presented. The best-described host emanation other than CO_2 is lactic acid.

Eiras and Jepson (1991) presented lactic acid, carbon dioxide and human sweat stimuli singly and in combination to female *Aedes aegypti* within a wind-tunnel system. For activation, a threshold CO_2 concentration of 0.03% above ambient was observed; lactic acid and human sweat did not elicit take-off when presented alone but enhanced the effect of CO_2. Kline et al. (1991) investigated the interactive effects of 1-octen-3-ol and CO_2 on biting insects in the Florida Everglades, trapping 300 234 mosquitoes of various species and 6588 deerflies within 36 h. A CO_2-release rate of 20 ml min^{-1} increased the catch against control by a factor of four, and 2000 ml min^{-1} increased the catch by a factor of 70. In comparison, octenol released at a rate of 3 mg h^{-1} increased catches by a factor of three. The effects of combined stimuli were additive in some species and synergistic in others. According to Kline et al. (1994), CO_2 is an effective attractant to three species of *Culicoides*, but each has a different response pattern. Octenol has a repellent effect in two species and an attractant effect in the other. Van Essen et al. (1994) observed that octenol increased catches of *Aedes vigilax* when presented in addition to CO_2, but no increases or even smaller catches were observed in two species of *Culex*.

Carlson et al. (1992) investigated whether CO_2 released by a human hand can be the factor that makes it attractive. They found that the release rate is in the order of 1 ml h^{-1}, which is negligible compared to the release rate in breath. No attractiveness of this release rate was observed for *Aedes aegypti*. On the other hand, Moskalyk and Friend (1994) observed that females of *Aedes aegypti* feed more readily, through an artificial membrane, from a diet containing ATP as a taste stimulus if that diet is warmed up to body temperature and also contains bicarbonate. This suggests that CO_2 released through the skin does not evoke attraction from a distance but that it might be involved in the release of feeding behaviour after landing, together with other sensory inputs.

The effects of CO_2 and other chemical stimuli on haematophagous flies are similar to those found in mosquitoes. In laboratory experiments on the stable fly, *Stomoxys calcitrans* (Warnes and Finlayson 1985a) an activation response occurs if the CO_2 concentration is raised by 0.01% above background; with increasing concentration, the response reaches a maximum at 1% and then declines. Air expired by a human elicited a larger response than would be predicted from its CO_2 content and it was shown that this is caused by breath odours other than CO_2. Breath odours without CO_2 elicited no response. The respective effects on the orientation response of *Stomoxys* (Warnes and Finlayson 1985b) are similar. In a wind tunnel (wind speed 0.3 m s^{-1}) flies tend to aggregate at the downwind end when left alone but tend to fly upwind when stimulated. An increase in CO_2 concentration by 0.01% or more elicited anemotactic responses which were enhanced by odours from human breath or a live guinea pig. Acetone also elicited a response.

The presence of tsetse flies, *Glossina* spp., the vectors of trypanosomiasis, is a major limiting factor to cattle breeding in tropical Africa leading to a considerable amount of research on host finding (Turner 1971; Vale and Hall 1985; Torr 1988, 1989, 1990; Brady et al. 1989; Warnes 1990; Gibson et al. 1991). The role of olfactory stimuli, including CO_2, was recently reviewed by Willemse and Takken (1994). As in the previous examples CO_2 is a key chemical attractant and its effect is enhanced by other host odours such as acetone, 1-octen-3-ol and phenols. For orientation in close vicinity of a target, shapes and contrasts of visual cues are important; striped targets can have a repellent effect, leading to the suggestion that the stripes of zebras are a defense against tsetse (Gibson 1992).

Warnes (1990) found in wind-tunnel experiments that the flight speed of *Glossina* spp. decreases in proportion to CO_2 concentration for the range 0.1–1%, but not for higher concentrations, and that the frequency of turns increases consistent with what would be the optimum strategy during closing in on a host. Bursell (1984) performed wind-tunnel experiments on the activation of *Glossina morsitans* by pulses of CO_2, acetone and cow odour. When the flies are adapted to air from which CO_2 has been removed, a pulse of 0.01% causes significant activation; the effect reaches a maximum at 0.04% above which it declines to statistical insignificance at 4%. The responses to a range of acetone concentrations also went through a maximum, but not the responses to a concentration series of cow odour.

The observation that CO_2 concentrations decrease in effectiveness if an optimum is exceeded has also been made for ticks (see below), suggesting that a mechanism exists whereby CO_2 (or a general olfactory stimulus such as acetone) is "ignored" unless the "expected" complement of other sensory stimuli is also present. Such a mechanism could ensure discrimination against non-host CO_2 sources such as decomposing animal or vegetable materials.

Data on the electrophysiological characteristics of CO_2-receptor neurons in haematophagous diptera are scarce. Kellogg (1970) identified CO_2-sensitive neurons on the maxillary palps of females of the yellow fever mosquito, *Aedes aegypti*. In CO_2-free air, the action potential rate was sometimes zero. Exposure to air containing 0.05% CO_2 evoked an excitatory phasic-tonic action potential response. Sudden increases above a background of air from outside the laboratory (quoted as 0.04%) caused an increase in the action potential rate; for an increase of 0.01%, which was the smallest increment tested, the action potential rate in the phasic part of the response was two to four times the rate of response to the background. With increasing concentration the response increased logarithmically and the phasic peak of the response became progressively more pronounced. Saturation of the phasic peak occurred at levels between 0.05 and 0.5% CO_2, with 10% giving little additional response.

The finding that an increase of 0.01% above background evokes a very pronounced response indicates that the threshold of detectability must be much lower. The subject is clearly one on which more precise information is required (Gillies 1980). Recent, more detailed observations seem to be lacking. However, the magnitude of the contrast response is comparable to or larger than that of lepidopteran CO_2 receptors (see below), where the contrast threshold has been determined using signal-to-noise ratio as a criterion. Extrapolation from these results suggests that it is not unlikely that the mosquito CO_2-receptor neurons can detect sudden changes of less than 0.001% or 10 ppm on a background of 350 ppm.

Stange (1975) identified CO_2-receptor neurons on the antenna of the sheep blowfly, *Lucilia cuprina*, with response properties similar to those found in honeybees. Bogner (1992) obtained single-cell recordings from CO_2-receptor neurons on the antenna of the tsetse fly *Glossina palpalis*. The neurons did not respond to chemical stimuli other than CO_2; the responses to CO_2 were phasic-tonic and in response to prolonged stimulation with 1% CO_2 the action potential rate remained above resting level indefinitely. Between stimuli, the preparation was kept in a current of air with approximately 0.03% CO_2; stimulation with CO_2-free air caused a decrease in action potential rate but responses to small increments were not tested. The intensity-response function is logarithmic over a range from ambient to 5%.

Den Otter and Van der Goes van Naters (1992) tested 182 receptor neurons on the antenna of *Glossina morsitans* for sensitivity to olfactory stimuli. Out of 141 cells which were olfactory receptors, 52% responded to 1-octen-3-ol and 13% were sensitive to CO_2 only. Given a total of 2000 receptor neurons per antenna, this suggests that the complement of CO_2-receptor neurons on a tsetse fly antenna is 260, a number similar to those found on the antennae of honeybees and in the labial palp organs of moths (see below).

3.3.2 Bugs

The behavioural responses of fifth instar nymphs of the fire bug, *Rhodnius prolixus*, to a range of host-related stimuli were determined by Núñez (1982). Both activation, in a wind tunnel, and orientation responses, in an Y-maze olfactometer, were examined. A natural food source (hamster or human forearm) is the most effective stimulus; neither the odour from a cage previously occupied by a hamster nor CO_2, released into the wind tunnel at a rate which raised the concentration in the test area by 0.017–0.03%, caused a difference in activity when presented alone, but a significant increase occurred in response to the combined stimulus. In the olfactometer, air to which 0.1% CO_2 had been added was significantly preferred to air alone.

3.3.3 Ticks

The arachnid ticks (Acarina: Ixodidae), in spite of being unable to fly, are strongly attracted to breath and CO_2; the use of CO_2-baited traps is an established method for monitoring tick densities (Mount and Dunn 1983). Stimulated ticks show questing behaviour, lifting their forelegs in the air to sample the surroundings as insects do with their antennae; a chemoreceptive sensory organ (Haller's organ) is located on the tarsi. In Y-maze choice experiments, Sauer et al. (1974) observed the behaviour of the lone star tick, *Amblyomma americanum*, in response to CO_2 concentrations ranging from ambient (0.03%) to 8%. When the concentration was below 2% but above atmospheric in one arm of the Y, ticks with extended forelegs and an increase in tick movement were observed but there was no preferential movement towards the arm with increased CO_2. However, for atmospheres of 2% and above such movement did occur in correlation with concentration. This indicates that ticks are stimulated by small concentrations of CO_2 but that directed responses do not occur until concentrations reach levels above 1%. Steullet and Guerin (1992a) found, in wind-tunnel experiments, that for the tropical bont tick, *Amblyomma variegatum*, a mean concentration of 0.15% CO_2 was the best locomotor stimulant and was more effective than higher concentrations. Perrit et al. (1993) observed questing behaviour in adult *Ambylomma americanum* and *Dermacentor variabilis* and found responses if CO_2 concentrations were increased by 9 ppm (0.0009%) above a mean ambient background of 396 ppm. It cannot be excluded that this extreme sensitivity is an artifact introduced by the method of stimulus control.

In *Amblyomma variegatum*, only two CO_2-sensitive receptor neurons are present in Haller's organ (Steullet and Guerin 1992a). One of them responds to CO_2 with excitation; it responds weakly to ambient air but its dynamic range extends beyond 5% CO_2, the concentration in breath at the source. The second neuron responds to CO_2 with inhibition. An increase from zero to 10–20 ppm causes detectable inhibition and concentrations above 0.5% completely suppressed spike activity. The firing rate of that receptor was most affected by modulation of about 100 ppm around 500 ppm, and less by modulation around 1200 ppm. By contrast, the CO_2-excited receptor poorly encoded changes at the lower concentration but became more efficient for changes at the higher level. Consequently, the CO_2-detection system of the tick involves range splitting between two different receptor neurons.

As in the case of the haematophagous diptera, ticks do not rely solely on CO_2 but also use other chemical stimuli (Steullet and Guerin 1992b): H_2S tested at a concentration of 0.02 ppm arouses 60% of resting ticks; some receptor neurons in Haller's organ are sensitive to H_2S, with a threshold of 0.1 ppb.

Ticks also respond to light and the responses are modulated by CO_2 (Werder and Gothe 1993).

3.4 Herbivores in the Open Atmosphere

Photosynthetically active parts of plants act as sinks for CO_2 if the rate of CO_2 assimilation exceeds the rate of respiration or as sources if respiration dominates. Flowers, developing fruits and rapidly growing tips of shoots are always sources. CO_2 assimilation occurs usually during daytime, simultaneously with the photolysis of water; however, the two processes can be decoupled: as an adaptation to arid conditions, some plants use the crassulacean acid metabolism (CAM) pathway, which fixes CO_2 at night in an intermediate metabolite (Osmond et al. 1979). The CO_2 gradients that can occur in the vicinity of plant materials are much smaller than those encountered in the vicinity of warm-blooded animals, because the metabolic rate of a plant is lower (per unit mass) and the total surface area is much larger. Most importantly, the bulk of the diffusion resistance between the enzymatic sites of metabolism inside leaves and the open atmosphere is located inside the leaf. This resistance is in the form of intra- and intercellular diffusion barriers as well as the stomatal resistance (Nobel 1991). Outside the leaf, most of the remaining diffusion resistance is located across the laminar boundary layer in the immediate vicinity of the leaf surface, where molecular diffusion is the limiting factor. However, small but measurable CO_2 gradients also occur at larger scales where turbulent diffusion dominates, ranging from the vicinity of the soil over the crop canopy to the planetary boundary layer (Fuller 1948; Monteith 1976; Desjardins et al. 1982, 1989; Denmead and Bradley 1987; Nobel 1991; Desjardins 1992).

3.4.1 Honeybees Revisited

In addition to the ventilation responses reported in Section 3.2.4, honeybees also show behavioural responses to experimental modifications of atmospheric CO_2 concentration when outside the hive. Within the context of foraging, the behavioural repertoire of honeybees includes the ability to learn the association of food sources with visual stimuli such as colours and shapes, and with olfactory stimuli. They can also be trained to use CO_2 as a cue (Lacher 1967): in choice experiments, 80% of free-flying bees, which have been rewarded with sugar solution in the vicinity of an air current containing 1% CO_2 or more, but not in the presence of air without raised CO_2, select the raised concentration. Below 1%, the preference becomes weaker and discrimination ceases below 0.45%. When the task consists of discriminating between two raised levels, flying bees can discriminate, for example, between 1 and 3% and 7 and 10%. The experimental conditions were such that the bees had to approach two different sources, 50 cm apart. Therefore, they had to traverse air at the atmospheric background. This implies that flying bees cannot only detect the existence of CO_2 gradients, but can also memorize their sizes.

It remains open to question whether this behavioural response to CO_2 reflects the ability of bees to generalize stimuli, even out of context, or whether CO_2 gradients of sufficient magnitude exist, under natural conditions, that provide cues about, for example, the quality of food sources to foraging bees. For example, the production of nectar in flowers could be associated with raised metabolic activity.

Lacher (1964) observed that the honeybee CO_2 receptors exhibited a firing rate of approximately 30 impulses s^{-1} if exposed to room air, and that resting activity was not, or was only insignificantly, modified on exposure to CO_2-free air. This would suggest that, unlike mosquitoes, bees cannot detect small changes around ambient concentration. This question was re-examined by Stange and Diesendorf (1973) with different results. When adapted to CO_2-free air, approximately half of the cells tested responded to stimulation by 300 ppm with an increase in action potential rate; in a receptor neuron adapted to a background of 300 ppm, stimulation with a pulse of 270 ppm (10% contrast) reduced the action potential rate from 10 impulses s^{-1} to zero during the time of stimulation. Under the same conditions, air samples from different parts of the laboratory and from outside influenced the rate in different ways, presumably due to slight variations in CO_2 content.

3.4.2 Caterpillars

First instar larvae of the polyphagous moth, *Helicoverpa armigera*, show a pronounced preferential response towards a CO_2 source (Rasch and Rembold 1994). Choice tests were performed in a Y-maze olfactometer; a current of CO_2-free air entered one arm and a range of CO_2-levels was presented in the other. Under those conditions, a significant proportion of caterpillars (40% more than random) preferred the side containing CO_2 if the concentration was 160 ppm; the response saturated at 800 ppm with a preference of 70%. This suggests that CO_2 gradients are used by the larvae to distinguish between photosynthetically active plant parts and those with net respiration. Correspondingly, field observations show consistently that flowers and developing fruits suffer most from attacks by *Helicoverpa*.

A CO_2-receptor neuron was identified electrophysiologically on the terminal segment of the maxillary palp of larvae of the small ermine moth *Yponomeuta cagnagellus* (Roessingh 1989), the same site at which CO_2 sensitivity in beetle larvae has been identified by amputation experiments (see Sect. 3.1.2). In this particular receptor neuron, responses to changes in CO_2 concentration were observed only at 0.2% or more, which is well above the concentrations occurring in the vicinity of plant parts. In contrast, Stange (unpubl.) observed that CO_2-receptor neurons of larvae of the moths *Helicoverpa* and *Cactoblastis* showed pronounced increases in action potential rate if stimuli of 30 ppm or more were presented on a background of CO_2-free air.

3.4.3 The Lepidopteran Labial Palp Organ

In adults of the Lepidoptera, the third, terminal segments of the labial palps consistently carry a particular type of sensilla (Fig. 2), ranging in numbers from a few in some species (non-herbivorous keratophagous moths) up to several thousand in others. In most species the segment is hollow with a pore at the tip and the sensilla form a dense lining of the inside of the cavity. The distinct sensory organ thus formed has been named the labial palp organ (LPO).

In a combined morphological and electrophysiological investigation, Lee et al. (1985) identified a field of about 80 club-shaped sensilla in the LPO of the cabbage white butterfly, *Pieris brassicae.* Most sensilla house single sensory cells of the same type; pore tubule structures in the walls of the sensilla identify them as olfactory. The distal section of the sensory dendrite is lamellated, in contrast to other olfactory dendrites which tend to be straight rods but similar to the CO_2-receptor neurons in mosquitoes and flies. The lamellation results in a large increase in the surface area which is 40 μm^2 for a single sensillum. In electrophysiological measurements, the cells exhibited strong responses to natural, living sources such as cabbage leaves, chamomille and carrot flowers and conspecific animals, both male and female. The odour of the cut wings of both sexes of conspecifics was ineffective.

Almost simultaneously, Kent et al. (1986) studied the anatomy of the LPO and its central nervous projections in the tobacco hornworm moth, *Manduca sexta.* The sensory organ is extremely well developed, containing 1750 sensilla with lamellated dendrites. The nerve fibres from the LPO project into the antennal lobe of the deutocerebrum, which is also the site of projection of predominantly olfactory antennal inputs.

Bogner et al. (1986) examined sensilla in the LPO of moths of several species in the genera *Rhodogastria* and *Achaea* and found anatomical structures closely similar to those in *Pieris* and *Manduca.* Using electrophysiological techniques, a wide range of pure volatile chemicals, temperature and humidity changes were tested, as was CO_2 from various sources. Although weak responses to some other stimuli were observed at high intensities, it became clear that CO_2 is by far the most effective stimulus. The overall response characteristics were similar to those in honeybees: the response to a step of increased CO_2 concentration consists of a pronounced phasic peak and a non-adapting plateau response. At the CO_2 concentration of room air (quoted as 0.03%), the cells are already in an excited condition. The intensity-response functions are logarithmic up to 1–10% CO_2. It was suggested that the biological function remains entirely open to speculation and that an obvious functional explanation would require more knowledge of spatial and/or temporal gradients in the CO_2 concentration of the microclimate.

The question as to whether those findings could be generalized for other lepidopteran species was examined by Bogner (1990). He found this to be the case and suggested that the CO_2 sensitivity of the labial palp organ is a common occurrence in the Lepidoptera. He also observed that, for the vast majority of Lepidoptera, small differences in the CO_2 content of the immediate environment would be measurable.

In electrophysiological recordings from the LPO of *Helicoverpa armigera* (Stange, unpublished), it turned out that local differences in atmospheric CO_2 are detected by individual receptor neurons: in an experimental situation where the organ was kept in a conditioning current of air pumped into the laboratory from the outside, air samples taken in syringes from different parts of the laboratory or from the outside atmosphere caused changes in action potential rate, either increases or decrease.

Actually, it turned out to be quite difficult to generate a stimulus which, when presented in alternation with a constant background, did *not* modulate the action potential rate. The best approximation to a null response to two switched stimulus currents was obtained with a setup designed to be as symmetrical as possible (Stange 1992): two currents of synthetic air from the same source contained 350 ppm CO_2. A small, adjustable fraction of one current was passed through a column of soda lime, to remove CO_2, and then recombined with the original current. With this arrangement, changes in action potential rate of a single cell in phase with the switching of the stimulus are clearly recognizable for a contrast of 2.5% or 9 ppm CO_2 (Fig. 4).

Each LPO of *Helicoverpa* contains approximately 250 receptor neurons. As they all appear to carry the same information, it is reasonable to assume that their outputs are simply pooled by higher-order neurons in the central nervous system. Assuming statistical independence between parallel channels, pooling leads to an improvement of detectability proportional to the square root of the number of channels. Recording from all cells simultaneously is not possible, but the situation can be approximated by pooling the appropriate number of repeated responses in a single channel. In recordings averaged over 256 repetitions of a switched stimulus, visualized as post-stimulus time histograms (Stange 1992), a response is visible for a contrast of 0.7% or 2.5 ppm CO_2.

To minimize the chance that this extreme level of sensitivity is an artifact, an alternative and indirect, but independent, experimental method was used to determine the minimum detectable contrast. As response magnitudes depend on the slope of temporal gradients, it was also desirable to determine the frequency dependence of the threshold. This required the generation of controlled waveforms of CO_2 modulation at a maximum frequency of 10 Hz. If it is assumed that the CO_2-receptor neurons measure the partial pressure of CO_2 rather than its molar fraction, modulation of overall atmospheric pressure should have the same effect as modulation of molar fraction. Methods of rapidly modulating pressure and, equally as important, of accurately measuring it, are more readily available than exact and rapid control and measurement of CO_2 concentration.

By applying sinusoidal modulation of pressure, the frequency dependence of the responses of the CO_2-receptor neurons was determined for a range of 0.05 to 12.8 Hz. Simultaneously, the amplitude spectrum of noise in the responses (spontaneous fluctuations in action potential rate) was measured in order to obtain an objective detectability criterion. Using a signal-to-noise ratio of 1 as the criterion, the minimum detectable contrast for the LPO of *Helicoverpa* as a whole is 0.5 ppm or 0.14% on a background of 350 ppm, over a bandwidth ranging from 0.3 to 3 Hz (Stange 1992).

This suggests that the lepidopteran CO_2 sense can readily detect the, presumably small, CO_2 gradients associated with plant metabolic activity, the question which follows is what is the actual magnitude of those gradients in a natural habitat and under conditions where interactions between an insect and its host are likely to occur. Stange (1992) examined the question as to whether or not hosts could be detected from a horizontal distance, analogous to the situation in haematophagous diptera, and came to the conclusion that a small shrub or tree is detectable from a distance of 10–100 m, provided that no other CO_2 sources or sinks are present.

The ecosystem consisting of the CAM plant, *Opuntia stricta*, and the moth, *Cactoblastis cactorum*, was selected as a model case to study interactions in a natural habitat (Stange et al. 1992, 1995). By 1920, *O. stricta*, indigenous to the south-east of the United States and to the Caribbean, had invaded large parts of Australia, and the subsequent successful introduction of *Cactoblastis* made this host-herbivore system one of the best-documented cases of biological control (Dodd 1940; Osmond and Monro 1981). In the woodland habitat of the plant, after dusk when the moths are active and when females are in search of oviposition sites, two types of CO_2 gradients occur that are detectable by the moth's sensors. The first gradient, associated with soil respiration at a time when the atmosphere is layered, is vertical and extends from the soil surface to an altitude of approximately 1 m. The concentration difference between ground level and an altitude of 1 m is in the order of 100 ppm, exceeding the detectability limit of the sensors by two orders of magnitude. The notion that this gradient provides, to a flying insect, a cue for the maintenance of a flight altitude favourable for host detection is supported by field observations of behaviour: females in search of oviposition sites miss plants which grow in elevated locations such as tree stumps and rock outcrops. On the other hand, and analogous to the situation outlined for nematodes (see Sect. 3.1.1), the size of the vertical gradient appears to rule out the possibility that horizontal gradients caused by plant activity are detectable.

A second gradient, associated with CO_2 fixation by the plant, extends from the surfaces of photosynthetic organs (cladodes) over a boundary layer distance of approximately 5 mm. Again, the concentration difference across this distance is in the order of 100 ppm. The notion that this gradient provides, to a walking insect, a cue to the physiological condition of the plant is supported by the observation that females of *Cactoblatis*, prior to oviposition, actively probe the plant surface by repeatedly approaching it with their labial palps (Myers et al. 1981): pronounced waving of the antennae accompanies this behaviour, indicating that other chemoreceptive cues are also examined. In a simulation of probing, pronounced responses of the CO_2-receptor neurons to the CO_2-fixing capacity of *Opuntia* were observed (Stange et al. 1995). This suggests that by probing the boundary layer, females of *Cactoblastis* can detect the healthiest, most active *Opuntia* cladodes, accounting for earlier observations that the most vigorous plants are preferred.

4 Implications of Elevated CO_2

CO_2 gradients are also generated by human activity. Gradients that fall within the temporal and spatial window that is detectable by the sensory organs are numerous and can be of considerable magnitude: for example, the rate of CO_2 release by a car exceeds the rate of release by a human by a factor of 10^2; it would appear that the occurrence of such signals must adversely affect the orientation of, for example, haematophagous insects. This question remains to be examined.

A much slower temporal gradient occurs because the present rate of CO_2 emission exceeds its sequestration by natural sinks. In the 10^6 years prior to the industrial revolution, the background fluctuated between 190 and 290 ppm (Barnola et al. 1987). Immediately before the industrial revolution the value was 280 ppm; a value of 300 ppm was measured in 1925 and 353 ppm as reached in 1990; the current rate of increase is 1.5 ppm $year^{-1}$. Estimates of the maximum concentration that will be reached and the time by which that will occur vary and depend on a number of variables including future political decisions. According to Gifford (1992), Siegenthaler and Sarmiento (1993) and Sundquist (1993), the total amount of carbon in the atmosphere is presently 7.4×10^2 Gt, while estimated reserves of easily accessible fossil carbon amount to 4×10^3 Gt, mainly in the form of coal. Consequently, instantaneous combustion of that material could raise the atmospheric CO_2 level by nearly an order of magnitude. The total amount is small compared to the hydrospheric (4×10^4 Gt) and lithospheric (6.5×10^7 Gt) pools, with the consequence that anthropogenic rises will be transient; however, equilibration with the oceans and the lithosphere is slow. A widely accepted estimate for the immediate future anticipates a value above 600 ppm before the end of the next century (Houghton et al. 1990). Nihous et al. (1994) evaluated strategies such as oceanic containment and reduction of hydrocarbon fuel consumption. In a worst-case scenario, atmospheric CO_2 could peak at 2800 ppm by 2110; in a more optimistic scenario, the maximum could be 1200 ppm encountered in 2350.

Within the scope of the present discussion, it is of interest to consider whether elevated CO_2 will affect the function of the CO_2 sense and the behavioural responses mediated by it; numerous other effects of elevated CO_2 on plants insects have been discussed, for example, by Bazzaz (1990), Bazzaz and Fajer (1992), Caulfield and Bunce (1994), Lincoln et al. (1993).

A CO_2 concentration of 2800 ppm or 0.28% in the open atmosphere is still small compared to concentrations in the soil and in hives; therefore, effects on the sensory performance of organisms in those habitats might be negligible. The situation is different for the case of animals relying on the detection of small CO_2 gradients in the open atmosphere. Many sensory organs including the CO_2 sense (e.g., Yamana et al. 1986) react to equal percentage changes of stimulus magnitude with equal response magnitude, independent of mean stimulus level (Weber-Fechner's law). This implies that the absolute value of the minimum detectable change increases proportionally if the background rises.

On the other hand, the CO_2 concentration in the air exhaled by a host is determined by the host's metabolic rate and will remain constant. At a given distance, for a raised background, a plume of breath released by a host will therefore continue to produce a given absolute change in concentration and hence a smaller fraction of background. Consequently, the range over which haematophagous arthropods can detect CO_2 released by their hosts will decrease. The notion that host detection is multimodal leaves it open to question as regards how significant the effect will be.

Similar considerations apply to gradients of relevance to phytophagous insects. However, the matter is more complicated because, unlike vertebrate hosts of haematophages, plants are expected to respond to elevated CO_2, leading to a change in absolute values of gradients generated.

In addition to these elementary considerations, there are other effects of elevated levels which influence the function of CO_2 receptors. For example, saturation has to be considered. Most of the concentration-response functions reported in the literature are still within the logarithmic range at ten times the present background, with the exception of *Cactoblastis*, where the slope decreases at 600 ppm and saturation is approached at 2400 ppm (Stange et al. 1995). Those and similar measurements were obtained by presenting short pulses on a zero background; however, it is more relevant what happens in response to prolonged exposure to an elevated level. It was observed for *Rhodogastria* (Bogner et al. 1986) that when exposed to a background as high as 1% receptor cells would fire at an increased but constant rate indefinitely, without signs of fatigue. The same observation was made for *Helicoverpa* (Stange and Wong 1993), where exposure to 900 ppm CO_2 led to a sustained rise in action potential rate. However, under identical experimental conditions, the situation was different for *Cactoblastis* (Stange et al. 1995): on prolonged exposure to elevated CO_2 (>600 ppm), a fraction of the receptor neurons initially responds with increased action potential rate, but then the rate continues to drop and firing ceases within a minute or so, resulting in loss of the ability to respond to small changes. This suggests that those receptor neurons are so highly specialized to the detection of CO_2 fluctuations around the present background that their function is impaired by elevated CO_2.

An impairment of the function of CO_2-receptor neurons by elevated CO_2 is also suggested by a further piece of evidence. Given that the lepidopteran receptor neurons are capable of detecting CO_2 contrasts of a fraction of a percent, the question re-emerges to what extent cross-sensitivity to other physical and chemical quantities negates this level of precision. One such quantity is temperature: numerous observations show that other arthropod cutaneous receptor neurons increase their action potential rate with an increase in temperature, possibly simply reflecting the fact that the rate of enzymatic reactions underlying the transduction process increases with temperature. Therefore, we would also expect that the CO_2-receptor firing rate should show a positive temperature coefficient. However, the sites of sensory transduction in and around cell membranes are surrounded by an aqueous phase in which CO_2 molecules must dissolve prior to reaching the transduction site. The temperature coefficient of the solubility of CO_2 in water is negative, meaning that the stimulus concentration at the transduction site decreases with an increase in temperature. This leads to the opposite expectation, namely that the temperature coefficient of the firing rate of CO_2 receptors is negative.

For the CO_2-receptor neurons in *Helicoverpa* (Stange and Wong 1993), it turns out that the effect of temperature on action potential rate varies with the CO_2 background to which they are exposed. At low CO_2 backgrounds, the action potential rate increases with an increase in temperature; at high backgrounds it decreases. The background at which the two effects cancel, and hence exact temperature compensation, occurs at 270 ppm which is close to the pre-industrial level in the open atmosphere, suggesting a functional adaptation to that level. At 600 ppm, a temperature fluctuation of 1 K causes a response of the same magnitude, with opposite sign, as a relative fluctuation in CO_2 of 0.8%. Sudden temperature fluctuations of several K are quite common in the microenvironment. Therefore, CO_2 gradients will be progressively confused with temperature gradients as the background rises.

References

Anderson JF, Ultsch GR (1987) Respiratory gas concentrations in the microhabitats of some Florida arthropods. Comp Biochem Physiol 88A:585–588

Barnola JM, Raynaud D, Korotkevich YS, Lorius C (1987) Vostok ice core provides 160000 year record of atmospheric CO_2. Nature 329:408–414

Bazzaz A (1990) The response of natural ecosystems to the rising global CO_2 levels. Annu Rev Ecol Syst 21:67–96

Bazzaz A, Fajer ED (1992) Plant life in a CO_2-rich world. Sci Am 266:18–25

Bengtsson G, Hedlund K, Rundgren S (1991) Selective odor perception in the soil collembola *Onychiurus armatus*. J Chem Ecol 17:2113–2125

Bjostadt LB, Hibbard BE (1992) 6-methoxy-2-benzoxazolinone: a semiochemical for host location by western corn rootworm larvae. J Chem Ecol 18:931–944

Bogner F (1990) Sensory physiological investigation of carbon dioxide receptors in Lepidoptera. J Insect Physiol 36:951–957

Bogner F (1992) Response properties of CO_2-sensitive receptors in tsetse flies (Diptera: *Glossina palpalis*). Physiol Entomol 17:19–24

Bogner F, Bopprè M, Ernst K-D, Boeckh J (1986) CO_2-sensitive receptors on labial palps of *Rhodogastria* moths (Lepidoptera: Arctiidae): physiology, fine structure and projection. J Comp Physiol A 158:741–749

Bowen MF (1991) The sensory physiology of host-seeking behavior in mosquitoes. Annu Rev Entomol 36:139–158

Boynthon D (1941) Soils in relation to fruit-growing in New York. Part XV. Seasonal and soil influences on oxygen and carbon-dioxide levels of New York orchard soils. Bulletin 763, Cornell Univ Agric Exp St Ithaca

Brady J, Gibson G, Packer MJ (1989) Odour movement, wind direction, and the problem of host-finding by tsetse flies. Physiol Entomol 14:369–380

Burkhardt JF (1991) The influence of different CO_2-concentrations on the behaviour of the ant, *Pheidole pallidula*. Verh Dtsch Zool Ges 84:303–304

Bursell E (1984) Effects of host odour on the behaviour of tsetse. Insect Sci Appl 5:345–349

Carlson DA, Schreck CE, Brenner RJ (1992) Carbon dioxide released from human skin: effect of temperature and insect repellents. J Med Entomol 29:165–170

Caulfield F, Bunce JA (1994) Elevated atmospheric carbon dioxide concentration affects interactions between *Spodoptera exigua* (Lepidoptera: Noctuidae) larvae and two host plant species outdoors. Physiol Chem Ecol 23:999–1005

David CT (1986) Mechanisms of directional flight in wind. In: Payne TL Birch MC Kennedy CEJ (eds) Mechanisms in insect olfaction. Clarendon Press, Oxford, pp 49–57

Davis EE, Bowen MF (1994) Sensory physiological basis for attraction in mosqitoes. J Am Mosquito Control Assoc 10:316–325

Denmead OT, Bradley EF (1987) On scalar transport in plant canopies. Irrig Sci 8:131–149

Den Otter CJ, Van der Goes van Naters WM (1992) Single cell recordings from tsetse (*Glossina m.morsitans*) antennae reveal olfactory, mechano- and cold receptors. Physiol Entomol 17:33–42

Desjardins RL (1992) Review of techniques to measure CO_2 flux densities from surface and airborne sensors. In: Stanbell G (ed) Advances in bioclimatology, vol 1. Springer, Berlin Heidelberg New York, pp 1–23

Desjardins RL, Brach EJ, Alvo P, Schuepp PH (1982) Aircraft monitoring of surface carbon dioxide exchange. Science 216:733–735

Desjardins RL, Macpherson JI, Schuepp PH, Karanja F (1989) An evaluation of aircraft flux measurements of CO_2, water vapor and sensible heat. Boundary-Layer Meteorol 47:55–69

Desser SS, Hong H (1992) Antennal and palpal sensilla of *Forcipomyia (Lasiohelea* Kieffer) *fairfaxensis* Wirth, 1951 (Diptera: Ceratopogonidae) from Algonquin Park, Ontario. Can J Zool 70:385–390

Dodd AP (1940) The biological campaign against prickly-pear. Queensland Government Printer, Brisbane

Dusenbery DB (1987) Theoretical range over which bacteria and nematodes locate plant roots using carbon dioxide. J Chem Ecol 13:1617–1624

Eiras AE, Jepson PC (1991) Host location by *Aedes aegypti* (Diptera: Culicidae): a wind tunnel study of chemical cues. Bull Entomol Res 81:151–160

Fuller HJ (1948) Carbon dioxide concentrations of the atmosphere above Illinois forest and grassland. Am Midl Nat 39:247–249

Fuzeau-Braesch S, Nicolas G (1981) Effect of carbon dioxide on subsocial insects. Comp Biochem Physiol 68A:289–297

Galbreath RA (1988) Orientation of grass grub *Costelytra zealandica* (Coleoptera: Scarabaeidae) to a carbon dioxide source. N Z Entomol 11:6–7

Galil J, Zeroni M, Bar Shalom D (1973) Carbon dioxide and ethylene effects in the co-ordination between the pollinator *Blastophaga quadraticeps* and the syconium in *Ficus religiosa*. New Phytol 72:1113–1127

Gibson G (1992) Do tsetse flies "see" zebras? A field study of the visual response of tsetse to striped targets. Physiol Entomol 17:141–147

Gibson G, Packer MJ, Steullet P, Brady J (1991) Orientation of tsetse flies to wind, within and outside host odour plumes in the field. Physiol Entomol 16:47–56

Gifford RM (1992) Interaction of carbon dioxide with growth-limiting environmental factors in vegetation productivity: implications for the global carbon cycle. In: Stanbell G (ed) Advances in bioclimatology, vol 1. Springer, Berlin Heidelberg New York, pp 24–58

Gillies MT (1980) The role of carbon dioxide in host-finding by mosquitoes (Diptera: Culicidae): a review. Bull Entomol Res 70:525–532
Hangartner W (1969) Carbon dioxide, a releaser for digging behaviour in *Solenopsis geminata* (Hymenoptera: Formicidae). Psyche (Cambridge) 76:58–67
Houghton JT, Jenkins GJ, Ephraums JJ (1990) Climate change: the IPCC scientific assessment. Cambridge University Press, Cambridge
Jones OT, Coaker TH (1977) Oriented responses of carrot fly larvae, *Psila rosae*, to plant odours, carbon dioxide and carrot root volatiles. Physiol Entomol 2:189–197
Jones OT, Coaker TH (1978) A basis for host plant finding in phytophagous larvae. Entomol Exp Appl 24:272–284
Kaib M, Ziesmann J, Wolfrum U (1993) Modulation of odour sensitivity by carbon dioxide in a termite sensillum: a possible mechanism for signal interpretation. In: Wiese K et al. (eds) Sensory systems of arthropods. Birkhaeuser, Basel, pp 481–488
Keil TA, Steinbrecht RA (1984) Mechanosensitive and olfactory sensilla of insects. In: King RC, Akai H (eds) Insect ultrastructure. Plenum Press, New York, pp 477–516
Kellogg FE (1970) Water vapour and carbon dioxide receptors in *Aedes aegypti*. J Insect Physiol 16:99–108
Kent KS, Harrow ID, Quartararo P, Hildebrand JG (1986) An accessory olfactory pathway in Lepidoptera: the labial pit organ and its central projections in *Manduca sexta* and certain other sphinx moths and silk moths. Cell Tissue Res 245:237–245
Kline DL, Wood JR, Cornell JR (1991) Interactive effects of 1-octen-3-ol and carbon dioxide on mosquito (Diptera: Culicidae) surveillance and control. J Med Entomol 28:254–258
Kline DL, Hagan DV, Wood JR (1994) *Culicoides* responses to 1-octen-3-ol and carbon dioxide in salt marshes near Sea Island, Georgia, U.S.A. Med Vet Entomol 8:25–30
Klingler J (1963) Die Orientierung von *Ditylenchus dipsaci* in gemessenen künstlichen und biologischen CO_2-Gradienten. Nematologica 9:185–199
Klingler J (1966) Über den Sitz der CO_2-Rezeptoren bei der Larve von *Otiorrhynchus sulcatus*. Entomol Exp Appl 9:271–277
Kramer E (1986) Turbulent diffusion and pheromone-triggered anemotaxis. In: Payne TL, Birch MC, Kennedy CEJ (eds) Mechanisms in insect olfaction. Clarendon Press, Oxford, pp 11–25
Lacher V (1964) Elektrophysiologische Untersuchungen an einzelnen Rezeptoren für Geruch, Kohlendioxyd, Luftfeuchtigkeit und Temperatur auf den Antennen der Arbeitsbiene und Drohne (*Apis mellifera* L.). Z Vergl Physiol 48:587–623
Lacher V (1967) Verhaltensreaktionen der Bienenarbeiterin bei Dressur auf Kohlendioxid. Z Vergl Physiol 54:75–84
Lance DR (1992) Odors influence choice of oviposition sites by *Diabrotica virgifera virgifera* (Coleoptera: Chrysomelidae). J Chem Ecol 18:1227–1237
Lee J-K, Selzer R, Altner H (1985) Lamellated outer dendritic segments of a chemoreceptor within wall-pore sensilla in the labial palp-pit organ of the butterfly, *Pieris rapae* L. (Insecta, Lepidoptera). Cell Tissue Res 240:333–342
Lee KE, Wood TG (1971) Termites and soils. Academic Press, London
Lewis CT (1971) Superficial sense organs of the antennae of the fly, *Stomoxys calcitrans*. J Insect Physiol 17:449–461
Lincoln DE, Fajer ED, Johnson RH (1993) Plant-insect herbivore interactions in elevated CO_2 environments. Trends Ecol Evol 8:64–68
Lüscher M (1961) Air conditioned termite nests. Sci Am 205:138–145
Mafra-Neto A, Cardé AT (1994) Fine-scale structure of pheromone plumes modulates upwind orientation of flying moths. Nature 369:142–144
McCallum ME, Dusenbery DB (1992) Computer tracking as a behavioral GC detector: Nematode responses to vapor of host roots. J Chem Ecol 18:585–592
McIver SB (1972) Fine structure of pegs on the palps of female culicine mosquitoes. Can J Zool 50:571–576

McIver SB, McElligott PE (1989) Effects of release rates on the range of attraction of carbon dioxide to some southwestern Ontario mosquito species. J Am Mosquito Control Assoc 5:6–9
Monteith JL (ed) (1976) Vegetation and the atmosphere, vol 2. Academic Press, London
Moskalyk LA, Friend WG (1994) Feeding behaviour of female *Aedes aegypti*: effects of diet temperature, bicarbonate and feeding technique on the response to ATP. Physiol Entomol 19:223–229
Mount GA, Dunn JE (1983) Economic thresholds for lone star ticks (Acari: Ixodidae) in recreational areas based on a relationship between CO_2 and human subject sampling. J Econ Entomol 76:327–329
Murlis J, Jones CD (1981) Fine-scale structure of odour plumes in relation to insect orientation to distant pheromone and other attractant sources. Physiol Entomol 6:71–86
Murlis J, Elkinton JS, Carde RT (1992) Odour plumes and how insects use them. Annu Rev Entomol 37:505–532
Myers JH, Monro J, Murray ND (1981) Egg clumping, host plant selection and population regulation in *Cactoblastis cactorum*. Oecologia (Berl) 51:7–13
Nelson RL (1965) Carbon dioxide as an attractant for *Culicoides*. J Med Entomol 2:56–57
Nicolas J, Sillans D (1989) Immediate and latent effects of carbon dioxide in insects. Annu Rev Entomol 34:97–116
Nihous GC, Masutani SM, Vega LA, Kinoshita CM (1994) Projected impact of deep ocean carbon dioxide discharge on atmospheric CO_2 concentrations. Clim Change 27:225–244
Nobel PS (1991) Introduction to biophysical plant physiology. Academic Press, San Diego
Núñez JA (1982) Food source orientation and activity in *Rhodnius prolixus* Stal (Hemiptera: Reduviidae). Bull Entomol Res 72:253–262
Omer SM (1979) Responses of females of *Anopheles arabiensis* and *Culex pipiens fatigans* to air currents, carbon dioxide and human hands in a flight-tunnel. Entomol Exp Appl 26:142–151
Osmond CB, Monro JM (1981) Prickly Pear. In: Carr DJ, Carr SGM (eds) Plants and man in Australia. Academic Press, Sydney, pp 194–222
Osmond CB, Nott DL, Firth PM (1979) Carbon assimilation patterns and growth of the introduced CAM plant *Opuntia inermis* in eastern Australia. Oecologia (Berl) 40:331–350
Paim U, Beckel WE (1963) Seasonal oxygen and carbon dioxide content of decaying wood as a component of the microenvironment of *Orthosoma brunneum* (Forster) (Coleoptera: Cerambycidae). Can J Zool 41:1133–1147
Paim U, Beckel WE (1964a) The carbon dioxide related behaviour of the adults of *Orthosoma brunneum* (Forster) (Coleoptera: Cerambycidae). Can J Zool 42:295–303
Paim U, Beckel WE (1964b) The behaviour of the larvae of *Orthosoma brunneum* (Forster) (Coleoptera: Cerambycidae) in relation to gases found in the logs inhabited by the larvae. Can J Zool 42:327–353
Perritt DW, Couger G, Barker RW (1993) Computer-controlled olfactometer system for studying behavioral responses of ticks to carbon dioxide. J Med Entomol 30:571–578
Pline M, Dusenbery DB (1987) Responses of plant-parasitic nematode *Meloidogyne incognita* to carbon dioxide determined by video camera-computer tracking. J Chem Ecol 13: 873–888
Rasch C, Rembold H (1994) Carbon-dioxide – highly attractive signal for larvae of *Helicoverpa armigera*. Naturwissenschaften 81:228–229
Roessingh P (1989) Trail marking and following by larvae of the small ermine moth *Yponomeuta cagnagellus*. PhD Thesis, Landbouwuniversiteit Wageningen, Netherlands
Sauer JR, Hair JA, Houts MS (1974) Chemo-attraction in the lone star tick (Acarinae: Ixodidae). 2. Responses to various concentrations of CO_2. Ann Entomol Soc Am 67:150–152
Seeley TD (1974) Atmospheric carbon dioxide regulation in honeybee colonies. J Insect Physiol 20:2301–2305
Seeley TD (1985) Honeybee ecology. Princeton University Press, Princeton
Siegenthaler U, Sarmiento JL (1993) Atmospheric carbon dioxide and the ocean. Nature 365:119–125

Snow WF (1970) The effect of a reduction in expired carbon dioxide on the attractiveness of human subjects to mosquitoes. Bull Entomol Res 60:43–48

Stange G (1975) Linear relation between stimulus concentration and primary transduction process in insect CO_2-receptors. In: Denton DA, Coghlan JP (eds) Olfaction and taste V. Academic Press, New York, pp 207–211

Stange G (1992) High resolution measurement of atmospheric carbon dioxide concentration changes by the labial palp organ of the moth *Heliothis armigera* (Lepidoptera: Noctuidae). J Comp Physiol A 171:317–324

Stange G, Diesendorf M (1973) The response of the honeybee antennal CO_2-receptors to N_2O and Xe. J Comp Physiol 86:139–158

Stange G, Wong C (1993) Moth response to climate. Nature 365:699

Stange G, Osmond CB, Stowe S (1992) Carbon dioxide receptors in moths as high resolution sensors of the metabolic performance of food plants. Chem Senses 17:881

Stange G, Monro J, Stowe S, Osmond CB (1995) The CO_2-sense of the moth *Cactoblastis cactorum* and its probable role in the biological control of the CAM plant *Opuntia stricta*. Oecologia (Berl) 102:341–352

Steullet P, Guerin PM (1992a) Perception of breath components in the tropical bont tick, *Amblyomma variegatum* Fabricius (Ixodidae). I. CO_2-excited and CO_2-inhibited receptors. J Comp Physiol A 170:665–676

Steullet P, Guerin PM (1992b) Perception of breath components in the tropical bont tick, *Amblyomma variegatum* Fabricius (Ixodidae). II. Sulfide-receptors. J Comp Physiol A 170:677–685

Sundquist ET (1993) The global carbon dioxide budget. Science 259:934–941

Takken W (1991) The role of olfaction in host-seeking of mosquitoes: a review. Insect Sci Appl 12:287–295

Torr SJ (1988) Behaviour of tsetse flies (*Glossina*) in host odour plumes in the field. Physiol Entomol 13:467–478

Torr SJ (1989) The host-orientated behaviour of tsetse flies (*Glossina*): the interaction of visual and olfactory stimuli. Physiol Entomol 14:325–340

Torr SJ (1990) Dose responses of tsetse flies (*Glossina*) to carbon dioxide, acetone and octenol in the field. Physiol Entomol 15:93–103

Turner DA (1971) Olfactory perception of live hosts and carbon dioxide by the tsetse fly *Glossina morsitans orientalis* Vanderplank. Bull Entomol Res 61:75–96

Vale GA, Hall DR (1985) The role of 1-octen-3-ol, acetone and carbon dioxide in the attraction of tsetse flies, *Glossina* spp. (Diptera: Glossinidae) to ox odour. Bull Entomol Res 75:209–217

Van Essen PHA, Kemme JA, Ritchie SA, Kay BA (1994) Differential responses of *Aedes* and *Culex* mosquitoes to octenol or light in combination with carbon dioxide in Queensland, Australia. Med Vet Entomol 8:63–67

Verron H (1963) Rôle des stimuli chimiques dans l'attraction sociale chez *Calotermes flavicollis* (Fabr). Insectes Sociaux 10:167–184, 185–296, 297–335

Warnes ML (1990) The effect of host odour and carbon dioxide on the flight of tsetse flies (*Glossina* spp.) in the laboratory. J Insect Physiol 36:607–611

Warnes ML, Finlayson LH (1985a) Responses of the stable fly, *Stomoxys calcitrans* (L.) (Diptera: Muscidae), to carbon dioxide and host odours. I. Activation. Bull Entomol Res 75:519–527

Warnes ML, Finlayson LH (1985b) Responses of the stable fly, *Stomoxys calcitrans* (L.) (Diptera: Muscidae), to carbon dioxide and host odours. II. Orientation. Bull Entomol Res 75:717–727

Werder J, Gothe R (1993) *Hyalomma truncatum* and *Rhipicephalus evertsi mimeticus* (Ixodidea, Ixodidae): reactions of adult ticks to vertically incident narrow- and wideband optical radiation with and without the influence of a CO_2-gradient. Exp Applic Acarol 17:489–501

White RA, Paim U, Seabrook WD (1974) Maxillary and labial sites of carbon dioxide sensitive receptors of larval *Orthosoma brunneum* (Forster) (Coleoptera, Cerambycidae). J Comp Physiol 88:235–246

Willemse LPM, Takken W (1994) Odor-induced host location in tsetse flies (Diptera: Glossinidae) J Med Entomol 31:775–794

Wilson EO (1962) Chemical communication among workers of the fire ant *Solenopsis saevissima* (Fr. Smith). Animal Behav 10:134–164

Wilson EO (1971) The insect societies. Harvard University Press, Cambridge

Wright RH, Kellogg FE (1962) Response of *Aedes aegypti* to moist convection currents. Nature 194:402–403

Yamana K, Toh Y (1987) Intracellular recording from receptor cells of the temporal organ of the Japanese house centipede, *Thereuonema hilgendorfi*: receptor potential and conductance changes. J Exp Biol 131:205–213

Yamana K, Toh Y (1990) Structure of the temporal organ of the Japanese house centipede, *Thereuonema hilgendorfi* Verhoeff. J Morphol 203:311–319

Yamana K, Toh Y, Tateda H (1986) Electrophysiological studies on the temporal organ of the Japanese house centipede, *Thereuonema hilgendorfi*. J Exp Biol 126:297–314

Subject Index